Ground
Improvement

Third Edition

Edited by Klaus Kirsch and Alan Bell

CRC Press
Taylor & Francis Group
Boca Raton London New York

CRC Press is an imprint of the
Taylor & Francis Group, an **informa** business

CRC Press
Taylor & Francis Group
6000 Broken Sound Parkway NW, Suite 300
Boca Raton, FL 33487-2742

First issued in paperback 2019

© 2013 by Klaus Kirsch and Alan Bell
CRC Press is an imprint of Taylor & Francis Group, an Informa business

No claim to original U.S. Government works

ISBN-13: 978-0-415-59921-4 (hbk)
ISBN-13: 978-0-367-86569-6 (pbk)

Visit the Taylor & Francis Web site at
http://www.taylorandfrancis.com

and the CRC Press Web site at
http://www.crcpress.com

Contents

Preface

Ground improvement techniques continue to progress in addressing ground engineering problems across the world, particularly in urban areas where land development and reuse need to be efficient not only in the geotechnical engineering but in time, cost, and energy used. As well as in expanding markets, recent growth has also been seen across a range of methods, in increasing productivity due to investment in plant and equipment, and in improvement in technical performance and quality due to electronic monitoring and control methods. Ground improvement methods are also frequently able to demonstrate low carbon impact and excellent sustainability credentials as these issues become more important.

The third edition of this well-known book provides a comprehensive overview of the major ground improvement techniques in use worldwide today. The chapters are fully updated with recent developments and have been written by recognised experts who bring a wealth of knowledge and experience to bear on their contributions.

Ground Improvement is written for civil and geotechnical engineers and for contractors involved in piling and ground engineering of any kind. Advanced graduate and postgraduate civil engineering and geotechnical students will find the book most helpful in guiding their studies.

Editors

Klaus Kirsch was a main board director of Keller Group plc, responsible for Group operations in Continental Europe and overseas until his retirement in 2001, and thereafter served the Group as adviser for its technical development. He has been involved with some fundamental breakthroughs in foundation engineering, notably the introduction of vibro stone column technology in the USA, and was instrumental in the development of a new generation of depth vibrators. He has authored many papers and also a book on vibratory deep compaction.

Alan Bell is technical consultant to the Keller Group plc. He was managing director of Keller in the UK until 2009 and has had close involvement in ground improvement techniques and their development at both national and international levels for over 30 years, including the introduction to the UK of vibrated concrete columns. He has contributed many technical publications, and is a visiting professor at the University of Strathclyde in Glasgow.

Contributors

George Burke, MSc
Hayward Baker Inc.
Odenton, Maryland

Jian Chu, PhD
Department of Civil, Construction,
 and Environmental Engineering
Iowa State University and
Ames, Iowa, USA and
Formerly School of Civil and Envir-
 onmental Engineering
Nanyang Technological University
Singapore

Marcus Dahlström, MSc
Inhouse Tech Geoteknik AB
Göteborg, Sweden

Eduard Falk, PhD
Keller Holding GmbH
Offenbach, Germany

James Hussin, MSc
Hayward Baker, Inc.
Tampa, Florida

Harald Krenn, MSc, PhD
Züblin Spezialtiefbau Ges.mbH
Vienna, Austria

Clemens Kummerer, PhD
Keller Grundbau GmbH
Söding, Austria

Venu Raju, Dr-Ing, PhD
Keller Asia
Singapore

Barry Slocombe, BSc, MSc
Keller Foundations
Coventry, United Kingdom

Wolfgang Sondermann, Dr-Ing
Keller Holding GmbH
Offenbach, Germany

Gert Stadler, Dr-Ing, Dr-mont
Graz University of Technology
Salzburg, Austria

Michał Topolnicki, PhD, DSc
Gdańsk University of Technology
Gdańsk, Poland
and
Keller Polska Sp. z o.o.,
Gdynia, Poland

Jimmy Wehr, Dr-Ing
Dresden University of Technology
Dresden, Germany
and
Keller Holding GmbH
Offenbach, Germany

Hiroshi Yoshida, DrEng
Chemical Grouting Co. Ltd.
Tokyo, Japan

Chapter 1

Introduction and background

Alan Bell and Klaus Kirsch

CONTENTS

1.1 PURPOSE OF GROUND IMPROVEMENT PROCESSES

When faced with difficult ground conditions at a project site, an engineer has a number of possible strategies to employ in order to achieve the project objectives. The most obvious is to find another site, but this is only very rarely practicable. Pressure on land, the need to use poor sites, and the location of many cities in estuaries or river situations make this option increasingly difficult. Another option is to redesign the building or structure to accommodate the prevailing difficulties arising from the ground, and where possible this is a good solution. Yet another possibility is to remove the troublesome ground and to replace it with more suitable material, and this can often be cost effective providing the depth to be addressed and the quantities concerned are relatively small.

If none of these avoiding strategies are technically or economically realistic, then the prevailing ground conditions must be addressed. A common potential solution is to adopt a system such as piling, in order to bypass the difficult ground and found in suitable material. However, this can be expensive and time consuming and may actually be difficult to achieve in very deep ground. In addition, for some classes of geotechnical problem such as tunnelling, piling may be unsuitable. For such reasons improving the ground to achieve an appropriate engineering performance is an increasingly successful approach worldwide when faced with problem ground conditions, partially evidenced by the two earlier editions of the present book.

Ground improvement is normally understood as the modification of the existing physical properties of the ground beneath a site to sufficient depth to enable effective, economic, and safe permanent or temporary construction in practical timescales. Typical objectives would be one or a combination of the following:

(1) An increase in shear strength or density to improve bearing capacity or to provide sufficient support for excavations or tunnels
(2) A reduction in compressibility to minimise total or differential settlements of buildings or structures, or other deformations in the ground arising from excavation or tunnelling
(3) A reduction in permeability to minimise flow of ground water to prevent inundation or water damage or to isolate zones of contaminated ground water
(4) Conversely, an improvement in deep drainage in order to assist pre-loading or surcharge techniques
(5) Controlled displacement of the ground in order to dispel previous differential settlements or ground distortions, or to compensate for ground movements arising from excavation or tunnelling
(6) Prevention of liquefaction or reduction in lateral spreading beneath or near both new and existing structures during earthquakes, employing densification, replacement with stronger materials, or deep drainage

The ground improvement processes used to deliver these objectives form the subject matter of this book, as set out in Section 1.2.

It should be noted that recent environmental legislation coupled with the need to recycle previously developed sites has led to considerable growth in a different concept of improving the ground—namely, to minimise or remove the hazardous effects of sites contaminated by toxic waste or chemical by-products from industrial processes. This subject is beyond the scope of this book and the reader should refer to the specialist literature on these topics.

1.2 WHAT THE BOOK COVERS

This third edition of *Ground Improvement* will provide the reader with a sound basis for understanding and further study of the most widely used processes for ground improvement.

Developments in equipment and methods have continued apace since the publication of the second edition, and where relevant are included in the ensuing chapters of this book. Indeed, the editors are grateful to the authors of the following chapters in the book, all of whom are recognised experts in their respective fields. Their contributions provide an overview of the processes concerned and the key geotechnical and design considerations involved, together with details of the equipment needed for successful execution. The methods are well illustrated with relevant case histories revealing applications in practice.

Since soil strength and compressibility are highly influenced by the particle packing or density in most engineering soils, densification is a useful approach. In granular soils this is frequently achieved most efficiently using vibratory methods to force particles into more closely packed configurations. Methods employing tools in which the vibrator can be taken deep into the ground are very efficient and are described in Chapter 2. Another important global technique is dynamic compaction, which employs large weights dropped from height to create the compactive energy needed, and is dealt with in Chapter 3.

In cohesive soils, an increase in shear strength and reduced compressibility can be achieved by consolidation, usually achieved by direct loading. The process is time dependent and can be hastened and better controlled using deep prefabricated drains and associated methods. These are comprehensively dealt with in Chapter 4.

The remaining six chapters in the book deal with various techniques involving the injection of materials into the ground in order to provide geotechnical improvement of various kinds. Chapter 5 describes permeation grouting, which involves the displacement of the ground water or air in soil pores or rock fissures, using an agent, usually termed grout, which is sufficiently fluid to permeate the ground. This agent subsequently hardens to create the intended improvements. Jet grouting is covered in Chapter 6. This technique uses powerful jets of grout, or grout with other fluids to displace or mix with the ground. In this way zones of ground with increased strength or stiffness, or barriers to flow can be formed. Soil fracture grouting is a displacement technique and employs finely controlled injection of relatively thin but multiple veins of grout to address excess building settlement, to lift structures, or to compensate for ground movement during tunnelling. It is described in Chapter 7. Compaction grouting, developed in the United States and now used around the world, is dealt with in Chapter 8. This is also a displacement method and can be used to compact or reinforce the ground with introduced grout. Recent changes in terminology originating in the United States are helpfully explained in detail. In-situ soil mixing processes continue to develop worldwide, and Chapter 9 provides comprehensive coverage of the main deep mixing processes in use across the world and describes several new techniques, such as the trench-mixing

TRD and the panel-mixing CSM approaches. Several new case histories are also included. Chapter 10 covers dry soil mixing using the Scandinavian approach, as this process has continued to see worldwide application.

In the remainder of the present chapter several topics which would not justify a separate chapter are included, such as the history of the two main means of creating improvement (see Section 1.3). Brief notes on health and safety for ground improvement sites are included in view of its common relevance (Section 1.4). The effects of ground improvement on greenhouse gas emissions have increasingly been addressed since the second edition and now warrant inclusion (Section 1.5). Overviews of two ground improvement techniques which, by their nature, have somewhat limited ranges of application have also been included. Blasting can be an effective means of improving granular soils by densification and is described in Section 1.6. The only reversible ground improvement process of ground freezing can be powerful where applicable, and is outlined in Section 1.7.

1.3 HISTORICAL DEVELOPMENT

Since earliest times humankind has found ways of dealing with poor ground in order to form pathways and later roads using such simple strategies as placing beds of reeds or saplings to support the weight of people and animals over soft ground. It is only relatively recently that the means of engineering difficult ground by compaction, consolidation, or by adding materials by permeation or mixing has seen significant advances. These processes developed during and after the Industrial Revolution, but mainly in the early twentieth century. In these years, better understanding of soil mechanical behaviour emerged through the work of Terzaghi and others; practical ground investigation became possible; and equipment and materials development reached the stage that significant volumes of soil could be treated. Two main approaches are in worldwide use today, namely deep vibratory treatment and injection or mixing of grouts. A brief historical review of these two topics follows in view of their importance.

Kirsch and Kirsch (2010) describe how depth vibrators were developed in Germany in the 1930s. Initially aimed at concrete densification, they soon were applied more effectively for sand compaction. Consequently, deep vibratory stabilisation for both natural and filled cohesionless soils was used widely for a range of applications in Germany and further afield, particularly after 1945. A further major area of application was added in 1956 when depth vibrators were employed to form stone columns in silty materials, leading to the application to cohesive soils more generally.

Further development and improvement of the special plant and equipment necessary for the execution of this ground improvement method together with the experience gained in practice considerably increased the range of application in foundation engineering, notably after the advantages of the method were recognised mitigating the liquefaction potential of soils in earthquake-prone regions by densification and/or drainage. Today the method can effectively be designed for its purpose and can be well controlled during its execution. It is interesting to note that first steps towards eventual process automation are already being used in practice.

The idea of injecting cement slurries, known as grouts, into the ground to improve their engineering characteristics also saw early development. It is believed that grouting of subsoil was first performed more than 200 years ago in France (about 1810) by the French engineer Charles Bérigny, using a suspension of pozzuolana cement in water to stabilise alluvial deposits forming the foundations of a bridge (Glossop 1961). Further development of the grouting process (procédé d'injection) in the nineteenth century was by the introduction of new hydraulic binders, particularly the invention of Portland cement in 1821. The method was already well developed at the outset of World War I with the use of pumps, pressure control, and the need for filtration all established.

Direct injection of simple cement grouts into the ground was often hampered by failure to permeate the ground, either because the pore size distribution of the soils were small in comparison to the grading of the cements or because the methods of injection, often from open-ended casings, were too crude. Low pressure permeation grouting using simple cement grouts is usually limited to gravel containing perhaps some coarse sand.

Important steps forward in addressing this limitation were accomplished by attention to the materials used for injection. Dutchman Hugo Joosten in Germany, by his invention of the Joosten system, used chemicals in the form of highly concentrated sodium silicates and calcium chloride as grout material to form precipitate silicates in situ to treat sandy cohesionless soils (Joosten 1926). These much finer-grained grouts could permeate more readily than simple cements. The method was then widely used in Berlin in the construction of the underground railway.

By the late 1950s a single shot approach was developed by mixing organic hardeners and sodium silicate before injection. Today various proprietary re-agents are available as hardeners, with widely differing properties. Another development involved creating, by fine grinding, so-called microfine or ultrafine cements, which also allow permeation into coarse sands or sometimes even finer soils, and these are also in use today. Various new chemical formulations were also developed in the 1960s and 1970s which enabled even finer grained cohesionless soils to be treated, but with limited application today due to concerns about toxicity.

Developments for dealing with some of the limitations of soil grouting also came from improvement in equipment. The invention of the tube à manchette pipe or TaM pipe (Ischy 1933), still very much employed in grouting processes today, was very significant. These pipes, consisting of grout ports with rubber sleeve valves, are placed in boreholes to attain the depths required, and are grouted in place using a relatively weak sleeve grout. One set of ports can be isolated at a time, and grout can be injected into the surrounding ground after it expands the surrounding seal and breaks the sleeve grout. This enables control of grout volume or pressure during injection at specific points in the subsoil. Littlejohn (1993) provides a useful summary of the history of injection processes.

Subsequent grouting development has concentrated more on the development of entirely new ground improvement processes using simple cement grouts, partly due to concerns over toxicity of some chemical grouts and partly to the desire for improved performance. By the early 1990s jet grouting; compaction grouting; soilfracture grouting; and soil mixing methods were all widely and successfully applied in addition to permeation grouting (e.g., Bell 1994). Since then there has been increased use of all of these, and soil mixing in particular is now more widely used. Further technical development of all grouting methods has continued, particularly in relation to electronic monitoring and control on site. Some history on these methods is included in Chapters 6 through 9.

1.4 HEALTH, SAFETY, AND ENVIRONMENTAL CONSIDERATIONS

In recent years the construction industry worldwide has seen significant improvements in the safety of construction workers and the public. Legislation, formal management systems, and motivational training have all played their part. Indeed, safety is a critical component of all construction in general and ground improvement in particular as it inevitably contains aspects that are potentially unsafe. The specialist piling and ground improvement industries are very committed to safety and minimising the environmental effects of construction. For example, the European Federation for Foundation Contractors (EFFC) holds the attainment of the highest standard of safety as a key objective and its members have established a health and safety charter, together with publications and advice on the subject.

Extensive procedures involving risk assessments, work instructions, method statements, and training, which together form site-specific project safety plans, are now commonly used in minimising safety and environmental hazards through general and specialist ground improvement contractors, as well as client and public bodies. The following provides some limited comment on the health, safety and environmental impact of ground improvement processes as a brief introduction to the subject in view of its consistent importance. However it cannot be a comprehensive presentation of the subject, and reference needs to be made to local regulations and written procedures, and the specialists in the particular processes.

1.4.1 Site mobilization and demobilization

In common with all site operations, health and safety considerations form an integral part of establishing construction activity and leaving after project completion. Clear delineation is needed for site entrances and exits, temporary roads for materials' supply trucks, pedestrian walkways, site boundaries with protection and exclusion of the general public, and clear storage and load/unload areas. The working surface, suitably lit, for all plant and equipment should be engineer-designed and capable of maintaining support in all weather. Simple site procedures can be used to ensure this work is done to the appropriate standard prior to commencement (e.g., by the use of a working platform certificate). If the platform is not integrated into the final works, a plan for its disposal is required. Overhead power lines and underground services need to be clearly identified and delineated, together with instructions as to avoidance or minimum clearances given by appropriate authorities prior to commencing work. Measures need to be introduced so that noise and dust are minimised and kept within agreed limits for the general public beyond the site boundaries, and for construction personnel on site.

Site operatives are required to be trained and experienced with certified skills, or if in training have adequate supervision from suitably experienced colleagues. Safety equipment supplied must be worn at all times and employed in accordance with training and advice given. Often the processes require physically lifting materials such as cement bags, or other heavy objects such as hoses or steel casing. Specific training in proper lifting procedures and specified max loads for any lift should be taken as a minimum approach. It is important to employ lifting devices such as winches and crane arms in all cases where limits are exceeded, and these may in some cases form part of the drilling equipment

Operational hazards on site must be identified in advance and plans put in place to deal with them. Some ground-improvement processes generate spoil from the ground and this needs to be controlled to prevent injury near drilling or boring equipment, or from flying debris necessarily generated on dynamic compaction sites. The spoil must be controlled on site to minimise deterioration of working and access areas, and measures taken for its safe re-use or disposal so as not to contaminate the immediate environment. Trip hazards such as open boreholes must be clearly marked on site.

The plant and equipment used to perform any ground improvement process needs to be in good operating condition. Since many pieces of equipment are often used together to enable the process to be efficiently executed on site, proper consideration needs to be given to all items including any attachments, not only the large plant. Safety precautions and operating instructions are generally published for each piece of equipment, including initial preparation on site, and should be reviewed prior to use and updated in the event of modifications or the introduction of revised items of equipment. Regular maintenance, inspection, and certification of all equipment at agreed intervals are needed to ensure continued safe operating equipment.

Measures including automatic cutoff devices or guarding should be in place to prevent injury to operatives near to rotating drilling or boring equipment. The correct mode for moving and operating equipment must be made clear, and in moving there must be clear guidance to ensure the safety of adjacent personnel.

One aspect particular to grouting processes is the condition of the grout hoses and hose connections. It is important that these be rated to safely withstand the pumping pressure, be operated properly and regularly checked and certified. Failure of the hoses or hose connections can result in the high-energy release of grout, potentially resulting in severe injury. Also, whipping high energy hoses for grout, air, or other fluids are highly dangerous, and whip checks at connections should be used. For both air and pressure grout equipment, clear procedures for pressure release in any circumstances including cleaning, must be identified and adhered to.

1.4.2 Hazardous materials

Material safety data sheets for all materials to be used on the project should be reviewed and training given prior to beginning work. Cements and other cementitious materials or chemicals are commonly used in grouting and soil mixing and other processes and are very caustic. Prolonged exposure to the skin or eyes can cause severe chemical burns and permanent injury. Consequently, risk assessments and the use and enforcement of safe working methods are vital. Hazards inherent from the design approach are also to be considered in minimising injury or illness in site personnel and the general public.

1.5 GREENHOUSE GAS EMISSIONS

In 1997 the Kyoto protocol was ratified by participating nations, and is a treaty aimed at stabilising greenhouse gas concentrations in the Earth's atmosphere at a level to limit anthropogenic (human) interference with the climate system. Indeed, there is now worldwide awareness, and acceptance in the scientific community, of the greenhouse gas effect on global climate. The construction industry, in common with other industries, is consequently looking at its own emissions, so that these can be better understood and consequently minimised or even eliminated. Methodology and databases are now available for identifying and calculating the key inputs. This is often done by employing the concept of carbon dioxide equivalent to the key greenhouse gases released by the process (for example, Hammond and Jones 2011).

Calculations can also be used to compare different ground improvement and other geotechnical processes such as piling methods, and this can be important in minimising the total carbon dioxide equivalent for a given project. Wintzingerode et al. (2011) list seven key potential sources of greenhouse gas emissions for ground improvement systems, namely:

- Raw materials
- Transport for materials
- Transport for personnel
- Transport for equipment
- Product manufacture
- Waste or spoil emissions
- Transport for waste or spoil

Such calculations can be used to examine different components of the construction process, and these clearly show the large impact of manufactured construction materials such as Portland cement and steel, with other inputs, notably the energy requirement for construction plant and equipment, usually much lower than for materials, as seen in comparative databases (e.g., EA 2010, GEMIS 2010). The other emissions are often very small for ground improvement projects. Nevertheless, each project must be studied separately. For example, the degradation of spoil consisting of peats or highly organic soils removed from below the water table can generate relatively large emissions (Hall 2006).

Several bases for comparison are possible and it is important to understand the implications of these. For example, Zöhrer et al. (2010) compare different methods using MJ/lineal metre for comparison. This provides a means of comparing a wide range of products, and is illustrated in Figure 2.18 in Chapter 2. The strong conclusion was the very low impact of the vibro processes by comparison with the others due to the use of quarried materials with this system.

Egan and Slocombe (2010) used several actual foundation projects for their basis of comparison and found that replacement or partial replacement of piles with vibro stone column ground improvement systems resulted in between 92.5%–96% reduction in embodied carbon dioxide, and where piles could only partially be replaced a reduction of 36.4% was noted.

Wintzingerode et al. (2011) draw attention to comparing the total emissions per unit load carried, and also the total emissions per square metre of final construction. Examples illustrate the large reductions of about a factor of 11 in carbon dioxide equivalents gained by switching from bored piles to vibro stone columns.

Pinske (2011) compared five different ground improvement methods (deep soil mixing, vibro replacement, vibro compaction, deep dynamic compaction, and earthquake drains). All were compared on the basis of a functional unit of treating 25,000 cubic meters (50 m × 50 m × 10 m) of loose, sandy hydraulic fill, intended for use at a specific site. Deep soil mixing was the most impactful method, mainly due to the use of Portland cement. However, introduction of slag cement reduced greenhouse gas emissions to nearly a quarter of the conventional method. A combination of vibro replacement stone columns, deep dynamic compaction, and vibro compaction resulted in the lowest environmental impact based on greenhouse gas emissions and life cycle energy, as with the previous examples primarily because these methods do not employ manufactured materials.

These studies indicate that ground improvement methods generally can offer environmental benefits as well as technical and cost advantages over other approaches. Systems such as vibro compaction, vibro stone columns, and dynamic compaction, which do not employ manufactured materials, have a clear advantage in many situations. Deep drains, soilfracture grouting, blasting, and freezing use relatively low amounts of manufactured materials and often develop relatively low emissions. Permeation grouting, jet grouting, compaction grouting, and deep soil mixing all use larger quantities of manufactured materials relative to other ground improvement methods, but can still show lower impacts than other geotechnical approaches such as piling, depending on the project. In addition, low impact cements, slag mixes, and other low impact materials are being developed and increasingly used to further enhance the sustainability of such methods.

However, studies also show that each site must be considered on its merits, with the technical requirements and the prevailing ground conditions primarily determining which systems are appropriate, at which stage the most sustainable from among these can be identified.

1.6 COMPACTION BY BLASTING

Densification of granular soils can be achieved by detonating explosives in the ground. Following field tests carried out in the former USSR in the 1930s, deep compaction by blasting became known as a method of ground improvement through publications of Abelev and Askalonov (1957) and Ivanov (1967). The treatment of loose granular soils by blasting is based upon a sudden dynamic load stimulating the grain structure to rearrange, reduce its porosity, and find a closer density. Not unlike the behaviour of granular soils when subjected to dynamic forces, deriving from the impact of heavy weights when dynamic compaction is carried out or during earthquakes, the prerequisite for efficient compaction is full water saturation.

To date the method of compacting loose granular soils below the water table by explosives has been used worldwide with positive results. It can be executed at relatively low costs generally for very large volumes of granular deposits. However, it does not yield such high densities as those achieved by vibro compaction with its scope of application being a moderate but rather homogeneous densification (Smoltczyk 1983). It should be noted that directly at the location of the charge, some heterogeneity or loosening can arise, particularly if the fines content in the soil is high.

The method can be used from irregular surface conditions and from ground unsuitable for heavy plant, but it is restricted to loose granular soils with low silt contents. Only occasionally the method has also been used to compact fine grained soils, such as loess. The depth range of the method is generally beyond the reach of vibro compaction, as has been reported by Solymar et al. (1984). The effective range of the explosion impact determines the placement of the charges in the ground. It is between 10 m for low charges of 10–15 kg of TNT equivalent per bore hole and up to 20 m for charges of 30 kg. Placement of the charges is generally in boreholes often supported by bentonite slurry.

Similar to vibro compaction and the effect of an earthquake, the explosive impact results in shear waves leading to partial or even total liquefaction of the granular deposit and ultimately to a densification of the soil when the pore water overpressure has dissipated. The density that can be achieved by the blasting method depends of course on the explosive energy, the distance of the bore holes containing the charges in the ground, and on the same soil characteristics determining the suitability of granular soils for vibro compaction. The improvement expressed by the increase in relative density D_r is generally ΔD_r = 15%–30%, in exceptional cases more, particularly when the original density is very low. The extension of the compaction reaches 20%–50% deeper than the installation depth of the explosive charge. The success of deep compaction by blasting also depends on the layout of the blast holes and the sequence of the ignitions in multiple blasting. Quality control measures are similar to those applicable for vibro compaction. In the absence of a reliable theory and simple design rules the method application relies on experience and on trial compactions ahead of any contract work. Gohl et al. (2000) have produced a promising theory based on cavity expansion theory and compare with experience on nine projects.

Densification arising from compaction blasting of loose water-saturated granular soils is strongest, whereas the effect on denser deposits tends to be less. This behaviour leads to an equalisation of density and homogenises the granular deposit. Although the method is regarded as an economical means of compaction, its application remains scarce, probably because the use of explosives for subsurface blasting requires special permissions that are not easy to obtain. In addition, the environmental impact is substantial: the emission of noise and far-reaching shock waves needs to be investigated and controlled throughout the project, as does the emission of gases and fumes from the explosion, which are injected into the ground and may contaminate the ground water.

The need to rehabilitate large deposits in the open brown coal mining areas in eastern Germany has resulted in an extensive use of the compaction blasting method in parallel to vibro compaction. It has triggered interesting field trials of blasting and intensive research work on this method of ground improvement (Kolymbas 1992, Raju 1994, Tamaskovics 2000). However, its application has been diminishing at this site due to safety considerations. General descriptions of the deep blasting ground improvement method can be found in Damitio (1970), Mitchell (1981), Kolymbas (1992), Gohl et al. (2000) and Gambin (2004).

1.7 GROUND FREEZING

The artificial freezing of soil has been known for well over 100 years. The method was patented in 1883 by Poetsch in Germany but previously, in 1862, practiced in the UK (Harris 1995, Jessberger and Jagow-Klaff 2003). By this method water-bearing soils are chilled to such an extent that the pore water freezes, providing the frozen soil with considerably higher strength than in its original state and rendering it at the same time impermeable to water. These changed soil conditions are transient and reversible as the soil returns to its original characteristics when thawing, provided that no changes in water content occur, which may happen in cohesive soils due to the development of ice lenses. Soil freezing is therefore the only reversible method of ground improvement, and it requires a continuous supply of energy during its application to maintain the necessary soil temperature and desired state (for example, its stiffness, strength, or impermeability).

The method was originally developed to sink large mining shafts through water-bearing soils. Only relatively recently soil freezing has also been used in tunnel construction and to resolve difficult problems in ground engineering, often as the method of last resort (closure of leakages in water barriers; retrieving of artefacts or valuable machinery from difficult ground conditions). It is also used in soil investigation measures to obtain undisturbed samples of saturated noncohesive soils to measure their density. Very recently the method has also been proposed in environmental engineering to freeze (encapsulate and immobilise) and subsequently safely remove hazardous soil. Its application is generally restricted in time.

Soil freezing is achieved by taking heat away from the ground, generally using either of the two methods:

- Freezing by brine ($CaCl_2$) circulation with refrigerators
- Freezing with liquid nitrogen (LN_2)

Both methods require special piping to be installed in the soil for the introduction of the coolant. Liquid nitrogen is generally used for short time, small volume applications (shock freezing) as the frost body builds up quickly at temperatures of −196°C of the coolant resulting in frozen soil temperatures between −20°C and −30°C, or even deeper. Brine freezing uses coolant temperatures of between −30°C and −40°C, allowing frost body temperatures of −10°C to −20°C to develop. Brine freezing is used in general for large-volume, long-duration projects since the installation costs of the refrigerating system are considerable and the time required to achieve the necessary frost body temperature is longer.

Soil freezing is generally used in water-bearing soils. Ground water flow strongly influences the development of the frozen soil body and needs to be considered in the analysis when it reaches values above 1 m/day, as a rule of thumb. The necessary energy to build up and maintain over time for the required frozen soil body dimensions can be calculated using the heat exchange formulae based upon the thermal characteristics of the soil (thermal capacity and conductivity). The design of these dimensions is based upon structural and heat transfer−related parameters of the frozen and unfrozen soil. Strength and deformation characteristics of frozen soil are time dependent and responsible for the creep developing and resulting from the visco-plastic behaviour of the frozen pore water. The dependency of the frozen soil strength from temperature and soil characteristics (mainly water content, density, grain size distribution, and salinity of the ground water) has been the subject of intensive research in recent years. Recommendations and special publications (such as from the International Symposium on Ground Freezing, ISGF) provide details of the method, including the necessary site and laboratory investigations prior to the design and execution of a soil-freezing project. The thermal and rheological computations to describe the time and temperature dependent stress strain behaviour of the frozen soil body are best carried out using the FE method (Kirsch and Borchert 2008).

Ground water conditions such as seepage flow velocity, temperature, and salinity are important factors of influence for maintaining the integrity of the frost body dimensions and are often decisive in determining the distances between the freeze tubes. Their proper knowledge is essential in optimising the energy cost of any soil freezing project. Quality-control measures are as with other methods of special foundations; they include temperature measurements indispensible to guarantee the integrity of the dimensions of the frost body throughout its maintenance.

It is a well-known phenomenon that freezing of the pore water is accompanied by a volume increase of 9%, which results in frost-susceptible soils, generally of fine contents in excess of 15% in unwanted soil heave. In addition, the development of ice lenses leads to continuous increase of heave and/or stresses. This heave is small with sufficient surcharge and generally does not occur in well-draining granular soils where unfrozen water is expelled by the ice front developing in the pores of the soil.

Soil freezing is a transient ground improvement method leaving behind after completion of the works very little disturbance in the ground (only coolant pipes if not extracted) and in this way can be regarded as almost reversible in restoring original ground conditions, albeit at relatively high energy cost and with a considerable carbon foot print. After thawing is completed, ground conditions return to their original state without any interference with or contamination of soil or aquifer. Access to the method can be found in special publications such as Jessberger and Jagow-Klaff (2003), Harris (1995), or Phukan (1985).

REFERENCES

Abelev, Y. M. and Askalonov, V. V. (1957). The stabilisation of foundations for structures on loess soils. *Proc. V. ICSMFE, Paris.*

Bell, A. L. (1994). *Proc. Conference Grouting in the Ground, Institution of Civil Engineers, London.* Ed. A. L. Bell, Thomas Telford.

Damitio, C. (1970). La consolidation des sols sans cohésion par explosion. *Construction* (France), 25:100–8.

EA. (2010). Carbon calculator for construction v 3.1.1. Environment agency UK spreadsheets and guidance. Jan.

Egan, D. and Slocombe, B. C. (2010). Demonstrating environmental benefits of ground improvement. *Proceedings of the Institution of Civil Engineers, Ground Improvement,* 163(1): 63–9.

Gambin, M. (2004). Densification des sables lâches par explosifs. In: ASEP-GI 2004, Vol. 2, 513–540. Ed. Magnan. Presses de l'ENPC/LCPC. Paris.

GEMIS. (2010). Database version 4.6, German Öko Institut for Applied Ecology, Berlin, Dec.

Glossop, R. (1961). The invention and development of injection processes. *Géotechnique* 10:91–100.

Gohl, W. B. , Jefferies, M. G., Howie, J. A. and Diggles, D. (2000). Explosive compaction design, implementation and effectiveness. *Géotechnique,* 50(6):657–65.

Hall, M. J. (2006). *A Guide to Calculating the Carbon Debt and Payback Time for Wind Farms.* London: Renewable Energy Foundation.

Hammond, G. and Jones C. (2011). Inventory of carbon and energy (ICE) version 2. Database: Sustainable Energy Research Team, University of Bath (UK) January.

Harris, J. S. (1995). *Ground Freezing in Practice.* Thomas Telford.

Ivanov, P. (1967). Compaction of non-cohesive soils by explosions. US Interior Dept. Report No. TT-70-57221.

Jessberger, H. L. and Jagow-Klaff, R. (2003). Ground freezing. In: Smotzcyk, U. (ed.) *Geotechnical Engineering Handbook,* Vol. 2. 117–168. Ernst & Sohn: Berlin.

Joosten, H. (1926). Verfahren zur Verfestigung von Gebirgsschichten. Deutsches Reichspatent Nr. 441622.

Kirsch, K. and Kirsch, F. (2010). *Ground Improvement by Deep Vibratory Methods,* Spon Press: London and New York.

Kirsch, F. and Borchert, K.-M. (2008). Bemessung von Vereisungskörpern bei Sicherungsmassnahmen im innerstädtischen Tunnelbau mit dem Teilsicherheitskonzept. 23rd Ch. Veder Koll., Schriftenreihe der Gruppe Geotechnik Graz. Heft 33.

Kolymbas, D. (1992). Sprengungen im Boden. *Bautechnik* 69(8):424–31.

Littlejohn, G. S. (1993). Chemical grouting. In *Ground Improvement*, Ed. Moseley. Blackie: London. 100–29.

Mitchell, J. K. (1981). Soil improvement. State of the art report. *Proc. X ICSMFE, Stockholm.*

Phukan, A. (1985). *Frozen Ground Engineering*. Upper Saddle River, NJ: Prentice Hall.

Pinske, M. A. (2011). Life cycle assessment of ground improvement methods. MSc Thesis, University of California, Davis.

Raju, V. R. (1994). Spontane Verflüssigung lockerer granularer Körper-Phänomene, Ursachen, Vermeidung. Diss. University Fridericiana Karlsruhe.

Smoltczyk, U. (1983). Deep compaction. General Report. *Proc. VIII ECSMFE, Helsinki*, Vol. 3.

Solymar, Z. V., et al. (1984). Earth foundation treatment at Jebba Dam site. *Journal of Geotechnical Engineering* 110(10):1415–30.

Tamaskovics, N. (2000). Beitrag zu Klärung der Mechanismen von Verdichtungssprengungen. Diss. Technische Universität Bergakademie Freiberg.

Wintzingerode, W., Zöhrer A., Bell, A. L. and Gisselmann, Q. (2011) .Calculations on greenhouse gas emissions from geotechnical construction processes. *Geotechnik* 3:218–21.

Zöhrer, A., Wehr, J. and Stelte, M. (2010). Is ground engineering environmentally friendly? Ecological balance of foundation engineering methods. 11th International EFFC-DFI conference, session 3: sustainability in the foundation industry, London, 26 to 28 May.

Chapter 2

Deep vibro techniques

Jimmy Wehr and Wolfgang Sondermann

CONTENTS

2.1 INTRODUCTION AND HISTORY

For over 70 years, depth vibrators have been used to improve the bearing capacity and settlement characteristics of weak soils. Vibro compaction is probably the oldest dynamic deep compaction method in existence. It was introduced and developed to maturity by the Johann Keller Company in 1936, which enabled the compaction of noncohesive soils to be performed with excellent results. A detailed description of the method from its beginnings up to the pre-war period is given by Schneider (1938) and by Greenwood (1976) and Kirsch (1993) for the period thereafter.

This original process, now referred to as vibro compaction, has since been applied successfully on numerous sites around the world. When carrying out compaction work using the vibro compaction method in water-saturated sands with high silt content, these sands, when lowering the depth vibrator and during subsequent compaction, are liquefied to such an extent that the compaction effect only occurs after a very long vibration period or it does not occur at all. In such soils, the vibro compaction method reaches its technical and economic limits.

In 1956, a technique to insert the vibrator into the soil without the aid of simultaneously flushing in water was developed to overcome the limitations of the vibro compaction method. After the vibrator is lifted, the temporarily stable cylindrical cavity is filled with coarse material, section by section. The coarse material is then compacted by repetitive use of the vibrator. This vibro replacement procedure came to be known as the conventional dry method. Such technical developments in dense stone column construction allowed for a greater range of treatable weak natural soils and man-made fills. Vibro replacement continues to be widely used in Europe to improve weak soil. It has a reputation for providing stable ground which allows for safe and economic construction of residential and light commercial and industrial structures.

The conventional dry method utilises the vibrator to displace the surrounding soil laterally rather than for primary compaction of the original soil. The crushed stone is pressed laterally into the soil during both the cavity-filling stage and compaction stage. This produces stone columns that are tightly interlocked with the surrounding soil. Groups of columns created in this manner can be used to support large loads. The conventional dry method reliably produces stone columns to depths of 8 m in cohesive soils that have a shear strength of at least 20 kN/m^2.

Bottom feed vibrators, which introduce the stones through the vibrator tip during lift, are used to overcome the disadvantage of possible cavity collapse that can occur with the conventional dry method in cohesive soils with high water content. During withdrawal of the vibrator, stone and compressed air are delivered through the vibrator tip, preventing cavity collapse. This method is known as dry vibro replacement. In 1972 this method was patented in Germany.

Reliable stone column production by vibro compaction in cohesive soils with high water content is achievable with the aid of a heavy water jet. Water is jetted from the vibrator tip as the vibrator is lowered to the desired depth. Mud flushes loosened soil and rises to the surface, stabilising the cavity. This is known as the wet vibro replacement method.

After the bottom feed system was developed in 1976, it was possible to install injected stone columns by means of an injection of a cement-bentonite suspension near the bottom of the vibrator (Jebe and Bartels 1983). The voids of the stone column skeleton are thereby filled with this suspension. Finally, vibro concrete columns were developed using a conventional concrete pump to deliver the concrete to the bottom of the vibrator via the tremie system.

In very soft nearly liquid soils, vibro replacement is not applicable due to the lack of lateral support of the soil. A geotextile coating may be used around the column to ensure filter stability and to activate tensile forces to avoid lateral spreading of the column. This method was developed in 1992 and first applied in early 1993 for a dam project in Austria (Keller 1993). A compilation of various projects with geotextile-coated columns may be found in Sidak et al. (2004).

These techniques have been chosen for many major structures in the United States and Europe, endorsing their value in promoting safe and economic foundations to a wide range of buildings and soil conditions. Probably the oldest recommendation on the use of vibro was issued by the German transport research society in 1979 (FGFS 1979). Later, the US Department of Transportation published the manual *Design and Construction of Stone Columns* (USDT 1983), followed by the British ICE *Specification for Ground Treatment* (ICE 1987) and the BRE publication *Specifying Stone Columns* (BRE 2000). The latest effort has been made by the European community to standardise the execution of vibro works in *Ground Treatment by Deep Vibration* (European Standard EN14731, 2005).

2.2 VIBRO PROCESSES

The operational sequence of the vibro compaction method is illustrated in Figure 2.1. During operation, the cylindrical, horizontally vibrating depth vibrator is usually suspended from a crane or like equipment. It weighs 15–40 kN, with a diameter of 30–50 cm and a length of 2–5 m. Details are provided by Kirsch and Kirsch (2010). The vibrator reaches application depth by means of extension tubes.

The vibrator shell is constructed of steel pipe, forming a cylinder. Eccentric weight(s) in the lower section are powered by a motor at the top end of a vertical shaft within the vibrator. Energy for the motor is supplied through the extension tubes. The rotational movement of the eccentric weights causes vibrations of the vibrator. The vibratory energy is transferred from the vibrator casing to the surrounding soil. This energy affects the surrounding soil without being dependent on the vibrator's depth of operation. A vibration damping device between the vibrator and extension tubes prevents the vibratory energy from being transmitted to the extension tubes. Supply pipes for water and air (optional) are also enclosed in the extension tubes. The pipes can deliver their payload through the vibrator tip as well as through special areas of the extension tubes to aid the ground penetration action of the vibrator.

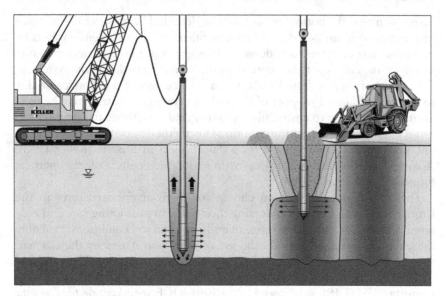

Figure 2.1 Vibro compaction method operating phases. (Courtesy of Keller Group.)

During vibro compaction, the motor runs as the depth vibrator is inserted into the soil (Figure 2.1). The insertion is aided by water flushing. Field experience has shown that penetration is more effective when a larger volume of water is used rather than a higher pressure. The water flow will expel some loosened sand through the annulus around the vibrator. The granular soil targeted for compaction sees a fast reduction in temporary excess pore water pressure. At compaction depths greater than 25 m, additional flushing lines and compressed air may need to be utilised.

The water and air flows are normally stopped or reduced after the vibrator arrives at its specified depth and the compaction process stages have been initiated. Field experience has determined that lifting the vibrator in stages of 0.5 m or 1.0 m after 30–60 seconds of application tends to produce the best results. During the compaction process, granular material adjacent to the vibrator sees a reduction in pore volume, which is compensated for by introducing sand via the annulus. It is possible for settlement of the surface to range from 5%–15% of the compaction depth. This range depends on the density prior compaction, as well as the targeted degree of compaction. After the initial insertion and compaction processes have been completed at a particular location, the vibrator is moved to the next location and lowered to the depth specified for compaction.

Compacted soil elements with specified diameters can be created by performing the compaction procedures in grid patterns. Open pit brown coal mining areas, such as those in the eastern part of Germany, have had vibro compaction performed at depths of 65 m. Typically, the layout of compaction probe centres is based on an equilateral triangle. A distance between 2.5 and 5.0 m usually separates the centres. This distance is determined by grain crushability (shell content), required density, vibrator capacity, and grain-size distribution of the sand. The production stage of extensive projects can be greatly enhanced if a comprehensive soil study is done, with the added benefit of a test programme prior to going out to tender/bidding on the project. Guideline values for the strength properties of sand, which can aid the design of such projects, are displayed in Table 2.1.

Currently, depth vibrators are used to produce vibro stone columns in cohesive soils that exhibit low water content. For this production variant to be successful, the soil consistency must be able to hold the form of the entire cavity after the vibrator has been removed. This allows for the subsequent repeated delivery and compaction of stone column material to proceed uninhibited by obstruction. With the dry or displacement method, the soil cavity is prevented from collapsing by the compressed air being released from the vibrator tip.

An alternative method to construct vibro stone columns in cohesive soils with high water content involves the use of a strong water jet that ejects water under high pressure from the vibrator tip. The cavity is stabilised by the mud that rises to the surface and flushes out loosened soil. The cavity is then filled in stages, through the annulus, with coarse fill, which surrounds the vibrator tip and is compacted into the stone column form as the vibrator is lifted. This is known as the wet/replacement method. A mud, or 'spoil,' containing high quantities of soil particles is transported to specially designed settling tanks, or ponds, by way of trenches. This procedure is complicated and can be messy, but it is important to separate the water and mud from the operations area, where it is easily accessed when the time comes to discharge it (Kirsch and Chambosse 1981).

Table 2.1 Guideline values for the strength properties of sand

Density	Very loose	Loose	Medium dense	Dense	Very dense
Relative density I_D [%]	<15	15–35	35–65	65–85	85–100
SPT [N/30 cm]	<4	4–10	10–30	30–50	>50
CPT q_c [MN/m²]	<5	5–10	10–15	15–20	>20
DPT (light) [N/10 cm]	<10	10–20	20–30	30–40	>40
DPT (heavy) [N/10 cm]	<5	5–10	10–15	15–20	>20
Dry density γ_d [kN/m³]	<14	14–16	16–18	18–20	>20
Modulus of deformation [MN/m²]	15–30	30–50	50–80	80–100	>100
Angle of internal friction [o]	<30	30–32.5	32.5–35	35–37.5	>37.5

Source: Kirsch, K. (1979). Geotechnik 1:21–32.

Note: After completion of the vibro compaction work, it may be necessary to re-compact the working surface down to a depth of about 0.5 m using surface compactors.

Grain diameters of the stones and gravel which comprise the fill material for the wet method range from 30 to 80 mm. Stone column installation to depths as great as 43 m has been reported (Wehr 2008). The wet method guarantees stone column continuity for a wide range of soft soils.

Grain diameters of the stones or gravel that comprise the fill material when using a bottom feed vibrator typically range from 10–40 mm. The fill is delivered to the vibrator tip by means of a pipe. After the vibrator arrives at the specified depth, compressed air is used to help deliver the fill as the vibrator is subsequently lifted in stages as it compacts the fill (Figure 2.2).

Carrier equipment typically consists of specially designed machines, known as vibrocats, which have vertical leaders. The vibrocats control the complex bottom feed vibrators, equipped with material lock and storage units, which deliver fill material to the vibrator by means of specialised mechanical or pneumatic feeding devices (Figure 2.3).

Vibro cats possess a particular feature which is an additional downward force (Figure 2.4). This so-called 'activation force' of approximately 150 kN causes a better vertical compaction of the column material and a repeated vertical loading. Furthermore, it is possible to increase the column diameter easily because the vertical action of the vibrator tip leads to a horizontal displacement with efficient compaction.

The installation of vibro mortar columns is similar to the dry vibro replacement method apart from the cement suspension filling the voids of the stone skeleton inside the column. This results in a much stiffer column compared to a conventional vibro stone column.

For the installation of vibro concrete columns, the tremie system is connected to a mobile concrete pump. Before penetrating, the system is charged with concrete. The vibrator then penetrates the soil until the required depth has been achieved. The founding layer, if granular, is further compacted by the vibrator. Concrete is pumped out from the base of the tremie at positive pressure. After raising the vibrator in steps, it re-enters the concrete shaft, displacing it into a bulb until a set resistance has been achieved. Once the bulb end is formed, the vibrator is withdrawn at a controlled rate from the soil while concrete continues to be pumped out at positive pressure. Once completed, the column can be trimmed and reinforcement placed as required.

Vibro geotextile columns consist of a sand or stone core with a geotextile coating. The advantage of a vibro geotextile column to other geotextile columns (Schüßler 2002) is the well-densified granular infill resulting only in small settlements of the soil-column system (Trunk et al. 2004).

The installation is usually performed in several steps in order not to damage the geotextile. First a hole is created with the vibrator to the required depth and the vibrator is extracted. In the next step the geotextile is mounted over the vibrator above the ground surface, and subsequently the penetration is repeated with the geotextile to the same depth as before.

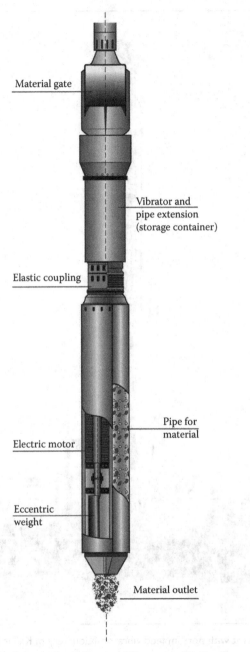

Material gate

Vibrator and
pipe extension
(storage container)

Elastic coupling

Pipe for
material

Electric motor

Eccentric
weight

Material outlet

Figure 2.2 Details of a bottom feed vibrator. (Courtesy of Keller Group.)

Figure 2.3 Vibro cat with bottom feed vibrator. (Courtesy of Keller Group.)

Figure 2.4 Vibro cat with activation force. (Courtesy of Keller Group.)

On the way up, it is preferable that stones are filled and densified inside the geotextile like in the usual dry bottom feed process.

If there is only one certain very soft layer it is possible to first build a vibro stone column below this layer, insert a vibro geotextile column or a vibro mortar column only in the very soft layer for economical reasons, and finish the upper part of the column as an ordinary vibro stone column.

2.3 VIBRO PLANT AND EQUIPMENT

The equipment developed for the vibro compaction and vibro replacement processes comprises four basic elements:

1. The vibrator, which is elastically suspended from extension tubes with air or water jetting systems
2. The crane or base machine, which supports the vibrator and extension tubes
3. The stone delivery system used in vibro replacement
4. The control and verification devices

The principal piece of equipment used to achieve compaction is the vibrator (Figure 2.5). The drive mechanism can be an electric motor or a hydraulic motor, with the associated generator or power pack usually positioned on the crawler rig in the form of a counter weight.

The typical power range in vibrators is 50–150 kW, and can go as high as 300 kW for the heaviest equipment. Rotational speeds of the eccentric weights in the cases of electric drives are determined by the frequency of the current and the polarity of the motor. For example, 3,000 rpm or 1,500 rpm vibrating frequency are obtainable from a 50 Hz power source, and 3,600 rpm or 1,800 rpm vibrating frequency from a 60 Hz power source with a single or double pole drive, respectively. A 5% reduction in the frequency applied to the ground occurs, corresponding to the magnitude of the 'slip' experienced with asynchronous motors. The use of frequency converters has recently become economical as a result of modern control technology. The frequency converters enable limited variation of the operating frequency of the electric motors.

During rotation, the eccentric weight generates horizontal force (see Figure 2.5). This horizontal force is transmitted to the ground through the vibrator casing and (depending on the vibrator type) ranges from 150–700 kN. Details are provided by Kirsch and Kirsch (2010). When the vibrator is freely suspended with a lack of lateral confinement, the vibration width (double amplitude) totals 10–50 mm. Acceleration values of up to 50 g are obtainable at the vibrator tip. It is practically complicated to measure crucial operational data during the compaction process. Therefore, any data

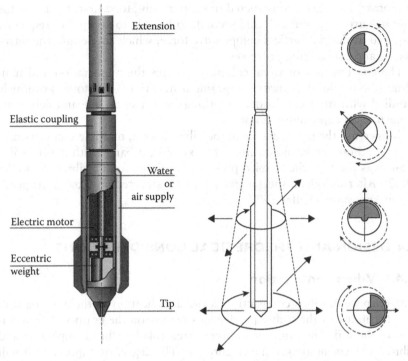

Extension

Elastic coupling

Water
or
air supply

Electric motor

Eccentric
weight

Tip

Figure 2.5 Depth vibrator and principle of vibro compaction. (Courtesy of Keller Group.)

given on vibrators apply to those which are freely suspended, lacking lateral confinement.

It is up to the designer to create a vibrator optimal for the specific application. One major challenge of design lies with keeping maintenance costs within standards that are economically tolerable. Based on field experience, the most effective compaction of sands and gravels is done by vibrating frequencies which approach the natural soil-vibrator system frequency, or 'resonance' for elastic systems, which ranges between 20–30 Hz (Wehr 2005).

Fellin (2000), who considered vibro compaction a 'plasto-dynamic problem,' has confirmed theoretically knowledge gained from practical vibro operation conditions. Fellin's goal, by constant analysis of information obtained on the vibrator movement during compaction performance, was to create 'on-line compaction control.' His work's theoretical results confirm the observation that when using a constant impact force, the vibration's effect range increases as the vibrator frequency decreases, whereas compaction increases when the impact force increases.

The thickness of soil depths to be treated determines the overall length of vibrator, extension tubes, and lifting equipment, which in turn determines the size of crane to be used. Purpose-built tracked base machines (vibrocats) have been constructed to support vibrators: first, to ensure the columns are truly vertical, and second, to be able to apply the frequently required or desired vertical compressive force, which accelerates the introducing and compacting processes.

The construction of stone columns requires the importation and handling of substantial quantities of granular material. This stone is routinely handled with front end loaders, working from a stone pile and delivering stone to each compaction point.

To increase the performance of the vibro system, multiple vibrators may be applied on one base machine. For example, a barge with a 120–150 t crane was used for the Seabird project in India with four vibrators (Keller 2002). Alternatively, a special frame was constructed on a barge suspending five vibrators (Keller 1997).

2.4 DESIGN AND THEORETICAL CONSIDERATIONS

2.4.1 Vibro compaction

The purpose of vibro compaction is the densification of the existing soil. The feasibility of the technique depends mainly on the grain-size distribution of the soil. The range of soil types treatable by vibro compaction and vibro replacement are given in Figure 2.6. The degree of improvement will depend on many more factors including soil conditions, type of equipment, procedures adopted, and skills of the site staff. Such variables do not permit

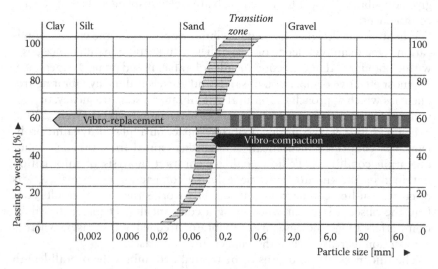

Figure 2.6 Range of soil types treatable by vibro compaction and vibro replacement (stone columns).

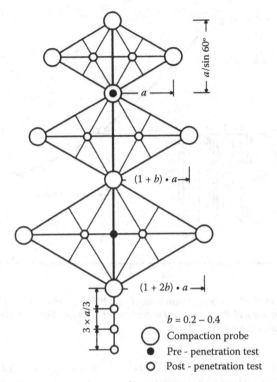

Figure 2.7 Trial arrangement for vibro compaction. (From Moseley, M.P. and Priebe, H.J. (1993). In: Moseley, M.P. (ed.) *Ground Improvement*, London, England: Blackie Academic & Professional.)

an optimum design to be established in advance but rather require the exercise of experience and judgement for their successful resolution.

For small projects, the design of vibro compaction work can be based on the experience of the contractor. For large projects it is preferable and advisable to conduct a trial in advance of contract works. A typical layout of vibro compaction probes for a trial is given in Figure 2.7. The trial allows for three sets of spacings between probes, together with pre- and post-compaction testing, often performed using cone penetration testing equipment. The degree of improvement achieved can be used to optimise the design, as shown in Figure 2.8.

The technical success of vibro compaction work is measured by the level of densification achieved against a specified target. The densification can be readily checked using standard penetration tests or, preferably, cone penetration tests. Comparisons can be made between pre- and post-compaction testing, and care should be taken to ensure that the same techniques of testing are used in each situation. Control of performance is a further important element in carrying out vibro compaction work. This is best achieved

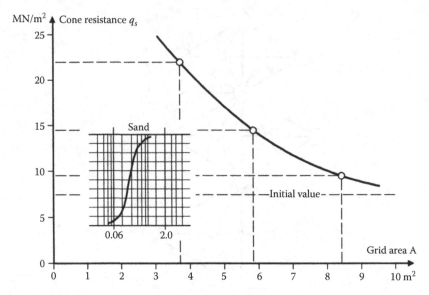

Figure 2.8 Results of vibro compaction trial. (From Moseley, M.P. and Priebe, H.J. (1993). In: Moseley, M.P. (ed.) *Ground Improvement*, London, England: Blackie Academic & Professional.)

by using a standardised procedure, established at the precontract trial, such as predetermined lifts of the vibrator at predetermined time intervals and/ or predetermined power consumptions. Only such a regular procedure can reveal whether variations in test results are due to the inherent inhomogeneity of the soils being treated or by insufficient compaction.

The soil being treated, the degree of densification required, the type of vibrator being used, and production rates all have an influence on the spacing of vibro compaction probes. Areas treated per probe vary commonly between 6–20 square metres. Vibrator development over the past decade has allowed considerable increases in the area treated by each insertion of the vibrator. This development continues and will enable further expansion of the treatment envelope.

Sands and gravels bearing negligible cohesion are compatible with vibro compaction. The silt (grain size <0.06 mm) percentage of such soils should be less than 5% for ideal performance. Compaction is substantially hindered by clay particles (grain size <0.002 mm) to the point that the procedure is unable to be performed without extra measures, including the introduction of coarse-grained fill. Reference to the grain-size distribution diagram (Figure 2.6) usually determines application limits. However, application limits for material that is very coarse are typically determined empirically, taking into consideration the penetration effectiveness of the respective vibrator. Static cone penetration tests can also serve to estimate values of soil compatibility for compaction methods. Given that the local skin friction-to-point resistance (friction ratio) falls between 0 and 1 and the point resistance is a minimum of 3 MPa, the soil can be considered to be compatible (Massarsch 1994).

The efficiency of compaction is also greatly influenced by the permeability of the soil. When permeability is too low ($<10^{-5}$ m/s), compaction effectiveness decreases as permeability decreases, whereas when permeability is too high ($>10^{-2}$ m/s), penetration of the soil by the vibrator becomes increasingly more difficult as the permeability increases (Greenwood and Kirsch 1983).

The carbonate or shell content is important for the densification of highly compressible soils with low cone resistance and high friction ratio. Cemented soils are not considered here.

Correlations between the CPT cone resistance and the relative density are well established for silica sand. Unfortunately there are not many references concerning this correlation for calcareous sands. Vesic (1965) added 10% of shells to quartz sand, which resulted in a decrease of the CPT cone resistance by a factor of 2.3. Bellotti and Jamiolkowski (1991) compared CPT cone resistances q_c(silica)/q_c(shells) = $1 + 0.015(D_r\text{-}20)$ yielding ratios between 1.3 and 2.2 increasing with relative density D_r. Almeida et al. (1992) compared normalised CPT cone resistances of calcareous Quiou sand and silica Ticino sand, which yielded ratios from 1.8–2.2 proportionate to increasing relative density. Foray et al. (1999) compared pressuremeter limit pressure of silica sand and carbonate sands, which resulted in ratios ranging from 2–3 proportionate to increasing initial vertical stress. Finally Cudmani (2001) looked at normalised cone resistances of seven sands yielding ratios between 1.4–3.5 depending on initial soil pressure and relative density. Meier (2009) executed systematic calibration chamber tests to investigate the influence of different silica/carbonate sand ratios including the influence of coarse material. This concept has been applied to the Palm Island projects in Dubai by Wehr (2005a).

2.4.2 Vibro replacement stone columns

The reduction of consolidation time and compressibility, and the increase of load-bearing capacity and shear strength, determine the effect of vibro replacement in soft fine-grained soils. The in-situ soil characteristics, the placement and geometry of the stone columns, and the soil-mechanical properties of the column composition are what determine the scale of ground improvement achieved. Aside from the settlement rate increase (generated by the stone column's drainage effect), the reduction of overall settlement is the goal of the vibro column installation. Quite simply, stone columns are effective in reducing settlement since they are stiffer than the surrounding soil. Between stone columns and the ground, the effective stiffness ratio relies considerably on lateral support provided by the surrounding soil when the stone columns have loads put upon them. In order to mobilise the lateral support and generate the interaction between the soil and columns, a horizontal deformation is required. This deformation inevitably causes settlement at the ground surface. Bell (1915) relays the most simplistic relationship for calculating load-bearing behaviour. A maximum lateral support of $\sigma_h = \gamma z + 2\,c_u$ can be provided by the adjacent cohesive soil possessing a cohesion c_u at depth z. If it is assumed that the passive earth pressure coefficient $K_p = \tan^2(\pi/4 + \phi/2)$ is used, then the above supporting pressure allows a maximum vertical column stress of $\sigma_o = K_p\,(\gamma z + c_u)$, with ϕ being the angle of the internal friction of the column material (Figure 2.9). This equation, while underestimating the column's load-bearing capacity, still conveys the significance of column and ground interaction. The equation also reveals the differences in load-bearing behaviours of stone columns when compared to load-carrying elements of greater stiffness.

Minimum shear strength of ground proposed for improvement used to be frequently given in the form of a very conservative c_u value of 15 kN/m² (AUFS 1979, Smoltczyk and Hilmer 1994). It must be noted that no attention is given to the positive effects of the three-dimensional behaviour, the influences of adjacent columns, the dilatation of column material (Van Impe and Madhav 1992), and most importantly, the rapid increase in the soil's shear strength owed to the stone column's drainage effect. As a consequence of these matters, the successful production of foundations in much softer soils with $c_u \geq 4$ kN/m² via vibro replacement has been achievable (Raju and Hoffmann 1996, Wehr 2006). Many model tests have been conducted in order to more clearly grasp the column/soil interactions and the influences of adjacent columns (Hu 1995). In qualitative terms, these tests show the failure mechanism on one side and the group effect on the other (Figure 2.10).

With ultimate vertical load, the failure of stone columns is a result of relatively low lateral support in the upper third (bulging), or the column toe

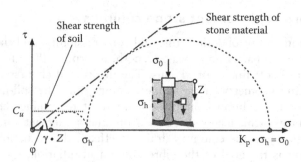

Figure 2.9 Influence of lateral support on column stress. (From Brauns, J. (1978). *Bautechnik* 55(8):263–271.)

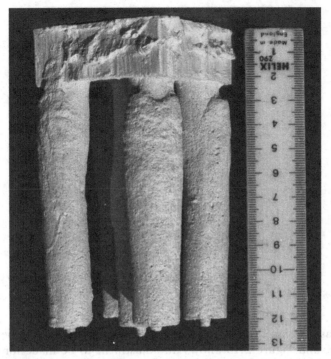

Figure 2.10 Failure mechanism of vibro replacement stone columns in the case of group effect. (From Hu, W. (1995). *Physical Modelling of Group Behaviour of Stone Column Foundations*, PhD Thesis, University of Glasgow.)

being punched into the underlying soil, such as with 'floating' foundations (Figure 2.11). However, such high rates of deformation precede the failure in every case that the column's serviceability is generally no longer provided. Therefore, we can conclude that the equations used to calculate the deformation, or 'serviceability state' of the discussed foundation, are much more relevant than the outcome of limit load assessment of stone columns.

Soyez (1987) and Bergado et al. (1994) have conducted a thorough overview of the various design methods. The authors show the distinction between calculating single columns and calculating column grid patterns. In Europe, Priebe's (1995) design method for vibro replacement stone columns has gained acceptance as a valid method (Figure 2.12).

Thus, in Figure 2.12, the improvement factor, depending on angles of internal friction of the stone column, is related to the ratio of the stone column area and the area being treated by the column. The improvement factor indicates how many times the compression modulus increases for a grid of stone columns and to what extent the settlement of a raft foundation will be reduced. Angles of internal friction are usually higher than 45° (Herle et al. 2008).

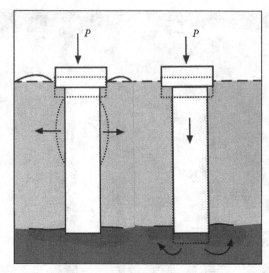

Figure 2.11 Failure mechanism of vibro replacement stone columns under vertical load. (From Brauns, J. (1978). *Bautechnik* 55(8):263–271.)

The basic design curves assume the stone column material to be incompressible, and Figure 2.13 allows an adjustment to be made for this by plotting a fictitious area ratio, which has to be added to the actual area ratio, against the compression modulus ratio for soil and stone column material.

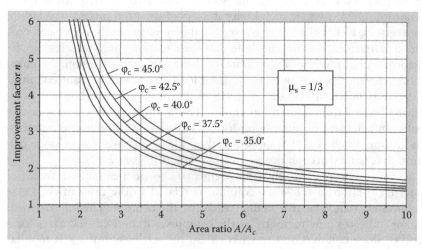

Figure 2.12 Design diagram for improving the ground by vibro replacement stone columns. (From Priebe, H.J. (1995). *Ground Engineering*, December, pp. 31–37.)

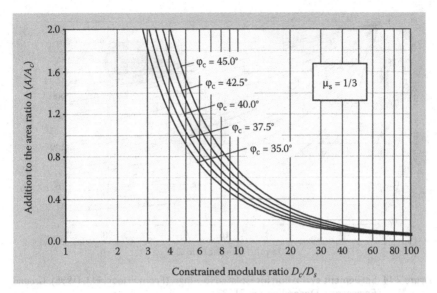

Figure 2.13 Area ratio addition. (From Priebe, H.J. (1995). *Ground Engineering*, December, pp. 31–37.)

With regard to settlement performance, theoretical approaches predominately refer to an infinite grid of columns. Load tests executed in practice on footings resting on small numbers of columns do not fulfill the assumptions. Accordingly, evaluations of settlement performance of a footing on a limited number of stone columns are only approximations.

Practical design charts that consider load distribution as well as reduced lateral support on columns situated underneath footing edges have been presented by Priebe (1995). These charts allow the estimation of settlement of a rigid foundation on a limited number of stone columns as a function of the settlement of an infinite raft supported by an infinite grid of columns, as outlined above.

The method presupposes that the footing area attributed to a stone column and the foundation pressure are identical. There exists an optimum layout for a given number of stone columns beneath a footing. However, in practical applications it is sufficient to determine the grid size required for the calculation by dividing the footing area by the number of columns. The main chart to use in the evaluation of load tests is shown in Figure 2.14. The application is relatively simple as the relevant settlement ratio depends on the number and diameter of the stone columns together with the treatment depth considered.

The United States has seen wider use of Goughnour and Bayuk's (1979) iteration method, even though it is generally considered much more complex. A great number of these calculations are derived from empirical or

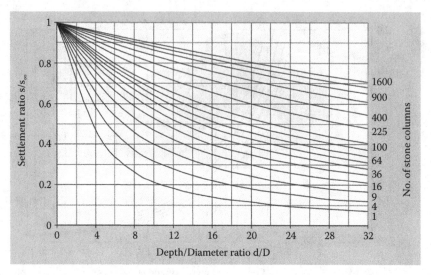

Figure 2.14 Settlement evaluation for isolated footings. (From Priebe, H.J. (1995). *Ground Engineering*, December, pp. 31–37.)

semi-empirical equations involving simplifying assumptions that do not effectively address the deformation behaviour's intricacies. There is currently a lack of an acceptable design method to adequately account for all mechanisms that take part in the load transfer process and is also simple enough for practical use. Therefore, it is best, before making final decisions for the execution of designs for sizeable ground improvement projects, to install test columns and use the achievable column diameters with the load test results to ensure an effective outcome (Chambosse and Kirsch 1995).

When determining the stress/deformation behaviour in the service load range, simulation calculations, such as the finite element method (FEM) often used in construction, are known to be highly effective. As for vibro replacement stone columns for ground improvement, Schweiger has proposed a method that utilises a homogenised model, the so-called 'ground/column matrix' (Schweiger 1990). Wehr (2004, 2006a) has produced noteworthy results regarding the simulation of the failure mechanisms of stone columns by use of his calculations for single columns and column groups for rigid and flexible footings. Brauns' proposed failure modes (1980) and Hu's (1995) model tests (Figures 2.11 and 2.10, respectively) have since been recalculated, resulting in the confirmation that shear zones dictate the settlement behaviour of columns. The influence of the different design parameters has been intensively investigated by Kirsch (2004). Currently, numerical analysis by means of FEM has gained acceptance as a valuable tool in designing stone column ground improvement when large projects are in the design phase or current concepts need optimisation.

The phenomenon of liquefaction of granular deposits during an earthquake has been well documented and increasingly studied. Engineering opinion has agreed that the role of liquefaction can be minimised by densifying the soils beyond their liquefaction potential for the site-specific design earthquake. A second method of minimising the role of liquefaction requires the provision of drainage paths, thus allowing rapid dissipation of pore pressure induced by an earthquake. The influence of the drainage capabilities of stone columns have been studied by Baez (1995), concluding that the Seed and Booker (1976) model is useful if allowable maximum pore pressure ratios are maintained below 0.6. Further investigations of in-situ stone columns composition indicated that in sands the columns generally have an 80/20 proportion (gravel to sand) due to the installation process.

A combination of vibro replacement and vibro compaction, where dense permeable stone columns are constructed and the density of the surrounding granular soil is increased, provides an excellent solution to liquefaction problems. Since its first application at Santa Barbara, California, in 1974 (Engelhardt and Golding 1975), it has been used many times. Perhaps the most significant are the documentations of the performance of the Santa Barbara project (Mitchell and Huber 1983) following a seismic event which induced ground accelerations equal to the design earthquake and the study of 15 sites in the San Francisco area (Mitchell and Wentz 1991) following the Loma Prieta earthquake. In the latter study, the sites treated by vibro techniques and the buildings founded on them were shown to have suffered no damage during the Loma Prieta earthquake.

The acceleration rates that affect the soil in the immediate vicinity of the depth vibrator greatly surpass those experienced in seismic events. On one project, peak ground accelerations of 1.7 g were detected 0.9 m from the stone column's centre (Baez and Martin 1992). As acceleration increases, the soil's shear strength is reduced. In saturated sand, complete liquefaction is possible in the event that the increase in pore water pressure generated by the vibrations surpasses the decrease in pore pressure which is naturally caused by filtration/dissipation (Greenwood and Kirsch 1983).

As long as the treatment medium consists of uniform coarse-grained sands and gravels with a minimum relative density of 80%, the following are attainable: acceptable load-bearing capacities, marginal settlement risk, and assurance against liquefaction induced by seismic events (Smoltczyk and Hilmer 1994). As the percentage of fines increases, higher densities become increasingly difficult to achieve. Therefore, when working in uniform fine-grained or silty sands, it is beneficial to install stone columns that enhance the drainage capacity. Cohesive soils appear to be more resistant to liquefaction than clean sands, but liquefaction is possible as well under seismic action of relatively long duration and high intensity. Many detailed site examples are given by Perlea (2000).

It is not possible to estimate, by statistical analyses, the extent to which the risk of liquefaction is reduced by vibro replacement. The key question is which part of the forces exerted by an earthquake is borne by the columns without any damages. The simple procedure for the design of vibro replacement by Priebe (1995) was modified to account for short-term seismic events (Priebe 1998). In this case it is more realistic to consider deformations of the soil with the volume remaining constant; that is, to calculate with a Poisson's ratio of 0.5, which also simplifies the formulae. In the above-mentioned procedure the improvement factor n_0, which is the basic value of improvement by vibro replacement, is determined initially using

$$n_0 = 1 + \frac{A_c}{A} \cdot \left[\frac{1}{K_{ac} \cdot (1 - A_c/A)} - 1 \right]$$

and

$$K_{ac} = \tan^2(45° - \varphi_c/2)$$

where

  A = attributable area within the compaction grid
  A_c = cross-section of stone columns
  ϕ_c = friction angle of column material.

The reciprocal value of this improvement factor is merely the ratio between the remaining stress on the soil between the columns p_s, and the total overburden pressure p taken as being uniformly distributed without soil improvement and, as such, can be used as a reduction factor $\alpha = 1/n_0$. On the understanding that the loads taken by the columns from both the structure and the soil do not contribute to liquefaction, it is proposed to use this factor to reduce the seismic stress ratio created by an earthquake and hence evaluate the remaining liquefaction potential according to Seed et al. (1983). A similar approach was proposed by Baez (1995) substituting the above K_{ac} with a ratio between the shear modulus of the soil and the stone column.

It is important to mention that excess pore water pressures play an important role in reducing the effective stresses but are neglected in the conventional above-mentioned design. A novel liquefaction approach including pore water pressures was applied by Cudmani et al. (2003) to two sites, one of them being Treasure Island influenced by the 1989 Loma Prieta earthquake. Liquefaction was predicted in a concentrated zone comprising both the bottom of a fine sand top layer and an underlying upper part of a silty sand layer. Mitchell and Wentz (1991) reported on the medical building in Treasure Island where the upper fine sand layer was improved with stone columns to a depth of 6.5 m, leaving the lower layer unimproved. This resulted in no liquefaction of the improved soil block but in liquefaction of the silty sand layer below 6.5 m, which was proved by the observation that the bottom 2.5 m of the 6.5-m-deep elevator shafts drilled prior to the earthquake were filled with silty sand. Furthermore, sand boils were clearly visible outside the improved area.

Another aspect is the design of the extent of soil improvement against liquefaction. An overview is given in Japanese Geotechnical Society (JGS 1998). There are two basic questions to be answered about the necessary width and depth of the soil improvement outside the loaded area.

Pore water pressures are transmitted from the liquefied area into the improved area of the ground. It is recommended (JGS 1998) to improve a lateral area corresponding to an angle of 30 degrees against the vertical axis starting from the edge of the foundation (point A in Figure 2.15). This shall be executed down to a nonliquefiable layer. The area ACD in the same Figure 2.15 exhibited particular unstable behaviour during model tests, and hence this part should be treated as liquefied in the soil improvement design.

Design guidelines for oil tanks in Japan (JGS 1998) recommend improving an area adjacent to the footing corresponding to 2/3 of the soil improvement depth, Figure 2.16. Recent research on sand drains for liquefaction remediation yields the lateral extent to be taken as the liquefiable depth (Brennan and Madabhushi 2002).

In many design codes and standards, a maximum treatment depth between 15–20 m is given according to experience. A special design chart

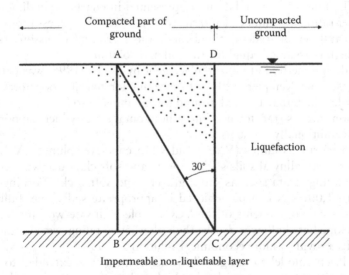

Figure 2.15 Stabilised area (ABCD) adjacent to foundation. (From Japanese Geotechnical Society. (1998). *Remedial Measures against Soil Liquefaction, from Investigation and Design to Implementation.* Rotterdam, Netherlands: A. A. Balkema.)

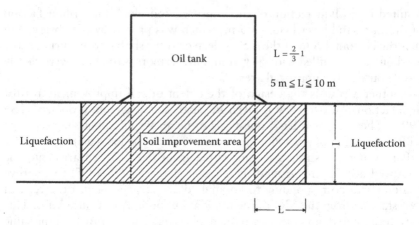

Figure 2.16 Soil improvement area for oil tanks. (From Japanese Geotechnical Society. (1998). *Remedial Measures against Soil Liquefaction, from Investigation and Design to Implementation.* Rotterdam, Netherlands: A. A. Balkema.)

is available for light-weight and small-scale structures to improve a limited depth leaving a liquefiable soil layer below (JGS 1998).

The time-dependent behaviour of sand or gravel drains may be analysed using charts proposed by Balaam and Booker (1981). This is an extension of the Barron solution for excess pore water pressure using the approximate diffusion theory for consolidation based on Biot's equation of consolidation. The rates of consolidation are presented in charts depending on the diameter ratio of the unit cell and the column d_e/d, the stiffness ratios of the column and the soil under drained conditions E_1/E_2 and a Poisson's ratio of 0.3, which is assumed equal for the soil and column.

The design method of vibro replacement by Priebe (1995) was extended by Raithel and Kempfert (2000) to account for tensile hoop forces in a geotextile which may be used around the columns in soils with $c_u < 4$ kPa. The hoop force is transformed into a horizontal stress which supports the column additionally to the soil.

Vibro mortar columns (VMC) and vibro concrete columns (VCC) are ideal for weak alluvial soils such as peats and soft clays overlying competent founding strata such as sand, gravels, and soft rock. Working loads of up to 1,000 kN can be achieved in appropriate soils. The 'bulb end' and frictional components of the VCC enable high safe working loads to be developed at shallower depths than alternative piling systems and thus generally provide a more economical solution.

The Priebe model to design vibro replacement was extended to allow also for stiff columns: if the load is higher than the inner strength of the columns, the conventional vibro replacement design by Priebe (1995) is executed. But if the column load is lower than the inner strength of the columns, the calculation is modified (Priebe 2003). At first the settlement of the soil below the bottom of the VCC is determined using the stress, which corresponds to the one of a shallow foundation in a homogeneous half space. This formulation is not on the safe side as the load distribution is smaller than in homogenous soil due to stiffer vibro columns. In a second step the settlement is determined from the difference to the increased stress below the bottom of the columns. This yields the punching effect of the column toe into the soil below. Because of the difference of the averaged stress, which has been assumed to be quite small, a certain compensation is given.

The value determined as column punching has to be added to the settlement of the soil below the columns. A similar model has been developed by Tomlinson (1980) for piled raft foundations.

Once completed, the columns exhibit stiffness 10–20 times greater than the adjacent soil. Construction of a supplementary layer of compacted material over the column heads is often performed in order to focus the surface load on the columns. The surface load is focused by means of an arching effect that occurs as this layer thickens. An alternative method involves using a horizontal geotextile. Suspended between the column heads, it prevents the columns from puncturing an attenuated load distribution layer (Kempfert 1995, Sondermann and Jebe 1996, Topolnicki 1996). For vibro stone columns, such a load distribution layer is not necessary.

2.5 APPLICATIONS AND LIMITATIONS

Vibro compaction is used to increase the bearing capacity of foundations and to reduce their settlements. Another application is the densification of sand for liquefaction mitigation. By reducing the amount of water which has to be pumped during groundwater lowering, sand can be compacted, which reduces the permeability. This solution is also possible for dams.

Vibro compaction is limited by the fines and carbonate content (see Section 2.4). Furthermore, a certain distance should be kept to existing buildings in order to limit settlements of new buildings (Achmus et al. 2010). Depths down to 65 m have been improved so far by vibro compaction.

Various ways of creating vibro stone columns have been developed in order to enhance load-bearing capacities of weak soils and limit settlement. For the support of individual or strip foundations, small groups of columns are employed. Large column grids are placed beneath rigid foundation slabs or load configurations that exhibit flexibility, as is the case with storage tanks and embankments. Due to inherent higher shear resistance, vibro replacement columns are a good choice for the enhancement of slope stability. When drainage takes precedence over bearing capacity, vibro drain columns can be employed to function as drains. This drainage type of sand or stone column is constructed simply by lifting the vibrator without compaction, leaving the sand or stone in a state ranging from loose to medium dense (European Standard EN 15237, 2007).

Vibro stone columns are not suitable in liquid soils with a very low und-rained cohesion because the lateral support is too small. However, vibro stone columns have been installed successfully in soil with 5 kPa $< c_u <$ 15 kPa, see Section 2.7. In case of very hard and/or cemented layers (i.e., cap-rock) or very well-compacted surface layers, pre-boring may be necessary to assist the penetration of the vibrator. Concerning the distance to build-ings the same applies as for vibro compaction (Achmus et al. 2010). Depths down to 43 m have been improved so far by vibro replacement.

2.6 MONITORING AND TESTING

Part of the state of the art methods is to monitor and record in great detail the operating parameters of any deep vibro work. Details are given in the European Standard 'Ground treatment by deep vibration' (European Standard EN 14731, 2005).

Vibro compaction is monitored online with devices that record, as a function of time, penetration depth, energy consumption of the motor and, if necessary, pressure and quantity of the flushing media used. If the vibra-tor frequency can be adjusted during the compaction process, this param-eter is also recorded.

For the vibro replacement method, all of the essential parameters of the production process (depth, up/down speed, activation force, vibrator energy, and stone/concrete consumption) are recorded continuously as a function of time, providing the user with visible and controllable data for producing a continuous stone column. A typical printout for stone column construction is given in Figure 2.17. Monitoring of the activation force is important to check that there is sufficient vertical compaction and hori-zontal displacement of the column material. Additionally, the monitoring of the fill level in the material supply tube indicates exactly how much material is inserted per linear metre column length. Monitoring the acti-vation force and the fill level together guarantees a high-quality column. Such instrumentation is available for leader-mounted, bottom feed vibra-tor systems and has been used in Europe since the 1980s (Slocombe and Moseley 1991).

Vibro compaction and vibro replacement are increasingly evaluated by means of 2D and 3D plots whereby areas of inhomogeneous soil can be detected. To monitor the installation quality covering a large area, the programme 'VibroScan' visualises automatically recorded vibro data, like depth, compaction energy, and activation force. Borelogs from core drill-ings, cone penetration tests, and standard penetration tests (SPT) represent only punctual explanations, which can never show the entire ground. After

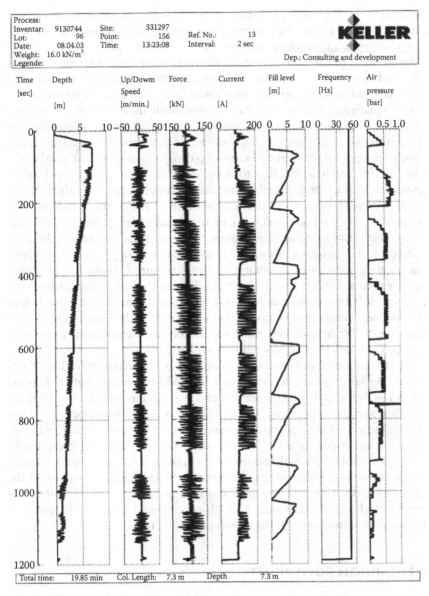

Figure 2.17 Typical printout for stone column construction. (Courtesy of Keller Group.)

collecting the data of all compaction points with VibroScan, a planar overview of the entire area of the ground improvement is displayed. Such areas
are highlighted whose soil conditions deviate from the remaining area, and
therefore a deviating quality of the ground after ground improvement measures is shown. The execution grid can be adapted to the respective soil

conditions, and auxiliary points can be arranged in areas of worse soil quality in order to receive a preferably constant quality and thus an optimum homogenisation of the improved ground at the end.

In addition to the online control, the final site records include the position and elevation of columns; the source, type, and quality of imported material; and, if necessary, environmental factors (noise, vibrations, etc.).

Cone penetration tests and SPTs are commonly used to verify the success of vibro compaction, with CPT being the better of the two. To compare the initial and final compressibility of the soil, pre- and post-tests undergo comparison. When evaluating post–deep compaction work test results, the ageing effect must be taken into account. This ageing effect on strength goes on for up to several weeks after the column has been installed. Many projects have demonstrated that the strength of compacted sands has the potential, over several weeks, to increase anywhere from 50%–100%. This substantial increase is attributed to pore water pressure reduction, sometimes in combination with the re-establishment of physical and chemical bonding forces to the column's grain structure (Mitchell et al. 1984, Schmertmann 1991, Massarsch 1991). Taking this strength increase into account, it is best to wait at least one week after compaction work before conducting formal compaction tests. The technical literature record contains much information regarding reports on compaction tests and monitoring (Covil et al. 1997, Slocombe et al. 2000).

The performance of vibro stone columns is monitored only for large projects using large plate load tests, which should be carried out by loading a rigid plate or cast in-situ concrete pad big enough to span one or more columns and the intervening ground. Zone load tests should be carried out by loading a large area of treated ground, usually by constructing and loading a full size foundation or placing earth fill to simulate widespread loads.

In case of soils with high sensitivity $S > 8$, the soil structure in the vicinity of the vibro stone columns may be disturbed resulting in a decrease of the initial undrained shear strength. It may take up to several months of recovery to reach the initial value. Enough time should therefore be foreseen between the installation of the columns and the tests.

2.7 CARBON FOOTPRINT

Beginning with the ratification of the Kyoto Protocol in 1997, the awareness of the global warming effect has risen all over the world. Approximately 1/5 of global greenhouse gas emissions are associated with the manufacturing and construction industries. In recent years, the awareness of the carbon footprint has reached the construction industry, leading to several certificates for buildings and engineering structures. The foundation of a building can have a significant impact on the total carbon footprint, depending on the local ground conditions. Ground improvement technologies can significantly reduce greenhouse gas emissions because the raw materials used have by far the biggest impact on the total emission of the foundation works, followed by the diesel consumption of the machinery used. By improving the existing soil, the use of concrete can be minimised or even avoided, resulting in a significant reduction of the foundation work's carbon footprint, and hence that of the total structure. Taking into account that concrete has a high input on carbon footprint emissions due to the manufacturing process, ground improvement technologies as vibro compaction or vibro stone columns are a sustainable and environmentally friendly solution.

Figure 2.18 shows the computed CO_2 equivalents in kg per linear m for different special foundation methods (Zöhrer et al. 2010). The vibro stone column method with gravel possesses the environmental friendly smallest CO_2 equivalent per linear m foundation element, followed by ready mixed mortar columns (VMC) and VCC. The large amount of cement used in CFA piles, deep soil mixing (DSM), and especially in bored piles (BP) leads to high or very high CO_2 equivalents. For foundation or ground improvement elements using binders, the employment of Portland cement instead of blast furnace cement leads to approximately 60%–70% higher CO_2 output.

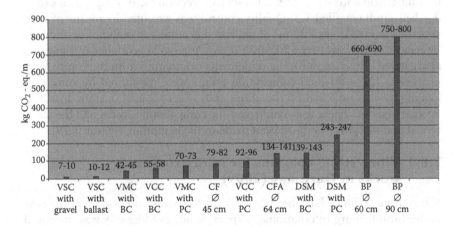

Figure 2.18 CO_2 equivalents for ground improvement and/or deep foundation methods. VSC = vibro stone columns, VMC = vibro mortar columns, VCC = vibro concrete columns, CFA = continuous flight auger piles, DSM = deep soil mixing, BP = bored piles, PC = Portland cement, BC = blast furnace cement. (From Zöhrer, A., Wehr, W., and Stelte, M. (2010). Is ground engineering environmentally friendly? *Proceedings of the 11th International EFFC-DFI Conference, Session 3: Sustainability in the Foundation Industry*, May 26–28, 2010, London, England.)

2.8 CASE HISTORIES

2.8.1 Vibro compaction for artificial islands in Dubai (2001–2008)

Extensive ground improvement using vibro compaction of reclaimed sand fill was carried out between 2001 and 2008 in Dubai for the 'Dubai Waterfront' project "with a total area of 130 km², consisting out of the 'Palm Jumeirah Island', the 'Palm Jebel Ali', the 'Palm Deira', and the 'World Island' (Wehr 2005a; Haß et al. 2010).

The material used to reclaim the islands originated from the seabed in front of the Dubai coast. The material was then put in place by means of a dredging process, leading to a loose state of density. Therefore, all that soil had to be compacted down to the bearing layer by means of deep vibro compaction. Design criteria allowed maximum settlements of 25 mm and a resistance against liquefaction in case of an earthquake with a magnitude M = 6.0. The allowed maximum in angular rotation was limited to 1:500. The design loads were up to 150 kN/m². Frequently, crane-hung twin vibrators were used for the vibro compaction works. To proof the achievement of the required compaction, a CPT reference diagram has been developed by means of Eurocode 8. Due to the fact that the material used had a high carbonate content originating from seashells and corals, the results of CPT soundings had to be adjusted to consider the modified strength of the reclaimed sand fill in comparison with silica sand. This shell correction factor $f = q_c$ (silica sand)/ q_c (shell sand) equalled 1.3. Quality control was executed by means of cone penetration tests every 900 m² of improved ground and material extractions by means of borings every 20,000 m². An area-wide geodetic levelling was also carried out to establish the settlements originating from the compaction processes. The individual compaction points were monitored by automatic logging devices attached on the crane units. Every 2 km², a static load test was carried out. These tests confirmed the compaction success and the theoretical assumptions that have been made in the beginning (Figure 2.19).

2.8.2 Vibro stone columns for infrastructure works in Germany (2009)

Vibro stone columns have been carried out for roads and bridges at the new Berlin Brandenburg International Airport. Sandy boulder clay was improved down to a depth of 8.0 m below ground surface, covering a total area of 90,000 m². The aim was to reduce the settlement behaviour of the soil by factor 2. The bearing layer has been identified as being sand with cone penetration resistance of $q_c > 10$ MN/m². The spacing and the vibro stone column's diameter based on results of field trials and static load tests carried out in representative locations with unfavourable soil conditions. These locations have been identified by additional CPTs prior to the execution of works. In

Figure 2.19 Compaction works at Palm Island.

addition to these field trials and load tests, numerical calculations have been executed to examine additional soil improvement effects by means of stiffness improvement of the soil surrounding the columns and stiffness improvement by means of displacement effects between columns. The static load tests and the numerical calculations resulted in a required stiffness modulus of 21 MN/m² and a stone column pattern of 2.75 m × 2.75 m. To be on a safe side, an execution point distance pattern of 2.5 m × 2.5 m was chosen. For some single footings, a reduction of the point distance to a smaller pattern was necessary to meet the settlement criteria. All compaction points were identified by means of GPS technology. The friction angle of the stone columns, used in the numerical calculations was obtained by means of laboratory shear tests prior to the design works, resulted in a friction angle $\varphi = 57°$. After finishing of the soil improvement works, the settlements have been observed during the ongoing construction of roads and bridges. The measured settlements stayed clearly below the forecasted values (Kirsch et al. 2009).

2.8.3 Vibro stone columns for shipyard infrastructure in India (2009)

Extensive vibro replacement works have been carried out in Pipavav, India, to improve the ground for approach roads and hardstand areas of a shipbuilding facility, consisting of making facilities for the fabrication of hull blocks, shiplifting facilities, and multiple land berths.

Soil investigation was carried out in and around the facility by means of boreholes. Soil profiles for approach roads and hard standing pavements showed marine clay down to 10 m, with SPT N = values of 2 to 4, underlain by weathered rock/bedrock with SPT N > 50.

Both the approach roads and hard standing pavements have to be designed for the heavy traffic loads during the transportation of the 400-ton shipbuilding hull blocks to the ship assembly area. The long term settlements were to be less than 200 mm and 150 mm for the hard standing pavements and approach road section, respectively.

Vibro stone columns were chosen as a foundation solution to meet these settlement limits. The design of the stone columns was carried out in accordance with Priebe's (1995) method. In order to meet the performance criteria, 900-mm-diameter columns were executed. The average installation depth was 12 m, with a triangular grid spacing of 2.5 m centre-to-centre. A section of stone columns along the approach road and hard standing pavement is shown in Figure 2.20.

Several vibro rigs with crane-hung vibrators were used to complete the installation works. About 144,000 linear metres of vibro replacement columns were installed in 2008 and 2009 to treat a total area of 57,500 m². To ensure quality and that the columns were consistently formed, the installation works were monitored and logged in real time by computers (Raj and Dikshith 2009).

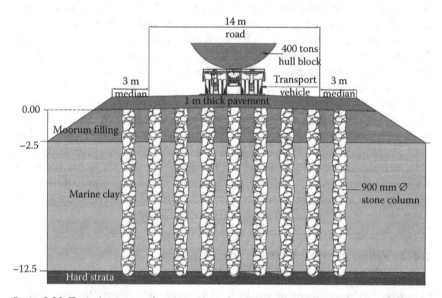

Figure 2.20 Typical section of approach road with vibro replacement columns. (From Vibro replacement columns for shipyard infrastructure at Pipavav, Gujarat, India. In: Leung C.F., Chu J., Shen R.F. (eds.) *Ground Improvement Technologies and Case Histories,* Singapore: Research Publishing Services, pp. 763–769.)

2.8.4 Vibro concrete columns for a sewage treatment plant in Malaysia (2008)

Ground improvement technology with VCC was utilised as the foundation system for the Jelutong Sewage Treatment Plant (JSTP) comprising 12 nos. of Sequential Batch Reactor (SBR) tanks and associated process tanks.

The subsoil conditions primarily consisted of 5-m-thick reclaimed fill underlain by about 5 m soft to firm silty clay. Stiff silt was found at a depth of about 10 m and a dense and hard stratum was encountered at a depth of more than 40 m. Figure 2.21 shows the extent of waste (domestic garbage) dump, demarcated from trial pits carried out at site, covering approximately one third of the site. Laboratory results indicated that plasticity index ranges between 20% and 40%; very low sulphate and chloride content (<0.2%) and average pH value to be around 8, so no additional protective measures for the cement-based foundation system were required. The SBR tanks were designed as twin tanks of approximately 90 m × 60 m × 7 m made out of reinforced concrete.

The foundation system was required to carry SBR tank loads up to 126 kPa with the total settlement of the structure to be less than 75 mm and differential settlement to be less than 1:360. VCCs were installed to support tank and ancillary structures within the garbage area, since soil mixing was inappropriate. The diameter of each VCC was about 0.6 m with working loads of 35 tons. Typical spacing of columns was 1.6 m c/c to support a foundation load of 126 kPa. The depth of columns was up to 14 m. The VCCs were designed to achieve an in-situ unconfined compressive strength (UCS) of around 10 MPa. The columns were installed using custom-built vibro

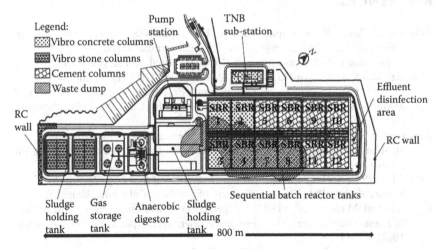

Figure 2.21 Different ground improvement methods for the sewage treatment plant (From Keller (2009). Brochure 10-65E: Foundation Works for a Sewage Treatment Plant Using Ground Improvement Methods in Malaysia.)

replacement machines where no water jetting was required. This method successfully installed the columns without removal of the existing garbage. To ensure quality, the installation works were monitored in real time by computers to ensure that the columns were consistently formed. Selected VCC were excavated for examination. It was proven that domestic waste material was displaced sideways during installation of VCC and did not contaminate the concrete. Some columns were tested up to 1.5 times the working load using plate load tests. A total 13 nos. VCC single column and 7 nos. VCC four-column group load tests were carried out to prove the performance of the constructed columns. All the load tests were successfully carried out. Concrete samples for VCC were collected for unconfined compression tests. The retrieved samples were subjected to UCS. Results of tests on VCC samples showed UCS in the acceptable range of 10–40 MPa (Yee et al. 2008).

2.9 CONCLUSIONS

Vibro systems have proven over 70 years to offer safe and economic methods of improving weak soils for a wide range of applications. Vibro compaction has been used to densify granular soils to significant depths (65 m), and the ability of this technique to reduce the risk of liquefaction during an earthquake is well documented. Vibro replacement is a widely accepted technique for improving cohesive and fine-grained soil to support a wide range of structures. Vibro concrete columns are a good alternative to piles.

REFERENCES

Achmus, M., Wehr, W., and Spannhoff, T. (2010). Building vibrations due to deep vibro processes. *Proceedings of the 7th International Conference on Ground Improvement Techniques,* June 23–25, 2010, Seoul, Korea, pp. 33–42.

Almeida, M.S.S, Jamiolkowski, M., and Peterson, R.W. (1991). Preliminary result of CPT tests in calcareous Quiou sand. *Proceedings of the International Symposium on Calibration Chamber Testing,* June 28–29, 1991, Potsdam, New York, United States, pp. 41–53.

Arbeitsgruppe Untergrund-Unterbau der Forschungsgesellschaft für das Straßenwesen (AUFS) (1979). Köln, Merkblatt für die Bodenverdichtung im Straßenbau.

Baez, J.I. (1995). *A Design Model for the Reduction of Soil Liquefaction by Vibro Stone columns,* PhD Thesis, University of Southern California, United States.

Baez, J.I. and Martin, G.R. (1992). Liquefaction observations during installation of stone columns using the vibro replacement technique, *Geotechnical News,* 10(3): 41–44.

Balaam, N.P. and Booker J.R. (1981). Analysis of rafts and granular piles, *International Journal for Numerical and Analytical Methods in Geomechanics,* 5: 379–403.

Bell, A.L. (1915). The lateral pressure and resistance of clay and the supporting power of clay foundations. In: *A Century of Soil Mechanics*, London, England: Institution Civil Engineers, pp. 93–134.

Bellotti, R. and Jamiolkowski, M. (1991). *Evaluation of CPT and DMT in Crushable and Silty Sands: Third Interim Report ENEL/CRIS*, Milano, Italy.

Bergado, D.T., Chai, J.C., Alfaro, M.C., and Balasubramaniam, A.S. (1994). *Improvement Techniques of Soft Ground in Subsiding and Lowland Environment.* Rotterdam, Netherlands: Balkema.

Brauns, J. (1978). Die Anfangstraglast von Schottersäulen in bindigem Untergrund. *Bautechnik* 55(8):263–271.

Brauns, J. (1980). Untergrundverbesserungen mittels Sandpfählen oder Schottersäulen. *Tiefbäu, Ingenieurbau, Straßenbau* 8:678–683.

Brennan, A.J. and Madabhushi, S.P.G. (2002). Liquefaction remediation by vertical drain groups, *Proceedings of the International Conference on Physical Modelling in Geotechnics, St. John's, Newfoundland*, July, pp. 533–538.

British Research Establishment. (2000). *Specifiying Vibro Stone Columns.*

Chambosse, G. and Kirsch, K. (1995). Beitrag zum Entwicklungsstand der Baugrundverbesserung, *Schriftenreihe des Lehrstuhls und Prüfamtes für Grundbau, Bodenmechanik und Felsmechanik der Technischen Universität München*, Munich, Germany, pp. 411–426.

Covil, C.S., Luk, M.C.W., and Pickles, A.R. (1997). Case history: ground treatment of the sand fill at the new airport at Chek Lap Kok, Hong Kong. *Proceedings of the 3rd International.Conference on Ground Improvement Geosystems*, London, England, pp. 149–156.

Cudmani, R.O. (2001). Statische, alternierende und dynamische Penetration in nichtbindigen Böden, Dissertation, Karlsruhe University.

Cudmani, R.O., Osinov, V.A., Bühler, M.M., and Gudehus, G. (2003). A model for the evaluation of liquefaction susceptibility in layered soils due to earthquakes, *12th Pan-American Conference on Soil Mechanics and Geotechnical Engineering*, Cambridge, Massachusetts, United States, Vol. II, pp. 969–976.

Engelhardt, K.and Golding, H.C. (1975). Field testing to evaluate stone column performance in a seismic area. *Géotechnique* 25(1):61–69.

European Standard EN 14731. (2005). *Ground Treatment by Deep Vibration.*

European Standard EN 15237. (2007). *Ground Treatment by Vertical Drainage.*

Fellin, W. (2000). *Advances in Geotechnical Engineering and Tunnelling: Rütteldruckverdichtung als plastodynamisches Problem, Vol. 3*, Rotterdam, Netherlands: A.A. Balkema.

Foray, P.Y., Nauroy, J.-F., and Colliat, J.L. (1999). Mechanisms governing the behaviour of carbonate sands and influence on the design of deep foundations, *Proceedings of the 2nd International Conference on Engineering for Calcareous Sediments Vol. 1*, February 21–24, 1999, Bahrain, pp. 55–68.

Forschungsgesellschaft für das Straßenwesen (FGFS). (1979). *Merkblatt für die Untergrundverbesserung durch Tiefenrüttler.*

German Patent: Nr. 22 GO 473.

Goughnour, R.R. and Bayuk, A.A. (1979). Analysis of stone column–soil matrix interaction under vertical load. *Paris Colloque International Renforcement des Sols: Terre Armée et Autres Méthodes, Vol. 1*, March 20–22, 1979, Paris, France, pp. 271–277.

Greenwood, D.A. (1976). Discussion. In: *Ground Treatment by Deep Compaction*. London, England: Thomas Telford for the Institution of Civil Engineers, p. 123.

Greenwood, D.A. and Kirsch, K. (1983). Specialist ground treatment by vibratory and dynamic methods. *Proceedings of the International Conference on Advances in Piling and Ground Treatment for Foundations*, March 2–4, 1983, London, England, pp. 17–45.

Haß, H., Wernecke, R., and Kessler, S. (2010). Tiefenverdichtung künstlicher Inseln in Dubai, *Vorträgen auf der 31st Baugrundtagung mit Fachausstellung Geotechnik*, Technischen Universität München, November 3–5, 2010, Munich, Germany, pp. 185–190.

Herle I., Wehr. W., and Arnold M. (2008). Soil improvement with vibrated stone columns – influence of pressure level and relative density on friction angle, *Proceedings of the 2nd International Workshop on Geotechnics of Soft Soils – Focus on Ground Improvement*, September 3–5, 2008, University of Strathclyde, Glasgow, Scotland, pp. 235–240.

Hu, W. (1995). *Physical Modelling of Group Behaviour of Stone Column Foundations*, PhD Thesis, University of Glasgow.

Institution of Civil Engineers. (1987). *Specification for Ground Treatment*. London: England: Thomas Telford.

Japanese Geotechnical Society. (1998). *Remedial Measures against Soil Liquefaction, from Investigation and Design to Implementation*. Rotterdam, Netherlands: A. A. Balkema.

Jebe, W. and Bartels, K. (1983). The development of compaction methods with vibrators from 1976 to 1982, *Proceedings of the 8th European Conference on Soil Mechanics and Foundation Engineering* organized by the Finnish Geotechnical Society, Helsinki, May 23–26, 1983, pp. 259–266.

Keller (1993). Brochure 13-21D: *Sanierung des Hochwasserdammes der Leitha bei Rohrau Pachfurt in Niederösterreich*.

Keller (1997). Brochure: *Rütteldruckverdichtung, Hamburg Elbtunnel, Vierte Röhre*.

Keller (2002). Brochure 11-31E: *Offshore Vibro, Compaction for Breakwater Construction, Seabird Project, Karwar, India*.

Keller (2009). Brochure 10-65E: *Foundation Works for a Sewage Treatment Plant Using Ground Improvement Methods in Malaysia*.

Kempfert, H.-G., and Stadel, M. (1995). Zum Tragverhalten geokunststoffbewehrter Erdbauwerke über pfahlähnlichen Traggliedern. *Geotechnik Sonderheft zur 4: Informations und Vortragsveranstaltung über Kunststoffe in der Geotechnik München*, Verlag Deutsche Gesellschaft für Geotechnik, March 16–17, 1995, Essen, Germany, pp. 146–152.

Kirsch, F. (2004). *Experimentelle und numerische Untersuchungen zum Tragverhalten von Rüttelstopfsäulengruppen, Mitteilung des Instituts für Grundbau und Bodenmechanik*, PhD Dissertation, Technische Universität Braunschweig.

Kirsch, F., Breitsprecher, G., and Rawitzer, S. (2009). Flughafen Berlin – Brandenburg International – Dimensionierung und Ausführung einer Baugrundverbesserung für Bauwerke und Verkehrsflächen der landseitigen Anbindung,' *Veröffentlichungen des Grundbauinstitutes der Technischen Universität Berlin, Hans Lorenz Symposium*, January 10, 2009, pp. 105–116.

Kirsch, K. (1979). Erfahrungen mit der Baugrundverbesserung durch Tiefenrüttler. *Geotechnik* 1:21–32.

Kirsch, K. (1993). *Baugrundverbesserung mit Tiefenrüttlern. 40 Jahre Spezialtiefbau: 1953–1993*, Düsseldorf, Germany: Technische und rechtliche Entwicklungen.

Kirsch, K. and Chambosse, G. (1981). Deep vibratory compaction provides foundations for two major overseas projects. *Ground Engineering* 14(8): 31–35.

Kirsch, K. and Kirsch, F. (2010). *Ground Improvement by Deep Vibratory Methods*, New York: Spon Press.

Massarsch, K.R. (1991). *Deep Soil Compaction Using Vibratory Probes in Deep Foundation Improvement*, STP 1089, ASTM.

Massarsch, K.R. (1994). Design aspects of deep soil compaction, *Proceedings of Seminar on Ground Improvement Methods*, Geotechnical Division, Hong Kong Institution of Engineers, May 19, 1994, pp. 61–74.

Meier, T. (2009). *Application of Hypoplastic and Viscohypoplastic Constitutive Models for Geotechnical Problems*, PhD Thesis, Veröffentlichungen des Institutes für Bodenmechanik und Felsmechanik der Universität Fridericiana in Karlsruhe, Heft 171.

Mitchell, J.K. and Huber, T.R. (1983). Stone columns foundation for a wastewater treatment plant—a case history. *Geotechnical Engineering* 14:165–185.

Mitchell, J.K. and Solmayr, Z.V. (1984). Time dependent strength gain in freshly deposited or densified sand, *Journal of ASCE*, GT11, 110:1559–1576.

Mitchell, J.K. and Wentz, J.R. (1991). *Performance of Improved Ground during Loma Prieta Earthquake*. Report No. UCB/EERC-91/12, Earthquake Engineering Research Center, University of California, Berkeley.

Moseley, M.P. and Priebe, H.J. (1993). Vibro techniques. In: Moseley, M.P. (ed.) *Ground Improvement*, London, England: Blackie Academic & Professional.

Perlea, V. (2000). Liquefaction of cohesive soils, *Soil Dynamics and Liquefaction 2000: Proceedings of Sessions of Geo-Denver 2000*, August 5–8, 2000, ASCE Geotechnical Special Publication No. 107, pp. 58–76.

Priebe, H.J. (1995). The design of vibro replacement, *Ground Engineering*, December, pp. 31–37.

Priebe, H. (1998). Vibro replacement to prevent earthquake induced liquefaction, *Ground Engineering*, September, pp. 30–33.

Priebe, H.J. (2003). Zur Bemessung von Rüttelstopfverdichtungen – Anwendung des Verfahrens bei extrem weichen Böden, *Bautechnik*, pp. 380–384.

Raithel, M. and Kempfert, H.-G. (2000). Calculation models for dam foundations with geotextile coated sand columns, *Proceedings of GeoEng 2000, an International Conference on Geotechnical & Geological Engineering*, November 19–24, 2000, Melbourne, Australia.

Raj, D. and Dikshith, C.V. (2009). Vibro replacement columns for shipyard infrastructure at Pipavav, Gujarat, India. In: Leung C.F., Chu J., Shen R.F. (eds.) *Ground Improvement Technologies and Case Histories*, Singapore: Research Publishing Services, pp. 763–769.

Raju, V.R. (2002). Vibro replacement for high earth embankments and bridge abutment slopes in Putrajaya, Malaysia, *Proceedings for the 4th International Conference on Ground Improvement Techniques*, March 26–28, 2002, Kuala Lumpur, Malaysia, pp. 607–614.

Raju, V.R. and Hoffmann, G. (1996). Treatment of tin mine tailings in Kuala Lumpur using vibro replacement. *Proceedings of the 12th Southeast Asian Geotechnical Conference*, May 6–10, 1996, Kuala Lumpur, Malaysia.

Schmertmann, J.H. (1991). The mechanical aging of soils. *Journal of Geotechnical Engineering* 117(9):1288–1330.

Schneider, H. (1938). Das Rütteldruckverfahren und seine Anwendungen im Erd- und Betonbau, *Beton und Eisen* 37(1).

Schüßler, M. (2002). Anwendung neuer, innovativer Gründungslösungen- ist das Risiko für den Auftraggeber überschaubar? *Deutsche Gesellschaft für Geotechnik Vorträge der Baugrundtagung*, September 25–28, 2002, Mainz, Germany, pp. 339–345.

Schweiger, H.F. (1990). Finite Element Berechnung von Rüttelstopfverdichtungen. *Vorträge des 5. Christian Veder Kolloquium*, Technischen Universität Graz, April 26-27, 1990, Graz, Austria, pp. 1–15.

Seed, H.B. and Booker, J.R. (1976). *Stabilisation of Potentially Liquefiable Sand Deposits Using Gravel Drain Systems*, Report No. EERC76-10, University of California–Berkeley, United States.

Seed, H.B., Idriss, I.M., and Arango, I. (1983). Evaluation of liquefaction potential using field performance data, ASCE, *Journal of Geotechnical Engineering* 109(3): 458–483.

Sidak, N., Strauch, G., and Wehr, J. (2004). Installation of geotextile covered stone columns by the Keller depth vibrator techniques, *Proceedings of ASEP-GI: International Symposium on Ground Improvement*, September 9–10, 2004, Paris, France, pp. 269–284.

Slocombe, B.C., Bell, A.L., and May, R.E. (2000). The in-situ densification of granular infill within two cofferdams for seismic resistance. *Workshop on Compaction of Soils, Granulates and Powders*, February 28–29, 2000, Innsbruck, Austria, pp. 33.

Slocombe, B.C, and Moseley, M.P. (1991). *The Testing and Instrumentation of Stone Columns*, ASTM STP 1089.

Smoltczyk, U. and Hilmer, K. (1994). Baugrundverbesserung. *Grundbau-Taschenbuch*, 5th ed., vol. 2.

Sondermann, W. and Jebe, W. (1996). Methoden zur Baugrundverbesserung für den Neu- und Ausbau von Bahnstrecken auf Hochgeschwindigkeitslinien. *Deutsche Gesellschaft für Geotechnik Vorträge der Baugrundtagung*, September 25–27, 1996, Berlin, Germany, pp. 259–279.

Soydemir, C., Swekowsky F., Baez, J.I., and Mooney, J. (1997). Ground improvement at Albany Airport. In: Schaefer V.R. (ed.) *Ground Improvement, Ground Reinforcement, Ground Treatment: Developments 1987–1997*, Geotechnical Special Publication No. 69, Logan, UT: American Society of Civil Engineers.

Soyez, B. (1987). Bemessung von Stopfverdichtungen. *BMT*. April, pp. 170–185.

Tomlinson, M.J. (1980). *Foundation Design and Construction*, New York: Pitman Publishing.

Topolnicki, M. (1996). Case history of a geogrid-reinforced embankment supported on vibro concrete columns. In: deGroot, M.B., den Hoedt, G., Termaat, R.J. (eds.) *Geosynthetics: Applications, Design and Construction*, Rotterdam, Netherlands: Balkema, pp. 333–340.

Trunk, U., Heerten, G., Paul, A., and Reuter, E. (2004). Geogrid wrapped vibro stone columns, *Proceedings of the Eurogeo 3, 3rd European Geosynthetics Conference*, March 1–3, 2004, Munich, Germany, pp. 289–294.

US Department of Transportation. (1983). *Design and Construction of Stone Columns*, Vol. 1.

Van Impe, W.F. and Madhav, M.R. (1992). Analysis and settlement of dilating stone column reinforced soil. *ÖIAZ: Österreichische Ingenieur- und Architekten-Zeitschrift* 137(3):114–121.

Vesic, A.S. (1965). Ultimate loads and settlements of deep foundations in sand. *Proceedings of the Symposium on Bearing Capacity and Settlement of Foundations in Sand*, Duke University, Durham, North Carolina, United States, pp. 53–68.

Wehr, J. and Raju, V.R. (2002). On- and offshore vibro compaction for a crude oil pipeline on Jurong Island, Singapore; *International Conference on Ground Improvement Techniques*, Malaysia, pp. 731–736.

Wehr, W. (1999). Schottersäulen – das Verhalten von einzelnen Säulen und Säulengruppen, *Geotechnik* 22(1):40–47.

Wehr, W. (2004). Stone columns—single columns and group behaviour, *Proceedings of the 5th International Conference on Ground Improvement Techniques*, March 22–23, 2004, Kuala Lumpur, Malaysia, pp. 329–340.

Wehr, W. (2005). Variation der Frequenz von Tiefenrüttlern zur Optimierung der Rütteldruckverdichtung, *Proceedings of the 1st Hans-Lorenz-Symposium*, October 13, 2005, Berlin, Germany, pp. 67–78.

Wehr, W. (2005a). Influence of the carbonate content of sand on vibro compaction, *Proceedings of the 6th International Conference on Ground Improvement Techniques*, July 18–19, 2005, Coimbra, Portugal, pp. 625–632.

Wehr, W. (2006). The undrained cohesion of the soil as criterion for the column installation effect with a depth vibrator, *Transvib 2006: International Symposium on Vibratory Pile Driving & Deep Soil Vibratory Compaction*, November 12, 2005, Paris, France, pp. 157–162.

Wehr, W. (2006a). Stone columns – group behaviour and influence of footing flexibility. *Proceedings of the Sixth European Conference on Numerical Methods in Geotechnical Engineering*, September 6–8, 2005, Graz, Austria, pp. 767–772.

Wehr, W. (2008). Offshore Schottersäulen in Singapur und Malaysia, *Darmstädter Geotechnik Kolloquium*, Technischen Universität Darmstädt, March 13, 2008, Darmstädt, Germany, pp. 85–93.

Yee, Y.W., Chua, C.G., and Yandamuri, H.K. (2009). Foundation works for a sewage treatment plant using ground improvement methods in Malaysia. In: Leung C.F., Chu J., Shen R.F. (eds.) *Ground Improvement Technologies and Case Histories*, Singapore: Research Publishing Services, pp. 677–684.

Zöhrer, A., Wehr, W., and Stelte M. (2010). Is ground engineering environmentally friendly? *Proceedings of the 11th International EFFC-DFI Conference, Session 3: Sustainability in the Foundation Industry*, May 26–28, 2010, London, England, on DVD.

Chapter 3

Dynamic compaction

Barry Slocombe

CONTENTS

3.1 INTRODUCTION

Dynamic compaction (DC) improves weak soils by *controlled* high-energy tamping where the applied energies can be greater than 100 times that of driven piling. The reaction of soils during dynamic compaction treatment varies with soil type and energy input. A comprehensive understanding of soil behaviour, combined with experience of the technique, is therefore vital to successful improvement of the ground. Dynamic compaction is capable of achieving significant improvement to substantial depth, often with considerable economy when compared to other geotechnical solutions.

3.2 HISTORY

The principle of dropping heavy weights on the ground surface to improve soils at depth has attracted many claims for its earliest use. Early Chinese drawings suggested the technique could be several centuries old (Menard and Broise, 1976). Kerisel (1985) reports that the Romans used it for construction, and Lundwall (1968) reports that an old war cannon was used to compact ground in 1871. In the twentieth century, compaction was provided to an airport in China and a port area in Dublin during the 1940s and to an oil tank in South Africa in 1955. However, the advent of large crawler cranes led to the current high-energy tamping levels first being performed on a regular basis in France in 1970 and subsequently in Britain in 1973 and in North America in 1975.

An extension of the concept of weights dropped onto the ground, rapid impact compaction (RIC), was developed in England in the late 1970s for the rapid repair of explosion damage to military airfield runways using modified (BSP) hydraulic piling hammers acting on a steel foot that remains in contact with the ground.

A further extension is for large three- and five-sided towed rollers, also called impact roller compaction, to compact ground. These and the RIC equipment apply energy 'from the top down' to limited depths that are easily achieved using lower than normal DC drop heights.

3.3 PLANT AND EQUIPMENT

At first sight, the physical performance of dynamic compaction would appear to be simple, using a crane of sufficient capacity to drop a suitable size of weight in virtual free-fall from a certain drop height. Most contracts are performed with standard crawler cranes, albeit slightly modified for safety reasons and productivity, with a single lifting rope attached to the top of the weight (Figure 3.1). Details such as crane counterbalance weights, jib flexure, torque convertors, line pull, drum size, type and diameter of ropes, clutch, brakes, as well as many other factors and methods of working have been subjected to rigorous analysis by the major specialist organisations to improve reliability and productivity. The operation must be performed

Figure 3.1 Typical crawler crane and equipment.

safely. As a result the Health and Safety Executive in Britain requires that a crane should operate at not more than 80% of its safe working load. Some cranes are better suited than others to the rigours of this type of work, even though on paper they appear to be of similar capacity.

Recent crane developments allow automation of the whole work cycle. This is controlled by a data processing unit that plots for each compaction point its location, number, weight size, drop height, number of blows, and measurement of imprint achieved. A particular feature of one European crane is the free-fall winch, which adjusts the rope length automatically after each blow. Some cranes include the ability for synchronous operation of two winches to lift larger weights than the conventional crane rating.

The majority of British and American contracts have utilised weights within the range of 6 to 20 tonnes dropped from heights of up to 20 m. The majority of UK work is now performed using 8-tonne weights dropped from heights of up to 12 m. Standard crawler cranes have also been used in America for weights of up to 33 tonnes and 30 m height. Specialist lifting frames with quick release mechanisms have been utilised to drop weights of up to 50 tonnes, and Menard built equipment to drop 170 tonnes from 22 m height in France. In America, and increasingly in Britain, the system is known as dynamic deep compaction.

Weights are typically constructed using toughened steel plate, box-steel and concrete, or suitably reinforced mass concrete where durability is the prime requirement. The effect of different sizes and shapes of the weight has also been extensively researched with narrower weights generally being used to specifically drive material down to depth to form dynamic replacement columns in peaty or Sabkha soils. Treatment has also been performed below water using barge-mounted cranes and more streamlined weights with holes cut out to reduce water resistance and increase impact velocity on the sea bed.

Within the UK RIC typically employs a 7-tonne weight dropped repeatedly through a 1.2 m height onto a 1.5 m diameter steel articulated compaction foot. Whilst the energy per blow is not large (typically 8.4 tonne-metres), the equipment permits a large number of impacts to be applied at a rate of about 40 blows per minute for typical treatment depth of up to 3.0 m. Weights from 5 to 12 tonnes are used worldwide. It is however less successful at treating the mixed soils generally encountered in the United Kingdom.

3.4 TERMINOLOGY

The original concept for dynamic compaction was to collapse voids, particularly for the treatment of natural sands plus granular, mixed, and cohesive fills. This was then extended to finer natural soils where the high impact energy effectively provided localised surcharge to squeeze water out of silts and clays, this being termed dynamic consolidation. Dynamic replacement was then developed to drive large-diameter columns of coarse imported materials through soft near-surface soils, particularly for peat and Sabkha strata.

The worldwide use of dynamic compaction has resulted in a large number of important terms, some of which can have different meanings to different nationalities or could be confused with other geotechnical descriptions. The following terms have been adopted in Britain:

(1) 'Effective depth of treatment' is the maximum depth at which significant improvement is measurable. The 'zone of major improvement' is typically 1/2 to 2/3 of this effective depth.
(2) 'Drop energy' is the energy per blow (i.e., mass multiplied by the drop height [tonne-metres]).
(3) 'Tamping pass' is the performance of each grid pattern over the whole treatment area.
(4) 'Total energy' is the summation of the energy of each tamping pass (i.e., number of drops multiplied by the drop energy divided by the respective grid areas [normally expressed in tonne-metre/m^2]). It is not the summation of the drop energy divided by the plan area of the DC weight.
(5) 'Grid area' is the treatment area per drop location for each individual treatment pass. The dimension of the first pass is often approximately equal to the target depth of treatment. Hence for 8 m target depth the grid area is about $8 \times 8 = 64$ m^2.

(6) 'Recovery period' is the time allowed between tamping passes to permit the excess pore pressures to dissipate to a low enough level for the next pass.
(7) 'Induced settlement' is the average reduction in general site levels as a result of the treatment.
(8) 'Threshold energy' is the energy input beyond which no further improvement can practically be achieved or where adverse response starts to develop.
(9) 'Overtamping' is a condition in which the threshold energy has been exceeded, sometimes deliberately, causing remoulding and dilation of the soil.
(10) 'Shape test' is the detailed measurement of a single or group of imprint volumes and surrounding heave or draw-down effect, which permits comparison of overall volumetric change with increasing energy input.
(11) 'Imprint' is the crater formed by the weight at a tamping location.

3.5 HOW DYNAMIC COMPACTION WORKS

In contrast to having constructed a vibro stone or concrete column, the treatment at that location then being completed, dynamic compaction (DC), whether to shallow or deep layers, improves the ground to the basal layers first and then progressively up to the upper layers in a series of tamping passes. In contrast, RIC and the three- and five-sided rollers improve the soils by first creating a 'plug' or surface layer of denser ground and then progressively driving this plug/layer to greater depth. The response of the ground to these two approaches is fundamentally different. Whereas the relatively lesser number of high-energy impacts at wide-grid centres of DC tends to initially bypass the upper layers and then by subsequent progressive treatment builds up the strength of the near-surface soils, the larger number of lower energy impacts of the RIC and roller require consideration of the possible generation of pore-water pressures that inhibit the required ground improvement in finer-grained soils.

There is then a fundamental difference between the responses of granular and cohesive soils when subjected to the high-energy impacts of the process. It is normal to visualise treatment as a series of heavy tamping passes with different combinations of energy levels designed to achieve improvement to specific layers within the depth to be treated. The most common approach is to consider the ground in three layers. The first tamping pass is aimed at treating the deepest layer by adopting a relatively wide grid pattern and a suitable number of drops from the full-height capability of the crane. The middle layer is then treated by an intermediate grid, often the midpoint of the first pass or half the initial grid, with a lesser number of drops and reduced drop height. The surface layers then receive a continual tamp of a small number of drops from low height on a continuous pattern. It is sometimes feasible to combine, and sometimes necessary to subdivide, the basic tamping passes for the reasons outlined in the subsequent discussion.

The performance of increasing correctly controlled total energy input will normally lead to better engineering performance of the treated ground. However, analysis of several hundred contracts where in-situ and large loading tests were performed has shown that this is not a linear relationship and that the post-treatment parameters are heavily dependent upon the characteristics of the soil. As a general rule, similar total energies, whether per m^2 of area or m^3 of treatment depth, provide better performance to granular than mixed soils. Mixed soils are then better treatable than cohesive, with refuse-contaminated soils generally offering the least performance.

For treatment using RIC the operator monitors and can record the number of impacts, the total energy input applied, the foot penetration per blow, and the cumulative penetration. When a specified parameter is reached, for example, foot penetration or set per blow, the RIC equipment is moved and positioned at the next treatment point. This primary treatment pass is normally performed on a closely spaced grid pattern, typically 1.5 to 2.5 m. Additional offset and/or lower energy passes, or conventional proof roller compaction, are occasionally performed to achieve better coverage.

3.5.1 Granular soils

In dry granular materials (i.e., sand, gravel, ash, brick, rock, slag, etc.), it is very easy to understand how tamping improves engineering properties. Physical displacement of particles and, to a lesser extent, low frequency excitation will reduce void ratio and increase relative density to provide improved load bearing and enhanced settlement characteristics. A feature that often develops when providing treatment to coarse fill materials is the formation of a hard 'plug' that inhibits penetration of stress impulses to the deeper layers but is very useful in providing superior settlement performance beneath isolated foundation bases. Dynamic compaction can also improve 'dirtier' sands with higher fines content than the vibro compaction technique when performed without the addition of stone, see Figure 3.9.

When granular materials extend below the water table, a high proportion of the dynamic impulse is transferred to the pore water which, after a suitable number of surface impacts, eventually rises in pressure to a sufficient level to induce liquefaction. This is the theory first proposed by Menard and is a phenomenon very similar to that occurring during earthquakes. Clearly the existing density and grading of the soils will be major factors in the speed at which this liquefied state will be achieved. Low frequency vibrations caused by further stress impulses will then reorganise the particles into a denser state. This is comparable to the response of sands from the vibro compaction technique for which D'Appolonia (1953) suggested that a vibrational acceleration in excess of 0.5 g was necessary to achieve such a densification effect.

Dissipation of the pore-water pressures, in conjunction with the effective surcharge of the liquefied layer by the soils above, results in further increase in relative density over a relatively short period of time. This can vary from 1 to 2 days for well-graded sand and gravel, to 1 to 2 weeks for sandy silts and varies with the applied energy. The testing programme should therefore recognise the time-dependent response for soils that are normally considered to be free-draining. Longer-term improvement, possibly as a result of chemical bonding or high residual lateral stresses within the soil matrix, has been reported by Mitchell and Solymar (1984).

There is, however, another school of thought in which the aim is to avoid the liquefied state. While it is recognised that liquefaction cannot be avoided in deep, loose, sandy deposits with a high water table, as are often encountered in parts of North America, the Middle and Far East,

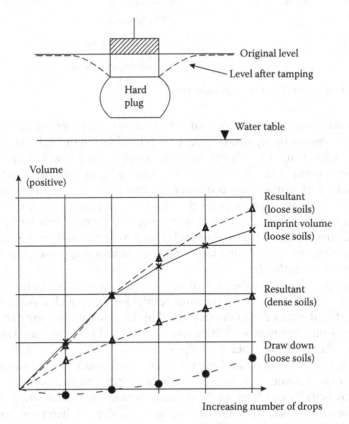

Figure 3.2 Volumetric response—granular soils.

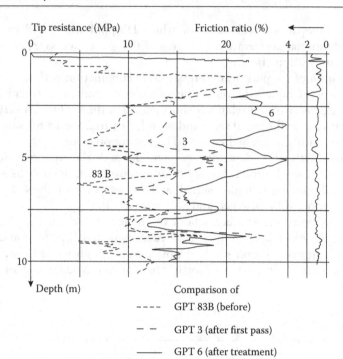

Figure 3.3 In-situ test results—granular soils.

such conditions are rare in Britain. The treatment is therefore designed to provide compaction by displacement without dilation or high excess pore pressures by using a smaller number of drops from a lower drop height. This method requires substantially lower energy input than the liquefaction approach, with consequent economies. Laboratory and in-situ tests have consistently shown that in order to achieve maximum density, the lowest number of stress impulses to attain the required energy input will provide the optimum result. Saturated granular soils will normally require higher treatment energy overall, in a larger number of tamping passes, than if the soils were essentially dry.

Figure 3.2 illustrates the typical volumetric response for granular soils and Figure 3.3 illustrates electric cone results for a site of clean sand with a water table at about 2.5 m depth treated by 15-tonne equipment. This also illustrates improvement with time since the second tamping pass was only capable of treating to about 4 m depth.

When the individual sand particles are weak, such as the calcareous sands of the Middle East, 'sugar' sands of North America, or the Thanet Sands of Britain, crushing tends to occur during the treatment. A similar response affects ash, clinker, and weak aerated slags. When these soils are dry, the effect of such particle breakdown is not particularly significant.

However, below the water table the higher proportion of fines developing with increasing energy input results in a rapid change from a granular to a pseudo-cohesive soil response.

The existence of very dense layers within the ground can cause anomalous results. Where, for example, cemented layers occur within natural sands, these tend to absorb the energy impulse and arch over the underlying stratum. A similar phenomenon can occur with vibro treatment where the cemented zones do not collapse around the vibrator to permit densification to occur. In these situations, where they occur at shallow depth, the dynamic compaction will break up the cemented layer. However, at greater depth, the energy levels required to break the stratum may be beyond the capabilities of the equipment on site. The presence of such layers is often not adequately revealed by normal site investigation.

In summary, excellent engineering performance can easily be achieved in dry granular soils using both DC and RIC equipment. However, care must be exercised for the treatment of soils with significant silt content, particularly below the water table.

3.5.2 Cohesive soils

The response of clays is more complex than that of granular soils. There is again the distinction between above and below the water table.

With conventional consolidation theory, a static surcharge loading will collapse voids within clay fills and expel water to induce consolidation and increase strength. The rate at which this occurs is dependent upon the imposed load, coefficient of consolidation, and length of drainage path. In contrast, dynamic compaction applies a virtually instantaneous localised surcharge that collapses voids and transfers energy to the pore water. This creates zones of positive water pressure gradient that induce water to drain rapidly from the soils matrix. This effect is further accelerated by the formation of additional drainage paths by shear and hydraulic fracture. Consolidation therefore occurs much more rapidly than would be the case with static loading. Dynamic compaction literally squeezes water out of the soil to effectively preload the ground. However, as with staged construction, the application of too high energy too soon can lead to problems. A typical volumetric response is illustrated in Figure 3.4.

Where the soils occur above the water table, the clays tend to be of relatively low moisture content, generally less than their plastic limit, where even a small reduction in water content can result in significant improvement in bearing capacity. As such, treatment is relatively straightforward and is mainly the collapse of air voids to provide a more intact soil structure. Care must however be exercised for the treatment of clays that are wetter than their plastic limit and higher plasticity clays.

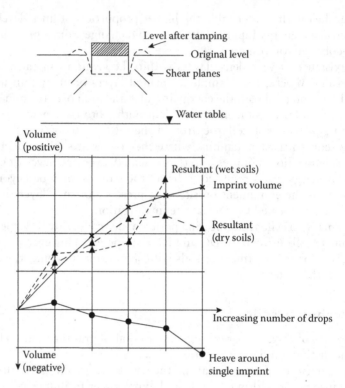

Figure 3.4 Volumetric response—cohesive soils.

Where the clays occur below the water table, a much larger reduction in moisture content is generally required in the presence of a smaller available pore-pressure gradient and a longer drainage path. These conditions can, if not properly controlled, result in the threshold energy being achieved much more rapidly and resulting in localised failure of the clay matrix. Control is then achieved by using greater numbers of tamping passes of lower energy input, requiring greatly extended contract periods in comparison to normal productivity.

To date, only nominal degrees of improvement have been achieved in thick layers of relatively weak saturated alluvial clays and silts, even with additional measures such as drainage trenches filled with sand or wick drains.

Where such layers are relatively thin and require treatment, a better speed of response is recorded due to the shorter drainage path. In some instances coarse granular material is driven into these materials to provide better grading that is more suited to treatment, or to displace from specific locations beneath part of a building area. It is, however, more common in the UK to adopt vibro stone columns in such soils to more critical locations, such as more heavily loaded foundations, and then perform dynamic compaction to preload the ground (see Slocombe 1989) with the benefit of the stiffer columns also acting as drains to control excess pore-water pressures.

For predominantly clay-type fill materials above the water table, the clay lumps can be considered as large weak particles of almost granular response. However, the major improvement is achieved by collapsing voids to provide a more intact structure. Clearly the strength of the lumps and sensitivity of the clay is of paramount importance in such soils. Differing degrees of weathering can also give rise to markedly variant responses on a site and experienced observation is required to define such locations. Mudstone and shale fragments can break down to a material of clayey response, particularly when heavy rainfall occurs.

For clay-type fills below the water table, the voided structure allows higher mobility of water causing lower excess pore pressures and shorter recovery periods in comparison to natural clays. The constituents would be of higher moisture content but again improvement would be achieved mainly by collapse of voids. Monitoring of excess pore-water pressures by means of piezometers is clearly useful but problematic above the water table.

It cannot be emphasised too strongly that the treatment of clay fills and clay soils requires experienced control on site. During treatment, after a small number of drops, heave starts to develop around the edges of each imprint. If tamping continues, the heave can build up to such an extent that it can exceed the volume of the imprint. Clearly this is the precise opposite of what is desired. Also, additive heave can occur by the performance of the adjacent tamping position at too narrow a grid dimension.

Particular care has to be exercised in the timing of successive tamping passes to permit adequate recovery of pore pressures to avoid excessive remoulding of the soils. Such approach can however be relatively slow and, in view of the emphasis placed these days on productivity, the vibro stone column in advance of dynamic compaction method described above is sometimes adopted.

If excessive heave around an individual imprint does start to occur, it is essential that the tamping at that position be stopped. This may only extend over a relatively confined area with better ground elsewhere. In soft areas it is better that twice the number of lighter energy input tamping passes be performed in a 'softly softly' approach.

Similar considerations apply when attempting to provide treatment to a significant depth where the surface layers are clayey. The strength of the surface soils can reduce in the short term and time has to be spent improving a disturbed matrix to reconstitute its original, let alone desired, properties. This is particularly difficult where thick crusts to, say, 2–3 m depth of stiff to very stiff clays overlay a granular deposit requiring treatment. In this situation even higher than normal energies are required to attain the deeper layers giving rise to even greater potential for virtually destroying the surface soils.

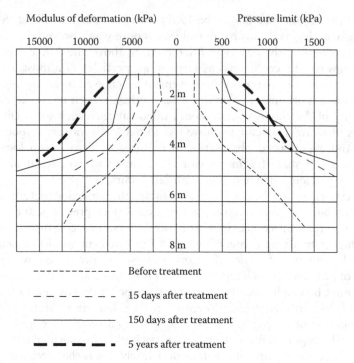

Figure 3.5 In-situ test results—cohesive soils.

The treatment of clayey soils will nearly always require a larger number of tamping passes when compared to a similar profile of predominantly granular constituents. Efficient treatment is achieved by attempting to provide as much improvement as quickly as possible while recognising that the response of the soils will dictate the speed of the treatment operations.

Clay soils will continue to improve for a significant period after treatment as reported by West (1976). Figure 3.5 illustrates further measurements on this site taken five years after treatment.

In summary, dry cohesive fills respond well to dynamic compaction. Care must be exercised in the treatment of weak natural clayey soils or clay fills below the water table. The prior performance of vibro stone columns to both stiffen the ground and enhance drainage has been successfully combined with dynamic compaction to weak clayey soils.

3.5.3 Landfills

The capability of dynamic compaction to treat every square metre of road and parking areas is increasingly used in the development of former landfill sites where, depending on their age, the original degradable constituents have decayed to create extensive voids. It has also been performed to reduce ground levels to avoid costly removal to specialist tip to permit development at the desired site level.

As a general rule, the older the landfill, the less the residual presence of matter susceptible to long-term decay and some older fills, particularly those of high ash content, have been compacted to also support structures that would normally be piled. However, the more recent fills generally contain significant proportions of organic matter and structures would normally be piled.

There is as yet little documented proof that dynamically compacted landfill can affect the rate of decay of residual degradable constituents, although a paper by Sharma and Anirban (2007) clearly records far better post-treatment performance at creep rate of 2% per log cycle than for static surcharge over a monitoring period of about 15 years. As there will be ongoing decay, when this technique is combined with piled structures, increasing differential settlements will become apparent with time and a degree of maintenance may be required at some future time.

The principle of treatment to landfills is comparable to the treatment of mixed clayey fills but with generally higher energy input than for inert fills. This is to collapse near-surface voids and to "overcompact" the remaining inert constituents. If a void then starts to develop due to localised long-term decay, the inert materials will tend to ravel into the void, bulk up, and spread the void effect rather than have a localised sharp deformity in the finished surface. Geogrids have also been used for a number of sites where the movements of heavy goods vehicles were critical to the development operations.

Many landfill sites have clay capping with basal clay liners to avoid downward migration of leachate into an aquifer. The DC drop energy should be limited to avoid shearing of the basal clay liner and care exercised in the design of the treatment operations to avoid the surface clays developing into a quagmire when attempting to apply the higher than for inert fill energy input to significant depth.

3.5.4 Collapsible soils

Rollins and Kim (2010) have reported on the successful treatment of natural cohesionless and low plasticity collapsible soils using typically higher compactive energy than would be used for noncollapsible soils. In Britain, there are many former opencast coal sites where the degree of control in backfilling has resulted in the presence of sufficient voidage to also provide potential for collapse settlement upon wetting. In such situations it is important when constructing vibro stone columns to 'seal' the columns to prevent the ingress of water into the collapsible fills via the free-draining stone columns.

Dynamic compaction to these sites reduces the near-surface voids to inhibit the collapse potential. It also reduces the permeability within the treatment depth to further inhibit the migration of any water into the ground. Consideration should however be given to how water is drained away from the development to avoid a possible plume of water extending beneath the development where there is greater thickness of collapsible fill than the treatment depth. Again, higher compactive energy than for non-collapsible soils would normally be performed.

The use of RIC to treat loess to nominal 3.0 m depth in Kazakhstan has been reported by Serridge and Synac (2006).

3.6 SITE INVESTIGATION

As with vibro designs, the extent of the site investigation should be appropriate to the type of development. For deep fill sites, desk studies to establish the locations of buried high walls plus age and degree of control of placement are essential data. Water contents for comparison with liquid and plastic limits should be performed for clayey constituents. Densities as revealed by SPT and CPTs plus the basal soils, whether clay, sand, and gravel or rock are also required. The presence of any overhead wires, buried services, or nearby structures should also be established.

3.7 DEPTH OF TREATMENT

Menard originally proposed that the effective depth of treatment was related to the metric energy input expression of $(WH)^{0.5}$ where W is the weight in tonnes and H the drop height in metres. This was modified by a factor of 0.5 by Leonards et al. (1980) for relatively coarse, predominantly granular soils, and factors of 0.375 to 0.7 by Mitchell and Katti (1981) for two soil types. The most exhaustive analysis yet published has been provided by Mayne et al. (1984). The author suggests that the range of treatment depths varies with initial strength, soil type and energy input as illustrated in Figure 3.6, as well as the depth to the groundwater table. Figure 3.6 suggests that factors as high as 0.9 could apply for shallow depths of loose soils and as low as 0.25 for deeper treatment.

There are many factors affecting this dimension, not least of which are the type and competence of the surface layers, position of the water table, and numbers of drops at each location. Assessment of in-situ results to determine such depths also tends to be subjective and will be affected by the recovery period after treatment. As noted in the previous section, a solid 'plug' of very dense material can form beneath the impact locations to inhibit the improvement to depth. Weak surface soils and a high water table

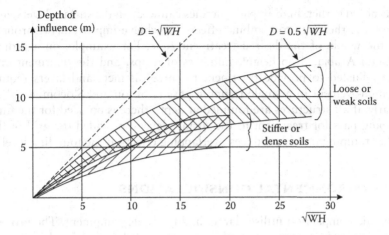

Figure 3.6 Depth of treatment.

can also limit the physical performance of a sufficient number of stress impulses to induce only a minor improvement to the basal layers. However, knowledge of the depth of any stress impulse is a vital factor in both the planning of the treatment operations and the potential for transmission of vibrations as discussed in the next section.

Kinetic energy at the point of impact is clearly a major factor in the depth of treatment and increasing the drop height will increase velocity. In Britain, high-speed photography has shown typical impact speeds of about 35 and 50 mph for 8 tonnes from 12 m and 12 tonnes from 15 m heights to achieve effective depths of treatment of about 5 to 6 m and 6 to 8 m, respectively.

For RIC, the BRE Report No 458 (2003) records depths of treatment of between about 2.0 and 4.0 m at total energy inputs ranging from 80 to 190 tonne-metre/metre^2 for granular fills and silty sands. Greater treatment depths using higher energy inputs in favourable conditions in Japan, Iran, and Canada, together with a range of applications, have been reported by Serridge and Synac (2006).

The shape of dynamic compaction improvement in the ground tends to be similar to the Boussinesq distribution of stresses for a square foundation. Modification of energy levels for each tamping pass can be used to custom-design the treatment scheme to the specific soils profile and engineering requirements. In contrast, the shape of vibro improvement tends to increase with depth. In earthquake areas the required density of soil from the Seed and Idriss (1971) analysis is often better provided at depth by the vibro technique, which has the added advantage of forming stone columns to act as drains in the finer soils. Stone columns can also be combined to reinforce weak cohesive soils at a depth that would be difficult to treat when using dynamic compaction for surface fill layers (Slocombe 1989).

As noted earlier, high-impact energies can weaken the surface layers, and the aim is therefore to combine effects to achieve improvement throughout the whole of the desired treatment zone. For example, on a project in Saudi Arabia, drop height, numbers of drops, and the treatment grid were adjusted to provide treatment to three distinct sand layers requiring improvement in a single tamping pass (Dobson and Slocombe 1982). Clearly, if all structures are founded at depth there is no need for the final tamping pass for treatment to the surface layers, provided the grid of the earlier tamping passes produces overlapping effects at the founding level.

3.8 ENVIRONMENTAL CONSIDERATIONS

Dynamic compaction utilises large, highly visible equipment. The process creates noise and vibration, both of which must be considered in Britain under the Control of Pollution Act, 1974. The standards listed in the reference section provide further details (BS 5228, 2009; BS 7385, 1990; BRE Digest 403, 1995).

Airborne noise levels are generated by a number of causes. Of these the point of impact is by far the highest noise level at typically 110 to 120dB at source. However, its duration only occupies about 0.5% of the lifting cycle. The considerably lower noise values during lifting and idling when combined with the impact noise using the L_{Aeq} calculation method will normally meet most environmental limitations at distances of greater than 50 m from the treatment operations. Lower than normal noise limits can be achieved by working within a specified zone for only a certain number of hours during the working day. Large plate glass windows can sometimes act as diaphragms to change the noise characteristics inside a property. Echoes, wind direction, and angle of crane exhaust are all factors that should also be considered.

By far the most important consideration, however, is ground vibration. In addition to the magnitude of the vibration, the typical frequency of about 5 to 15 Hz is potentially damaging to structures and services, and particularly noticeable to human beings. It is suggested that there are three vibration levels that will influence the design of the treatment scheme. Guide values of resultant peak particle velocity at foundation level for buildings in good condition are

Structural damage	40 mm/s
Minor architectural damage	15 mm/s
Annoyance to occupants	2.5 mm/s

Lower values must be adopted for buildings in poor condition or environmentally sensitive situations such as schools, hospitals, and computer installations. Certain major computer companies recognise the importance of the vibration frequency by requiring more onerous limits for frequencies below 14 Hz than above. It should be noted that some amplification can occur as the vibration rises up certain types of structure; for example, that 1.0 mm/s at ground level could be 2.5 mm/s at the third floor. Services and utilities must be considered on an individual basis depending upon their age, condition, and importance with values of 15 to 20 mm/s normally being considered acceptable, except for higher pressure gas mains.

The level of vibration transmitted through the ground is an imprecise science because of the variable nature of the characteristics of soils. Field measurements of vibrations at ground level have revealed a number of trends, which are illustrated in Figure 3.7. The upper dynamic compaction limits tend to occur in the presence of granular or refuse-type soils and the lower limit in cohesive strata. A high water table will also tend towards the higher limit. The upper vibro limit is for vibrators operating at a frequency of 30 Hz and the lower for 50 Hz.

Careful assessment is required where the soil being treated is directly underlain by relatively dense sand, gravel, or rock which will tend to transmit vibrations to larger than normal distances with comparatively little attenuation. Pre-existing dense surface or buried layers can have a similar

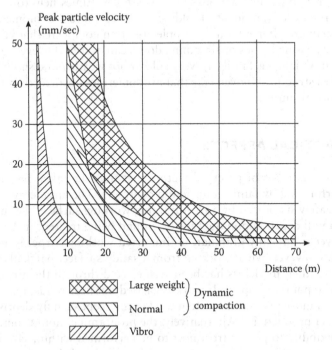

Figure 3.7 Vibrations.

effect of causing the transmission of higher than anticipated vibration levels. The physical performance of the treatment work improves interparticle contact of the soils and, as such, vibration levels can sometimes increase towards the end of the treatment operations even though the final impact energy levels are substantially lower than those performed for the initial tamping passes.

When vibrations become a problem, there are three main methods of reducing their effect. The first is to simply reduce the height of drop and compensate by increasing the number of drops per imprint. This reduces both the impact energy and penetration of the stress impulse that may have attained an underlying dense stratum. The second method of reduction is to utilise a smaller weight and the third is to excavate a cut-off trench to sufficient depth to intercept the surface wave. Cut-off trenches have been found to reduce the transmitted vibration levels by about 50% and are most effective when located near to the structure or service requiring protection.

As the RIC equipment is based on lower impact energy but greater numbers of drops, this method has been employed as close as about 10 to 15 m from an existing structure that was to be extended.

However, human beings are particularly sensitive at detecting vibrations and have a psychological reaction in believing that damage is caused even though the values are far below the well-established damage threshold levels. A thorough public relations exercise can sometimes help to overcome concern among local residents. Building surveys prior to the commencement of treatment are often advisable. People are often not aware that vibrations caused by passing lorries or slamming doors can exceed the levels of minor architectural damage. Similarly, very cold or hot weather and snow loading can lead to structural movements that are sometimes incorrectly attributed to vibrational causes.

3.9 PRACTICAL ASPECTS

There are a number of practical factors that must be taken into account when performing dynamic compaction contracts. The large crawler crane must be safely supported by a free-draining working surface, the thickness of which will depend upon the type of ground being treated. If the surface 1.0 m layer is basically granular, no imported working carpet is generally required. However, when working from a sandy surface, particularly during wet weather, fly-debris has been seen ejected through the air up to 60 m from the point of impact. If work is carried out near roads, railways, or property, a moveable screen is often used to intercept such fly-debris, albeit these affect productivity. Alternatively, the programme should contain sufficient flexibility to permit treatment to be performed within, say, 50 m of such features only when the surface conditions permit its safe operation.

As the RIC foot remains in contact with the ground, there tends not to be fly-debris issues with this equipment.

Where cohesive surface conditions exist, a free-draining granular working carpet is normally required. The thickness can be as little as 150 mm for light energy treatment in reasonably competent soils up to 1.5 m when treating heavily voided refuse fills. When aiming for substantial depth of treatment, thick working carpets of 1.0 m or more have been found to inhibit the stress impulse. A more efficient operation, which also provides greater control of backfill quantities, is to start with only 0.5 m thickness and to backfill imprints directly, thus preserving the working carpet for successive tamping passes. In such cases it is useful to compare the imported quantities to the assessed volume of the directly infilled imprints. The most commonly adopted approach is to blade these preferably coarse (up to 200 mm or single brick size) granular materials into the localised deep imprints using a large dozer. In such cases it is useful to perform a grid of levels before and after treatment to assess the induced settlements.

Winter working will place more onerous requirements on the adequacy of the working surface. The general rule is to increase the depth of the granular working carpet by 25% in comparison to summer thickness. When working in arid climates, there is often no need for any working surface, even for clayey soils.

As the performance of dynamic compaction tends to induce increases in water pressures, a pre-existing groundwater table within about 1.0 to 1.5 m of the working level can inhibit the productivity of the technique. In such cases bottom-feed vibro stone columns may be the preferred approach.

Safe working is a prime consideration. If more than one rig unit is to be used, they should be separated by at least 30 m. Similarly, subsequent operations by the main contractor may have to be delayed until the treatment operations are sufficiently remote. Whilst dynamic compaction can be performed over areas of vibro stone columns it has to be performed before any adjacent piling operations to avoid possibly damaging the constructed piles.

3.10 INDUCED SETTLEMENT

The general densification and collapse of voids will induce general reduction in site levels, the induced settlement being dependent on the total energy input and the manner in which it is applied. Initial shape tests are performed when the soils are loosest. As such, simple extrapolation of these results will overestimate the amount of induced settlement. Mayne et al. (1984), as part of his survey of 124 different sites, reported that the magnitude of induced settlement depended on the applied total energy input, also stating that the thickness of the layer was probably an important factor for

Table 3.1 Approximate induced settlement as percentage of treatment depth

Soil type	% depth
Natural clays	1–3
Clay fills	3–5
Natural sands	3–10
Granular fills	5–15
Refuse and peat	7–20

six soil types. This analysis does not, however, take into account either the initial softness/density of the soils or the proportion of total energy applied by the high velocity initial passes or low velocity final tamping pass that numerically is very significant in determining total energy input. Also, the application of too high an energy in clayey soils will result in less than optimum induced settlement occurring in practice.

A convenient simple approach is to adopt approximate percentages of the target treatment depth for 8-tonne (50 to 100 tonne-metres/m^2) and 12-tonne equipment (100 to 200 tonne-metres/m^2), the total energies with the 12-tonne energy applying to greater depth of treatment.

Higher percentages can be induced. However, the increase in energy input will not be linear (e.g., to increase from 10%–15% induced volume in refuse would require 200%–250% of the normal energy because during the treatment the material becomes progressively stronger and there is less and less potential void reduction available). Care has to be exercised to avoid overtreatment and possible loss in strength in these situations, especially since refuse tips tend to be capped by clay soils.

Loose materials will obviously settle more than denser soils. As noted earlier, ash and certain types of slag also tend to break down during treatment to produce induced movements towards the higher value for granular fills given above (see Table 3.1).

3.11 ADDITIONAL COMMENTS

Dynamic compaction is a highly sustainable technique since it does not use cement or quarried stone, normally only requiring suitable inert free-draining granular waste as a working platform and to infill localised deep imprints. It is also an area treatment technique that permits changes in foundation layouts and localised loadings, for example, mezzanine support foundations, anywhere within the treatment area. A number of dynamic compaction contracts have permitted the rebuilding of developments where fire destroyed the original by simply performing a number of loading tests upon the treated ground.

Pulverised fuel ash (PFA) comprises relatively fine single-sized particles that exhibit pozzolanic properties when properly compacted by conventional methods. When tipped into water, the materials tend to flocculate with little self-compaction occurring with time. Dynamic compaction has been attempted on a number of settlement lagoons with little success. However, dry PFA is considered suitable for treatment, albeit with a number of controls. Similarly, weak chalk tends to crush upon high-energy impact and, when wet, rapidly loses strength. Extreme caution is recommended when considering the feasibility of treating this weak rock.

Peaty soils can be treated in many different ways, depending on the required end result. High energy can be applied to physically displace the material wholescale from beneath the line of a major road. Discrete columns of sand- to cobble-sized fill can be driven into the peat in a manner similar to stone column theory or normal treatment methods applied to simply squeeze out some of the water and preload the ground. The basic fact that must be considered throughout is that peat tends to be a very weak material of high moisture content. As such, the pre-loading method will take time to perform.

Dynamic compaction has been performed to collapse shallow solution cavities. It is important that the extent of these voids be accurately determined prior to treatment so that the crane is positioned sufficiently remote when the cavity caves in to avoid falling into the void. Considerable care has to be exercised during the dropping of the weight to avoid the rope pulling off the crane drum or the weight punching into the void and becoming trapped beneath the surface. Long, narrow weights are better suited to this operation.

Many sites of former heavy industry are now being reclaimed in Britain. These often present the designer with the problems of deep fill and massive obstructions from old basements and foundations. Any technique that has to make a hole in the ground will experience difficulty in gaining adequate penetration, and large excavations often have to be performed to remove the obstructions. However, with dynamic compaction such features can be left in place provided they occur at sufficient depth to avoid excessive differential performance. For most industrial or low-rise housing developments, this depth would be a minimum of 1.0 m below the underside of new foundations or floor slabs, sometimes a depth equal to the width of the buried foundation. The sequence of operation would be for advance earthworks to remove all known features down to a specific level, then perform treatment and normal construction operations. In choosing this excavation level, the designer should take into account the type of structure and its tolerance to some degree of differential performance and the fact that the treatment will induce a reduction in site level that is slightly higher than normal as a result of the loose nature of the surface materials after the pre-excavation operations. It is also normal to apply a slightly higher than normal energy input to reduce the differential performance between the areas of massive foundations and abutting weaker soils which, if weak clay fills, could be significant. Where the backfill comprises large concrete posts, slabs, or waste that could arch over a void, care must be taken in the design of the foundations since the treatment may not necessarily cause the cavity to wholly collapse and could weaken the member forming the roof to the void. In these circumstances, higher than normal energies are preferred to break down such potential.

Brownfield sites with minor contamination are well suited to treatment since the technique does not create a bore that could permit the migration of leachate. Care should be exercised to ensure that the impact energies do not shear any underlying clay layer or basal liner to former landfill that may prevent the downward migration of contaminants into an aquifer. Similarly, if there is a high water level that is contaminated, attention should be paid to the possibility of a rise in level during the treatment and its effect on adjacent property or features. In some cases installation of drains lined with HDPE membrane or monitoring may be required.

Areas contaminated with chemicals or asbestos now have to be developed. The major advantage that dynamic compaction has over alternative methods is that it can be controlled by infilling deep imprints before they penetrate through the working carpet into the contaminated soils, to avoid exposure of hazardous material to the atmosphere while still compacting soils at depth.

Sites of former quarries are prime situations for treatment in view of the potential for piles to glance off the buried subvertical face between fill and rock, and the possibility of constructing piles to inadequate depth as a result of false readings from boulders or inaccurate historical information on the depth of the quarry.

Similarly, former opencast coal mining sites are now being treated, even when the fills are placed to better than 95% maximum dry density, to further reduce voids for better settlement control. One such site involving load testing recorded typical pre-treatment short-term moduli within the range of 16–18 MPa being increased to 30–40 MPa using 8-tonne weights and 50–60 MPa with 12-tonne weights.

3.12 TESTING

Many contracts have simply involved the measurement of depth of first pass imprints and monitoring of site levels. Post-treatment in-situ (SPT and CPT) and loading tests are often performed, and since the technique provides treatment to large areas very quickly, the speed at which such tests provide the necessary information is important, particularly if testing between tamping passes. It is rare, therefore, to recover samples for

Table 3.2 Suitability for testing dynamic compaction

Test	Granular	Cohesive	Comments
Dynamic cone	**	*	Too insensitive to reveal soil type. Has difficulty penetrating densely compacted ground.
Electric cone	***	*	Particle size important. Can be affected by lateral earth pressures generated by treatment. Best test for seismic liquefaction evaluation in sands.
Boreholes and SPT	***	**	Efficiency of test important. Recovers samples.
Small plate	*	*	Poor confinement to zone being tested. Affected by pore-water pressures.
Large plate	**	*	Better confining action.
Skip	**	**	Can maintain for extended period.
Zone loading	****	****	Best test for realistic comparison with foundations.
Full-scale	*****	*****	Rare.

Key: * least suitable, ***** most suitable

laboratory testing. In clayey soils, as with the performance of the treatment, it is essential that sufficient recovery period be allowed to avoid ambiguous results.

It is common for dynamic compaction to be performed for sites underlain by coarse fills or including obstructions that would cause penetration problems for vibro or piling methods. Similar problems could therefore be reasonably anticipated in attempting to perform in-situ tests. Air drills have been used on a small number of contracts to predrill a test location to below the level of the potential obstruction. However, surface loading tests only are more normally performed in this situation.

Table 3.2 describes the relative merits of various test methods. Additional comments on the advantages and limitations of certain in-situ tests follow.

3.12.1 Standard penetration test

This is probably the most useful in-situ test as it is applicable to both granular and cohesive soils. However, being of a dynamic nature it is particularly sensitive to the presence of residual pore-water pressures, quickly liquefying the stratum being tested and producing lower than expected results. A sample is normally recovered and the speed of provision of information is adequate for most contracts. The main drawback is that a considerable amount of time and money can be spent chiselling to penetrate the very dense surface layers normally provided by the treatment.

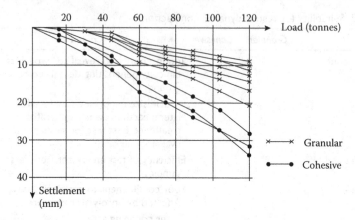

Figure 3.8 Post-treatment zone loading test results.

Tests normally reveal superior performance being achieved when treating granular materials in comparison to clayey soils; see Figure 3.8 for identical total energy input and zone test size for different site areas of granular and cohesive fills.

3.12.2 Pressuremeter

The dynamic compaction technique has been historically associated with this test. While this method is often used in mainland Europe, it is now rarely used in Britain.

3.12.3 Dynamic cone test

This is relatively cheap and robust but is limited by the inability to determine, without the performance of alternative parallel testing, whether a zone of low blow count is caused by loose zones or cohesive layers that would be expected to respond differently to the treatment. Since the majority of ground improvement contracts performed in Britain require treatment to variable fill sites this method has been found to be of limited value.

3.12.4 Static cone penetration test

These tests are ideally suited to the testing of sands because they illustrate the soil type by means of the friction ratio. They are considered less successful in clayey soils since experience has shown that this test is particularly affected by the presence of residual pore-water pressures. Being relatively sophisticated it is not recommended that these tests be used in the presence of coarse fill materials.

3.12.5 Dilatometer

This method would appear to have potential. However, no information has yet been published for its evaluation of the treatment of the soils and fills of Britain.

3.13 CASE HISTORIES

As part of a larger vibro compaction contract to densify clean sand fills to up to 18 m depth, dynamic compaction was performed to areas of surface sands to depths of up to 4 m where vibro was not performed. The sand fills in the DC areas were by design of significantly higher fine contents than the vibro areas with up to 40% total fines content, typical CPT friction ratios being up to 2.0% plus local pockets and thin layers of clay. The specified requirement was for post-treatment CPT tests with minimum relative density of 75% using the Jamiolkowski et al. (1985) method and clay layers to be for inferred undrained cohesion of at least 40 kPa.

As the groundwater was at 1.5 m depth, advance trials were performed to confirm that the required performance could be achieved. The trials were successful and the treatment proceeded using typically two, or locally three depending upon visual monitoring, tamping passes with 8-tonne weight dropped from heights of up to 8 m. Post-treatment CPT tests confirmed the required sand density and that all weak clay pockets and layers had been improved to minimum 60 kPa, typically 100 kPa. This success with the clay layers is considered to be due their thicknesses being less than 200 mm, this resulting in very short drainage paths for dissipation of the generated excess pore-water pressures, combined with very careful attention to the recovery periods between the tamping passes. The soils were considered unsuitable for treatment by RIC.

The second site has received a number of visits to combine dynamic compaction to up to 6 m depth, with compaction grouting to chalk solution features to up to 15 m depth, to ensure the integrity of the proposed basal liner and leachate extraction system beneath substantial depth of landfill.

The works were performed in the base of a former sand quarry where the exposed solution features were clearly visible in their local extents. These were then investigated by CPTs on nominal 3 m grid to assess whether dynamic compaction alone was required or whether compaction grouting had to be performed prior to the dynamic compaction. Some treatment areas were as small as 10 m^2, some as large as 1200 m^2. The solution features had been infilled by variable mixtures of sandy clays varying in strength between very loose/soft and stiff, sandy silt, sand and medium-dense gravelly sands. Areas of intact chalk were not subjected to treatment.

Post-treatment testing of the compaction grouting was performed by CPT prior to the DC treatment. The dynamic compaction was then performed using three tamping passes with 8-tonne weight and drop heights of up to 12 m to achieve an undrained Young's modulus of 50 MPa, as proven by 1.5 × 1.5 m large plate test loaded to 150 kPa.

The construction of a football stadium in Poland required treatment to depths of 15 m and 10 m beneath the stadium and pitch areas, respectively. A minimum CPT q_c of 15 MPa was specified below 1.0 m depth to permit the adoption of foundation design pressures of up to 300 kPa.

The upper 3–4 m depth comprised peat and organic silt. These were to be excavated and replaced by sandy soils. The soils to about 10 m depth then comprised fine silty sands of about 10% silt content with pre-treatment qc of about 5 MPa. The sands then became slightly coarser with up to 5% silt content and qc varying between about 6 and 15 MPa. The sands also contained thin silty layers and occasional traces of organics. Groundwater was at about 1.5 m depth. As there were existing structures on several site boundaries, the ground improvement design was based on vibro compaction to the deeper sands followed by dynamic compaction to the upper layers. Extensive vibro trials were first performed, followed by dynamic compaction trials. These revealed that wide vibro compaction grids alone could achieve the specified q_c > 15 MPa at depth, but not to the shallower layers. Closer and/or secondary vibro compaction grids to suitable depth could then be performed or the dynamic compaction be designed to treat to greater depth. Comparative costings revealed the optimum approach to be relatively wide vibro compaction grids followed by dynamic compaction using a 16-tonne weight dropped in free-fall from almost 20 m height to improve the upper 8 m depth of the sands. The high groundwater table would, however, have inhibited the efficient performance of such DC treatment, with water entering the deep imprints before the full drop numbers could be performed, thereby requiring phased treatment passes with suitable recovery periods between. Site levels were therefore raised by about 1.0 m to permit just two DC treatment passes for faster programme.

Post-treatment testing by CPT confirmed all specified criteria had been achieved (see Figure 3.9) with recorded settlements induced by dynamic compaction, performed after the vibro compaction, of about 200 to 300 mm.

3.14 CONCLUDING REMARKS

The dynamic compaction method is a powerful tool when applied to suitable sites. A large database has been collated over the years to define its limitations and, more importantly, its capabilities, which can be utilised with confidence. As with every specialist technique, the designer and contractor performing this method of ground improvement must understand these capabilities and limitations. Such understanding can only arise by experience.

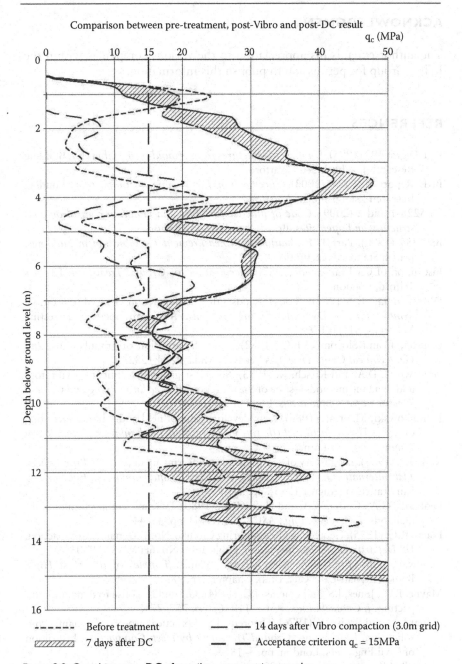

Comparison between pre-treatment, post-Vibro and post-DC result

Figure 3.9 Case history—DC after vibro compaction results.

ACKNOWLEDGMENT

The author extends his appreciation to the various companies within the Keller Group for permission to publish this information.

REFERENCES

BRE Digest 403 (1995). *Damage to structures from ground-borne vibration.* Building Research Establishment, Watford.
BRE Report BR 458 (2003). *Specification for Dynamic Compaction.* Building Research Establishment, Watford.
BS 5228-1 and 2 (2009). *Code of practice for noise and vibration control on construction and open sites.* British Standards Institution.
BS 7385 (1990). Part 1. *Evaluation and measurement for vibration in buildings.* British Standards Institution.
Institution of Civil Engineers (1987). *Specification for Ground Treatment.* Thomas Telford, London.
D'Appolonia, E. (1953). Loose sands—their compaction by vibroflotation. *Symposium on Dynamic Testing of Soils,* American Society of Testing Materials, STP, p.156.
Dobson, T. and Slocombe, B.C. (1982). Deep densification of granular fills. *2nd Geotechnical Conference,* Las Vegas, Nevada. April 1982.
Greenwood, D.A. and Kirsch, K. (1983). Specialist ground treatment by vibratory and dynamic methods—state of the art report. *Proceedings, Piling and Ground Treatment for Foundations,* London, pp. 17–45.
Jamiolkowski, M. et al. (1985). New developments in field and laboratory testing of soils. *Theme lecture, 11th Int. Conf. on Soil Mechanics and Foundation Engineering,* San Francisco.
Kerisel, J. (1985). The history of geotechnical engineering up until 1700. *Proceedings, 11th International Conference on Soil Mechanics and Foundation Engineering,* San Francisco (August 1985), pp. 3–93.
Leonards, G.A., Cutter, W.A. and Holtz, R.D. (1980). Dynamic compaction of granular soils. *J. Geotech. Eng., ASCE* 106 (GT1), pp. 35–44.
Lucas, R.G. (1995). Geotechnical Engineering Circular No. 1: Dynamic Compaction, *US Department of Transportation,* Publication No. FHWA-SA-95-037.
Lundwall, N.B. (1968). The Saint George Temple. *Temples of the Most High,* Bookcraft, Salt Lake City, Utah, Chapter 3, p. 78.
Mayne, P.W., Jones, J.S. and Dumas, J.C. (1984). Ground response to dynamic compaction. *J. Geotech. Eng., ASCE* 110 (6), pp. 757–774.
Menard, L. and Broise, Y. (1976). Theoretical and practical aspects of dynamic consolidation. *Proceedings, Ground Treatment by Deep Compaction,* Institution of Civil Engineers, London, pp. 3–18.
Mitchell, J.K. and Katti, R.K. (1981). Soil improvement – state of the art report. *Proceedings, 10th International Conference on Soil Mechanics and Foundation Engineering,* Stockholm (June 1981), pp. 509–565.

Mitchell, J.K. and Solymar, Z.V. (1984). Time-dependent strength gain in freshly deposited or densified sand. *J. Geotech. Eng., ASCE* 110 (11), pp. 1559–1576.

Rollins, K.M. and Kim, J. (2010). Dynamic compaction of collapsible soils based on US case histories. *J. Geotech. Geoenviron. Eng., ASCE* 133 (9), pp. 1178–1186.

Seed, H.B. and Idriss, I.M. (1971). Simplified procedure for evaluating soil liquefaction potential. *J. Geotech. Eng., ASCE* 97 (SM9), pp. 458–482.

Serridge, C.J. and Slocombe, B. (2012). Ground improvement. *ICE manual of geotechnical engineering*. Eds. Burland, Chapman, Skinner and Brown. Chapter 84, pp. 1247–1269.

Serridge, C.J. and Synac, O. (2006). Application of the Rapid Impact Compaction (RIC) technique for risk mitigation in problematic soils. *Engineering Geology for Tomorrow's Cities*. Eds. Culshaw, Reeves, Jefferson and Spink. *Engineering Geology Special Publication* 22, Paper 294 (CD-ROM), Geological Society of London.

Sharma, H.D. and Anirban, D. (2007). Municipal solid waste landfill settlement: Postclosure perspectives. *Journal of Geotechnical and Geoenvironmental Engineering, ASCE* 133 (6), pp. 619–629.

Slocombe, B.C. (1989). Thornton Road, Listerhills, Bradford. *Proceedings, International Conference on Piling and Deep Foundations*, London (May 1989), pp. 131–142.

West, J.M. (1976). The role of ground improvement in foundation engineering. *Proceedings, Ground Treatment by Deep Compaction*, Institution of Civil Engineers, London, pp. 71–78.

Chapter 4

Prefabricated vertical drains

Jian Chu and Venu Raju

CONTENTS

4.1 INTRODUCTION

Preloading is one of the most common ground improvement techniques for soft clay. The modern application of the preloading method is combined almost always with the use of vertical drains or prefabricated vertical drains (PVDs). PVDs have been used successfully in many soil improvement and land reclamation projects in the world (Hansbo, 1979, 2005; Holtz, 1987; Holtz et al., 1991; Balasubramaniam et al., 1995; Bergado et al., 1990, 1993a, 1993b, 1996, 2002; Li and Rowe, 2001; Chu et al., 2004; 2009a; Choa et al., 2005; Bo et al., 2003; 2005; Arulrajah et al., 2004; Indraratna et al., 2005; Seah, 2006; Kitazume, 2007; Varaksin and Yee, 2007). Therefore, the theories, design, and construction methods for PVDs have become the core technical issues in the preloading or consolidation method. In recent years, PVDs or their variations have also been used for other purposes such as for dissipation of pore water pressures for liquefiable sand (Towhata, 2008; Chu et al., 2009a) or in environmental engineering for vapour extraction system (Schaefer et al., 1997; Collazos et al., 2002).

Depending on how a preload is applied, the preloading methods can be subdivided into preloading using fill, preloading using vacuum pressure, and combined fill and vacuum preloading methods, as described in Table 4.1. In addition to preloading, PVDs have also been used for some other relatively new methods such as dynamic consolidation for clays, which are also listed in Table 4.1. In most of the applications, the main purpose of using PVDs is to reduce the drainage path so that the time taken for the consolidation of soft soil or the dissipation of excess pore water pressure can be substantially reduced.

The practice of using vertical drains started with sand drains and then evolved into PVDs. According to Hansbo (2004), the use of vertical sand drains was first proposed in 1925, and patented in 1926 by Daniel D. Moran. A sand drain is formed in situ by placing sand directly into a borehole or into a "sock" made of geosynthetic fabric or geotextile in a borehole. One of the most well-documented case histories for the use of sand drains is the test field at Skå-Edeby in Sweden (Hansbo, 1960). However, there are construction constraints in the installation of sand drains. PVDs, also named band drains or wick drains, were introduced as a better alternative. The first type of PVD was developed by Walter Kjellman in 1947 (Kjellman, 1948). It was made of wood and cardboard. Nowadays most PVDs are made of corrugated plastic cores surrounded by geotextile filters.

The size of the PVDs has been standardised to a width of around 100 mm and a thickness of 3 to 6 mm. Although PVDs are used for almost all the projects, sand drains are still in use for some special projects. One example is the Kansai International Airport where 400 mm diameter sand drains were installed from an offshore barge into the soft seabed soils (Kitazume, 2007). PVDs made of natural products such as jute or coconut coir (Lee et al., 2003) are also available.

A great deal of experience and development in both research and construction have been accumulated in the past as a result of the extensive use of PVDs. Two books devoted exclusively to the use of PVDs for soil improvement have been published (Holtz et al., 1991; Bo et al., 2003). A number of other books and reports with good coverage of PVDs have also been published. These include Mitchell and Katti (1981), ICE (1982), Jamiolkowski (1983), Akagi (1994), Bergado (1996), Moseley and Kirsch (2004), Raison (2004), Hansbo (2005), Indraratna and Chu (2005), Chu et al (2009a), and Chu et al. (2012).

This chapter intends to provide a practical guide to researchers and practicing engineers who have to deal with design and construction issues related to PVDs. It also tries to introduce briefly the latest development and technologies in the use of PVDs. Two case studies are also given to illustrate the real operation of soil improvement using PVDs.

4.2 CONSOLIDATION THEORIES AND ANALYSIS

The design of PVDs provides solutions to questions such as (1) what is the drain spacing required to achieve a required degree of consolidation within a given time; or (2) how long will it take to achieve a required degree of consolidation for a given drain spacing and duration? To answer these questions, consolidation theories are required.

The most fundamental consolidation theory is Terzaghi's one-dimensional consolidation equation, which can be written as

$$\frac{\partial u_e}{\partial t} = c_v \frac{\partial^2 u_e}{\partial z^2} \quad \text{and} \quad c_v = \frac{k_v}{m_v \gamma_w} \tag{4.1}$$

where c_v is the coefficient of consolidation of soil in the vertical direction, u_e is the excess pore water pressure, t is the real time, z is the position of the soil element, k_v is the coefficient of permeability of soil in the vertical direction, $m_v = \Delta\varepsilon_v/\Delta\sigma_v'$ is the coefficient of vertical compression, and γ_w is the unit weight of water.

Equation 4.1 is derived under the assumptions that the dissipation of water is only in the vertical direction (i.e., it can only be applied for consolidation with vertical flow). However, with the use of PVDs, water flows

Table 4.1 Ground improvement methods where PVDs may be used

Method		Description/Mechanisms	Typical applications	Advantages	Limitations
Preloading methods	A. Preloading using fill with vertical drains	Preloading is a process to apply surcharge load on to ground prior to the placement of structure or external loads to consolidate the soil until most of primary settlement has occurred. Vertical drains are used to provide radial drainage and accelerate the rate of consolidation by reducing the drainage paths.	The method is applicable to soils having low permeability or when the compressible soil layer is thick.	Rate of consolidation can be greatly accelerated. The construction time can be controlled by adjusting the spacing of the drain.	The method may not be applicable when the construction schedule is very tight or when the ground is so soft that vertical drains cannot be installed.
	B. Vacuum preloading with vertical drains	The method is the same as A, except the surcharge is applied using vacuum pressure. The vacuum pressure is usually distributed through vertical drains. It also provides immediate stability to the system. The treated soil is enclosed by an air- and watertight barrier in all directions.	The method is applicable to ground and consists of mainly saturated, low permeability soils. The method can be used when there is a stability problem with fill surcharge. This method can also be used to extract polluted ground pore water, if required.	1). The method does not require fill material; 2). The construction period can be shorter, as no stage loading is required; 3). It may be more economical than using fill surcharge; 4). The vacuum brings immediate stability to the system.	1). This method causes inward lateral movement and cracks on the ground surface, which may affect surrounding buildings or structures; 2). The vacuum pressure is limited to 50–90 kPa, depending on the system adopted.
	C. Combined fill and vacuum preloading with vertical drains	The method is a combination of A and B when a surcharge more than the limit of vacuum pressure is required.	The same as for A and B.	1). Construction time can be much reduced as compared to staged loading using fill surcharge alone;	1). It is technically more demanding than A and B; 2). Data interpretation is also more complicated.

Category	Method	Principle	Applications	Advantages	Limitations
				2). The lateral movement of soil can be controlled by balancing the amount of vacuum and fill surcharge used; 3). The vacuum brings immediate stability to the system.	
Dynamic consolidation methods	D. Drainage enhanced dynamic consolidation (DC)	This method improves the soil properties by combining the DC method with vertical drains which facilitates the dissipation of pore water pressure generated during DC.	This method can be used to improve the bearing capacity of soft soil with low permeability.	1). This method makes the application of DC possible to fine-grained soil; 2). The duration of soil improvement can be reduced.	1). The method may only work for cohesive soil with relatively low plasticity index; 2). The compaction energy applied has to be within a certain limit, so that the depth of improvement is limited; 3). The technique has not been fully developed. Thus, the success of the method cannot always be guaranteed.
	E. Dynamic consolidation (DC) combined with vacuum or de-watering	This method improves the soil properties by conducting DC and applying vacuum or de-watering alternately for a number of times. The vacuum facilitates a quick dissipation of water pressure generated by DC.	This method can be used to improve soft clay or soft ground with mixed soil.	1). The soil improvement time can be reduced; 2). Can be applicable to most types of soil.	1). The method has not been fully established. Thus, the success of the method cannot always be guaranteed; 2). The depth of improvement is normally limited to 8 m.

Source: Modified from Chu et al. 2009.

mainly in the horizontal direction. In this case, the following radial consolidation theory is required (Barron, 1948):

$$c_h\left(\frac{\partial^2 u_e}{\partial r^2} + \frac{1}{r}\frac{\partial u_e}{\partial r}\right) = \frac{\partial u_e}{\partial t} \quad \text{and} \quad c_h = \frac{k_h}{m_v\gamma_w} \tag{4.2}$$

where c_h is the coefficient of consolidation of soil in the horizontal direction, k_h is the coefficient of permeability of soil in the horizontal direction.

Equation 4.2 was derived under the following two assumptions (Barron, 1948): (1) All vertical loads are initially carried by excess pore water pressure; and (2) All compressive strains within the soil occur in the vertical directions.

When combining the vertical and horizontal flow, the consolidation equation becomes (Carillo, 1942):

$$c_v\frac{\partial^2 u_e}{\partial z^2} + c_h\left(\frac{\partial^2 u_e}{\partial r^2} + \frac{1}{r}\frac{\partial u_e}{\partial r}\right) = \frac{\partial u_e}{\partial t}. \tag{4.3}$$

For ground improvement, the progress of consolidation of soil is evaluated using the average degree of consolidation, U, of the compressible clay layer:

$$U = \frac{S_c(t)}{(S_c)_{ult}}100\% \tag{4.4}$$

where $S_c(t)$ is the consolidation settlement at a given time and $(S_c)_{ult}$ is the ultimate consolidation settlement.

Using Terzaghi's consolidation equation, Equation 4.1, and by assuming the form of initial pore water pressure distribution, a relationship between the average degree of consolidation and the time factor $T_v = c_v t/H_d^2$ can be established as shown in Figure 4.1 or expressed approximately in equations by curve fitting. One of the closed-form equations is given by Sivaram and Swamee (1977):

$$U_v = \frac{S_c(t)}{(S_c)_{ult}}100\% = \frac{(4T_v/\pi)^{0.5}}{[1 + (4T_v/\pi)^{2.8}]^{0.179}} \tag{4.5}$$

where U_v is the average degree of consolidation due to vertical flow.

For radial consolidation, the following solution was given by Barron (1948) by assuming equal strain and all the other assumptions adopted for Equation 4.1:

$$U_h = 1 - \exp\left[\frac{-8T_h}{F(n)}\right] \tag{4.6a}$$

$$F(n) = \frac{n^2}{(n^2-1)}\ln(n) - \frac{(3n^2-1)}{4n^2} \approx \ln(n) - 0.75 \tag{4.6b}$$

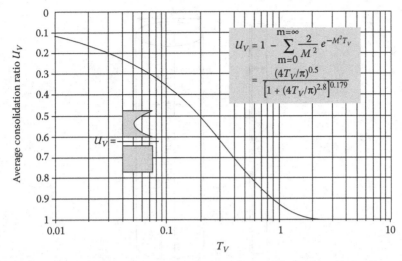

Figure 4.1 Relationship between average degree of consolidation U_v and time factor T_v derived based on solutions to Equation 4.1. (Redrawn from Bo, M.W., Chu, J., Low, B.K. and Choa, V. (2003). *Soil Improvement: Prefabricated Vertical Drain Technique*, Thomson Learning, Singapore.)

$$T_h = \frac{c_h t}{d_e^2} \tag{4.7}$$

$$n = \frac{d_e}{d_w} \tag{4.8}$$

where T_h is the time factor, n is the ratio between the diameter of soil discharging water into a vertical drain, d_e, and the diameter of the drainage well or equivalent diameter of a vertical drain, d_w, as shown in Figure 4 2. In this figure, a and b are the thickness and width of band drain.

The solutions given in Equations 4.6 through 4.8 were developed for a unit cell (i.e., a cylindrical column of soil surrounding a circular well, as shown in Figure 4.2). A graphical illustration of the relationships between U_h and T_h for different n values is shown in Figure 4.3. The curve shown in Figure 4.1 for vertical flow is also plotted in Figure 4.3 for comparison.

However, it should be noted that a PVD band drain is not circular. It has a typical width of 100 mm and thickness ranging from 3–6 mm. To use Equations 4.6 through 4.8, a conversion to compute the equivalent drain diameter, d_w, is thus necessary. One conversion method based on an equal perimeter has been proposed by Hansbo (1979) as:

$$d_w = \frac{2(a + b)}{\pi} \tag{4.9}$$

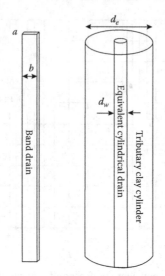

Figure 4.2 Barron's equal-strain solution for radial drainage in a cylindrical cell. (Redrawn from Bo, M.W., Chu, J., Low, B.K. and Choa, V. (2003). *Soil Improvement: Prefabricated Vertical Drain Technique*, Thomson Learning, Singapore.)

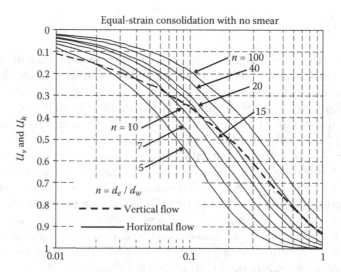

Figure 4.3 Relationships between U_h and T_h for different n according to Equation 4.6 and U_v and T_v according to Equation 4.5. (Redrawn from Bo, M.W., Chu, J., Low, B.K. and Choa, V. (2003). *Soil Improvement: Prefabricated Vertical Drain Technique*, Thomson Learning, Singapore.)

The diameter of soil discharging water into a vertical drain, d_e, is dependent on the drain spacing and the drain installation pattern. The d_e is calculated based on equivalent cross-section area. For PVDs installed in a square grid pattern with a spacing s as shown in Figure 4.4a, d_e can be calculated as:

$$s^2 = \frac{\pi d_e^2}{4}, \quad \text{i.e.,} \quad d_e = 1.128s \tag{4.10}$$

For PVDs installed in a triangle grid as shown in Figure 4.4b, d_e can be calculated as:

$$s^2 \sin 60° = \frac{\pi d_e^2}{4}, \quad \text{i.e.,} \quad d_e = 1.05s \tag{4.11}$$

For relatively long PVDs, consolidation of clay is controlled by horizontal drainage. However, for relatively short PVDs, both vertical and horizontal drainage may contribute a fair proportion. In this case, the combined degree of consolidation U_{vh} can be calculated using Carillo's equation (Carillo, 1942):

$$\left(1 - U_{vh}\right) = \left(1 - U_v\right) \times \left(1 - U_h\right) \tag{4.12}$$

In this case, U_v and U_h should be calculated separately using Equations 4.5 and 4.6.

It should be pointed out that Equation 4.6 was derived by assuming that the well resistance of the PVD can be ignored. If the well resistance has to be considered, the $F(n)$ equation in Equation 4.6b will have to be changed into (Hansbo, 1981):

$$F(n) \approx \ln(n) - 0.75 + \pi z(2l - z)\frac{k_h}{q_w} \tag{4.13}$$

where l = longest drainage path along vertical drain; z = depth; and q_w = discharge capacity of PVD.

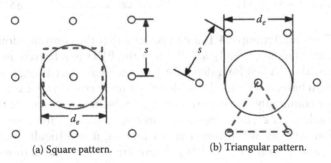

(a) Square pattern.　　　　　(b) Triangular pattern.

Figure 4.4 Patterns of PVD installations.

It can be seen from Equation 4.13 that when q_w is sufficiently large, the last term in Equation 4.13 will become very small so that the effect of well resistance can be ignored. Some modern high-quality PVD products can provide sufficiently large q_w. Therefore, well resistance may be ignored in the design when PVDs with sufficient large q_w are used. Practically this requires the q_w value of PVDs to be specifically checked as part of the quality-control process during the construction. It is thus important to ensure that the PVDs are selected properly and the quality of the PVD products is checked whenever PVDs are used for soil improvement (Chu et al., 2004).

Working Example 4.1

PVDs were installed in a compressible clay layer of 10 m thickness in a square pattern with a spacing of 2 m. The PVD used is 100 mm wide and 4 mm thick. The coefficients of permeability of the clay in the vertical and horizontal directions is 2.0 m²/year and 3.0 m²/year, respectively. The boundary below the clay was impervious. Calculate the degree of consolidation achieved in one year's time.

Solution:
$d_e = 1.128s = 1.128 \times 2 = 2.256$ m $= 2256$ mm
$d_w = 2(a + b)/\pi = 2(100 + 4)/3.14 = 66$ mm
$n = d_e/d_w = 2256/66 = 34$
$F(n) = \ln(n) - 0.75 = \ln(34) - 0.75 = 2.78$
Time factor due to radial drainage:
$T_h = c_h t/d_e^2 = 3 \times 1/2.256^2 = 0.589$
Degree of consolidation due to radial drainage:
$U_h = 1 - \exp[-8T_h/F(n)] \times 100\% = 1 - \exp[-8 \times 0.589/2.78] = 82\%$
Time factor due to vertical drainage:
$T_v = c_v t/H_d^2 = 2 \times 1/10^2 = 0.02$
Degree of consolidation due to vertical drainage:
$U_v = (4T_v/\pi)^{0.5}/[1 + (4T_v/\pi)^{2.8}]^{0.179} \, 100\% = 16\%$
The combined degree of consolidation is:
From $(1 - U_{vh}) = (1 - U_h)(1 - U_v)$, we can calculate $U_{vh} = 85\%$

From Working Example 4.1, it can be seen that the consolidation due to vertical drainage is normally small when the PVD is relatively long (say, more than 10 m). Therefore, depending on the design situation, it is possible to design based on radial drain alone as a more conservative estimation.

Another commonly encountered design problem is to calculate the time taken to achieve a certain degree of consolidation, U_{vh}, for a given PVD installation scheme. When time t is not known, it is difficult to calculate T_v and T_h and thus U_v, U_h, and U_{vh}. There are three methods to solve this problem.

(1) Take $U_h = U_{vh}$ and calculate t using Equation 4.6 by ignoring the contribution of vertical consolidation.

(2) By a trial and error method. The t calculated in method 1 can be used as the first estimate to calculate U_{vh}. If U_{vh} is greater than the assumed, then a smaller t can be used to calculate U_{vh} again until the U_{vh} value matches the assumed.

(3) Use a special function "GoalSeek" in the Microsoft Excel to carry out iterations automatically to obtain the time required to achieve the required degree of consolidation. The detail of the third method and Excel code that can be used for this purpose are provided in Bo et al. (2003).

It should be pointed out that these solutions are provided for perfect drain conditions (i.e., the installation of PVDs does not affect the soil properties). However, during the PVD installation process, the soil around the PVDs is disturbed or smeared. The 'smear' effect comes from the compressibility of soil and the disturbance to the soil structure during the insertion and removal of the mandrel (see Section 4.3 for details). The zone in which the soil is disturbed or smeared is called the smear zone, as shown in Figure 4.5. The diameter of the smear zone, d_s, varies from soil to soil and is also affected by the size of the mandrel. Based on past studies (e.g., Holtz and Holm, 1972; Hansbo, 1981, 1983; Indraratna and Redana, 1998; Onoue et al., 1991; Hird and Moseley, 2000; Xiao, 2002), Bo et al. (2003) have proposed that d_s to be estimated as:

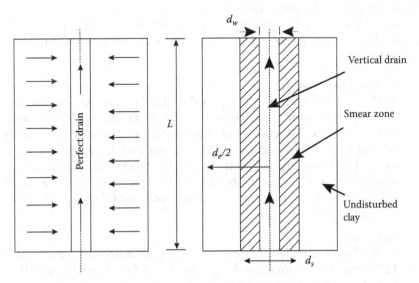

Figure 4.5 Analytical model of smear zone around vertical drain.

$$d_s = (4 \text{ to } 7) \, d_w \qquad \text{or} \qquad d_s = (3 \text{ to } 4) \, d_m \qquad\qquad (4.14)$$

where d_m = equivalent diameter of the mandrel, d_w = equivalent drainage diameter.

As a result of smear or sample disturbance, the coefficient of permeability or coefficient of consolidation in the smeared zone is greatly reduced compared to the intact soil. The reduction in permeability can be normally taken as 2 or 3 (i.e., the permeability of the smeared soil, k_s, is 2 to 3 times smaller than that of the intact soil, k_h):

$$k_h = (2 \text{ to } 3) \, k_s \qquad\qquad (4.15)$$

However, the study of Bo et al. (2003) and Chu et al. (2004) indicates that the reduction in permeability can be as large as 2 to 10 times depending on the sensitivity of the soil.

By taking the smear effect into consideration, Barron (1948) and Hansbo (1979, 1981) have derived another set of consolidation equations by assuming an annulus of the smeared clay around the drain with a diameter of d_s and a permeability of k_s:

$$U_h = 1 - \exp\left[\frac{-8T_h}{F_s(n)}\right]$$

$$F_s(n) \approx \ln(n) - 0.75 + \ln(s)\left(\frac{k_h}{k_s} - 1\right) \qquad\qquad (4.16b)$$

where $s = d_s/d_w$ is the smear ratio.

It can be seen from Equation 4.16 that when $s = 1$ and $k_h = k_s$, so when there is no smear effect, Equation 4.16 becomes identical to Equation 4.6.

Working Example 4.2

PVDs were installed in a compressible clay layer of 10 m thick in a square pattern with a spacing of 2 m. The PVD used is 100 mm wide and 4 mm thick. The coefficient of permeability of the clay in the vertical and horizontal directions is 2.0 m²/year and 3.0 m²/year, respectively. The boundary below the clay was impervious. Assuming the smeared ratio is 3 and the coefficient of permeability of the smeared soil is 1.5 m²/year. Calculate the degree of consolidation achieved in one year's time.

Solution:

In Working Example 4.1, we have already calculated $d_e = 2.256$ m = 2256 mm, $d_w = 66$ mm, $n = d_e/d_w = 34$, $T_h = 0.589$, $T_v = 0.02$, and $U_v = 16\%$. Using Equation 4.16b:

$F_s(n) = \ln(n) - 0.75 + \ln(s)(k_h/k_s - 1) = \ln(34) - 0.75 + \ln(3)(2 - 1) = 3.88$

Degree of consolidation due to radial drainage:

$U_h = 1 - \exp[-8T_h/F_s(n)]\ 100\% = 1 - \exp[-8 \times 0.589/3.88] = 70\%$

The combined degree of consolidation can be calculated using Equation 4.12:

From $(1 - U_{vh}) = (1 - 0.7)(1 - 0.16)$, we have $U_{vh} = 75\%$.

Comparing the answers for the two working examples, it can be seen that the degree of consolidation U_{vh} has reduced by 10% from 85% to 75% due to smear effect.

It should be pointed out that all the above analytical methods were established based on Darcian flow. For non-Darcian flow, solutions have also been provided by Hansbo (1997; 2001; 2004). However, the non-Darcian flow consolidation theory has not been widely used in practice. Furthermore, the effect of non-Darcian flow on one-dimensional consolidation is negligible in the beginning of the consolidation process (Hansbo, 2004). As the non-Darcian flow consolidation equations have been presented elsewhere (Hansbo, 1997; 2001; 2004; 2005), it will not be elaborated in this chapter.

Note that the use of the analytical solutions presented above is restricted to the assumptions of one-dimensional, linear-elastic, small strain behaviour of soil. The spatial variation of soil properties is also not taken into consideration. For the modelling of two- or three-dimensional boundary value problems with spatial variation of soil properties, such as consolidation under an embankment, the finite element analysis should be used. Several numerical procedures have been developed for this purpose. However, the coverage of these topics is beyond the scope of this chapter. Interested readers are referred to Hird et al. (1992); Bergado et al. (1993b); Chai et al. (1995; 2001); Indraratna and Redana (1997; 2000); Indraratna et al. (2005); Rujikiatkanjorn et al. (2008); and Chu et al. (2012).

4.3 DESIGN

For the design of a soil improvement project using preloading and PVDs, the following design and construction procedure can be adopted:

1. Conduct proper site investigation to establish the soil profile on site, characterise the geotechnical properties of the soil, and determine the design parameters.
2. Determine the depths that PVDs need to be installed and the pattern of installation.
3. Select PVDs that meet the design specifications and design requirements.
4. Calculate the drain spacing required to achieve the required design specification.

5. Estimate the ground settlement and draw the surcharge placement plan.
6. Install PVDs and carry out quality-control tests and inspections during the PVD installation at predetermined intervals.
7. Design a field instrumentation scheme, install instruments, collect field monitoring data, and monitor the soil improvement process.
8. Calculate the degree of consolidation and other design parameters used for design and check whether design specifications have been met.

Some of these steps will be explained in detail in the following sections. The flow chart illustrating the above procedure is also shown in Figure 4.6.

4.3.1 Determination of design parameters

Once the consolidation theories and methods of analysis are in place, the next step in the design process is to obtain soil parameters to feed into the equations or computer software for analysis. This is not a simple task as the determination of soil parameters is still one of the most challenging tasks facing geotechnical engineers. We need to obtain a value for each soil parameter, but few soil parameters are constant. For example, the coefficient of consolidation is assumed to be a constant in either Terzaghi's or Barron's consolidation theory (i.e., Equation 4.1 or 4.2). However, in practice, the coefficient of consolidation for soft soil is not a constant. Its value is affected by many factors, such as the overconsolidation ratio, the stress state, the fabric of the soil, and even the method of determination (Holtz and Kovacs, 1981; Chu et al., 2002). Therefore, the so-called engineering judgment is sometimes required in deciding which value would be the most appropriate. Good engineering judgment comes from good understanding of soil behaviour and the past experience in dealing with similar types of soil and geotechnical problems. The coefficient of permeability is another key parameter required for vertical drain design. However, it happens that the coefficient of permeability of soil is one of the most difficult soil parameters to be determined. This is partially because the coefficient of permeability of the soil has the widest range of variation among all the soil parameters. Its value can vary from 10^{-11} m/s for soft clay to 10^{-3} m/s for sand and gravel, a change of 10^8 times. Although the permeability of the soil that needs to be treated with vertical drains is normally low, the error involved in the permeability estimation can still range from 10 to 100 times. This is not unusual as the permeability of the same soil can change 10 to 100 times during the process of consolidation. An error of one order of magnitude in permeability can result in an error of the same order of magnitude in the time taken to achieve a specific degree of consolidation based on Terzaghi's

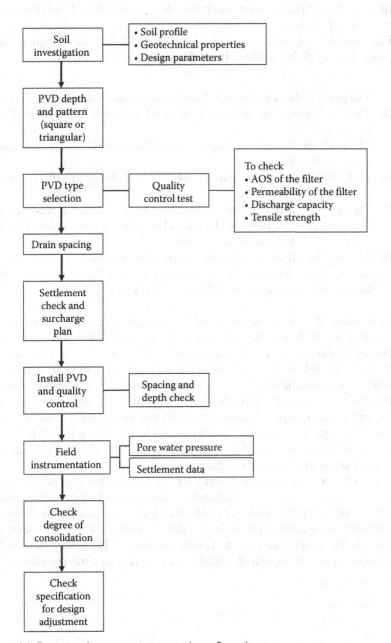

Figure 4.6 Design and construction procedures flow chart.

consolidation theory.* Therefore, it makes sense economically to conduct some proper site investigation work and determine the soil parameters as accurately as possible.

The specific soil parameters that are required for the design of soil improvement work involving vertical drains in soft clay include:

(1) The preconsolidation stress, σ_p', and the overconsolidation ratio (OCR).
(2) The coefficient of consolidation in both horizontal and vertical directions, c_h and c_v.
(3) The coefficient of permeability in both horizontal and vertical directions, k_h and k_v.
(4) The diameter of the smeared zone, d_s, and the permeability of the smeared soil, k_s.
(5) The coefficient of compressibility, C_c, the coefficient of recompressibility, C_r, and sometimes the secondary compression index, C_α, are required for settlement estimation.
(6) The undrained shear strength, c_u, and the undrained Young's modulus, E_u, may also be required for analysing the stability of a dike or the stability of a drain installation rig on soft clay.

As mentioned before, the values of c_h and c_v, or k_h and k_v change with stress state or OCR. One has to accept the fact that neither the coefficient of consolidation nor the coefficient of permeability of soil is a constant. As such, the selection of those parameters has to be based on its in-situ stress conditions and the anticipated stress changes. Therefore, it is also necessary to establish relationships between the coefficient of permeability and void ratio, and relationships between the coefficient of consolidation, and the stress state. A proper site investigation should be planned not only to determine the soil parameters but also to understand how the soil parameters vary with stress and loading conditions.

Generally the consolidation parameters of soil can be determined using laboratory tests, in-situ tests, back-calculation from field measurements, or a combination of them. In laboratory tests, the stress states and drainage conditions can be defined precisely and the variation of soil parameter with stress and consolidation process can be evaluated. However, the results are usually affected by sample disturbance. It is also time consuming to

---

* $$t = \frac{TH_d^2}{c_v} = TH_d^2 \frac{m_v \gamma_w}{k_v}, \dots$$

$$dt = TH_d^2 m_v \gamma_w \frac{-dk_v}{k_v^2},$$

$$\frac{dt}{t} = \frac{TH_d^2}{t} \frac{m_v \gamma_w}{k_v} \frac{-dk_v}{k_v} = c_v \frac{1}{c_v} \frac{-dk_v}{k_v} = \frac{-dk_v}{k_v}$$

conduct laboratory consolidation tests. In-situ tests are normally relatively quick to conduct and therefore are more useful than laboratory tests in identifying the soil profile and characterising the soil behaviour over a large extent. However, in in-situ tests, the stress and drainage conditions are generally not well defined. The data interpretation from physical measurements to soil parameters are sometimes based on arbitrary assumptions or correlations which are established for a specific type of soil only. Therefore, when in-situ tests are adopted, laboratory tests may still be required to verify the assumptions and check the correlation relationships. The back-calculation from field measurements can provide a good check on the selection of design parameters. However, the back-calculated value is only a factored parameter. It reflects not only the soil property, but also other factors, such as the disturbance to the soil during construction.

The types of laboratory and in-situ tests that are suitable to the determination of consolidation properties are summarised in Table 4.2. The

Table 4.2 Types of tests for measurement of consolidation properties

Type of test	Name of test	Parameter determined	Remarks
Laboratory tests	Oedometer test	c_v, k_v (indirect measurement[i]), C_c, C_r, σ_p', and C_α[ii]	Need high-quality 'undisturbed' samples.
	Rowe cell test	c_h and k_h (direct[iii] or indirect measurement)	
	Other consolidometers	c_h and k_h (direct or indirect measurement)	
In-situ tests	Piezocone dissipation test (CPTU)	c_h and k_h (indirect measurement)	Based on pore water pressure dissipation.
	Pressuremeter or self-boring pressuremeter (SBPM) test	c_h and k_h (indirect measurement)	Based on lateral pressure change or pore water pressure dissipation.
	Flat dilatometer test (DMT)	c_h and k_h (indirect measurement)	Based on lateral stress change.
	Field permeability test (e.g., BAT permemeter)	k_h (direct measurement)	Using a piezometer.
Back-analysis	Based on pore water pressure measurements	c_h (factored value)	Using piezometers.
	Based on settlement measurements.	c_h (factored value)	Using settlement gauges.

[i] In this case, k_v is calculated based on the value of c_v.
[ii] When secondary consolidation is measured.
[iii] k_h is measured directly as part of the consolidation test.

settlement prediction for projects using vertical drains is the same as those without the use of vertical drains. Those methods are covered in many textbooks (e.g., Holtz and Kovacs, 1981). As far as land reclamation or the other similar types of geotechnical problems (where the extent of load is much greater than the thickness of the compressible layer) are concerned, the settlement predicted using the one-dimensional analysis and parameters determined by laboratory tests is reasonable although not always reliable. Ground settlement should always be monitored as part of the soil improvement works.

Different types of laboratory and in-situ tests that are suitable for the determination of consolidation properties are discussed in detail in Chu et al. (2002) and Bo et al. (2003). One example for the determination of the c_h values for the intact Singapore marine clay at Changi by different methods is presented in Figure 4.7. Among the tests shown in Figure 4.7, the in-situ tests and the Rowe cell test measure the c_h values, whereas the conventional oedometer test measures c_v. The c_h back-calculated based on the settlements measured at different elevations are also presented in Figure 4.6. In the back-calculation, the ultimate settlement was estimated based on Asaoka's method (Asaoka, 1978). As shown in Figure 4.6, the back-calculated c_h values were lower than the c_h values determined by either laboratory or in-situ tests. Similar observations have been made at other sites in the Singapore marine clay (Chu et al., 2002) and by Balasubramaniam et al. (1995) for the Bangkok clay. Prefabricated vertical drains were installed at those sites at a close spacing. It implies that when vertical drains are used in soft marine clay, the overall c_h value of the soil will be lower than the c_h value of the intact soil. This could be due to the smear effect to soil induced by the installation of vertical drains. The effect of disturbance can be relatively large particularly when the drains are installed at a close spacing (Chu et al., 2002).

It has been generally observed from the comparisons made in Figure 4.7 and the other cases that:

(a) The c_h of soft clay determined by the Rowe cell test is generally 2–4 times larger than the c_v by the conventional oedometer test, reflecting the anisotropic nature of the soil.

(b) The CPTU dissipation test tends to agree reasonably well with that from the Rowe cell tests. Therefore, CPTU can be a good tool for c_h determination if it is calibrated properly.

4.3.2 Properties of smeared soil

With the substantial reduction in recent years in the costs involved in vertical drain products and installation, there is a tendency to use closer drain spacing. However, when the drains are installed too close to each other,

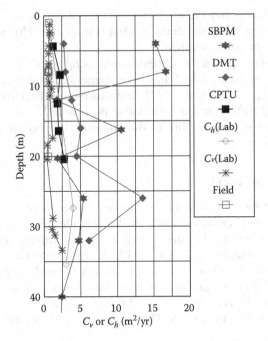

Figure 4.7 Comparison of c_h profile of Singapore marine clay measured by different methods.

the benefit resulting from use of more PVDs may be overshadowed by the increased smear effect due to PVD installation. Although Equation 4.14 has been proposed to estimate the diameter of the smear zone, d_s, it is difficult to quantify d_s accurately as the value d_s is affected by many factors, such as the shape, the size of the mandrel, the type of soil, and the sensitivity of soil. Several studies have been conducted to determine d_s and the effect of smear zone on the consolidation of soil. Hansbo (1981, 1997) estimated $d_s = (1.5 \sim 3.0)d_w$. This relation has been commonly used in design. Based on a laboratory study and back analyses, Bergado et al. (1991) proposed that $d_s = 2d_w$ could be assumed. Indraratna and Redana (1998) observed from some model tests that the smear zone could be as large as $d_s = (4 \sim 5)d_w$. Holtz and Holm (1972) also suggested that d_s be equal to two times the equivalent diameter of the mandrel. The studies of Xiao (2002) and Bo et al. (2003) indicated that the smear zone could be as large as 4 times of the size of the mandrel or 5–7 times the equivalent diameter of drain. The dimension of a small rectangular mandrel is 120 mm by 60 mm. This is equivalent to a diameter of 115 mm using Equation 4.9. The equivalent diameter of a drain is 66 mm. Therefore, the size of the smear zone can be as large as 460 mm. If the drain spacing is 1 m, it means almost everywhere the soil is disturbed. As smear can cause a significant reduction in the permeability and the coefficient of consolidation of soil, it can be counterproductive when the drain spacing is too close. This is particularly the case when the soil is sensitive.

The parameter, k_s, is normally estimated using a reduction ratio, k_s/k_h. A value in the range of 2 to 6 has been suggested by various researchers (Hansbo, 1981; Onoue et al., 1991; Indraratna and Redana, 1998; Hird and Moseley, 2000). Hansbo (1997) also proposed the ratio be put equal to the ratio of the permeability in the horizontal direction to that in the vertical direction.

4.3.3 Types of PVDs

PVD products have become quite standard in terms of size and performance. The typical cross-section is 100 mm wide and 3–6 mm thick. A PVD normally consists of a core and filter made of different types of materials. Classified in terms of design, the PVDs available in the market can be classified into three different types. The first is the ordinary type of band drains where the core and filter are fitted loosely and are separable. Some typical forms of PVDs are shown in Table 4.3. The photos of three PVD samples are shown in Figure 4.8. The second type is the so-called integrated

Table 4.3 Ordinary types of PVD

Core		Filter type	Method of assembly	
			Separated filter and sleeve	Filter jointed to core
Cross-section	Description			
⊓⊓	Corrugated groove	Nonwoven fabric	Mebradrain MD 7007	
┼┼┼	Ribbed groove	Synthetic fiber	Mebradrain MD88	
⋀⋁⋀	Monofilament	Needle-punched nonwoven fabric		Colbond CX1000
⋁⋀	Double Cuspated	Nonwoven fabric	Flodrain	
▮▪▮	Studded (one side)	Nonwoven fabric	Alidrain ST	
◆▮◆▮	Studded (two sides)	Nonwoven fabric	Alidrain DC	

Figure 4.8 Samples of ordinary types of PVDs.

PVD where the core is adhered to the filter, as shown in Figure 4.9. This type of PVD offers a number of advantages over the ordinary type of PVDs. For examples, it offers a higher discharge capacity and tensile strength as discussed in detail by Liu and Chu (2009). The third type is circular PVDs. A picture of it is shown in Figure 4.10. The circular PVD has better resistance to buckling and has been used exclusively for vacuum preloading projects (Chu et al., 2009).

It should be mentioned that there are different types of drains, such as electric vertical drain with a metal foil embedded in the drains as anodes and cathodes for electro-osmosis (Shang, 1998; Bergado et al., 2000; Karunaratne, 2011). There are also PVDs for geoenvironmental use. For example, PVDs have been used for a vapour extraction system (Schaefer et al., 1997; Collazos et al., 2002). There have also been

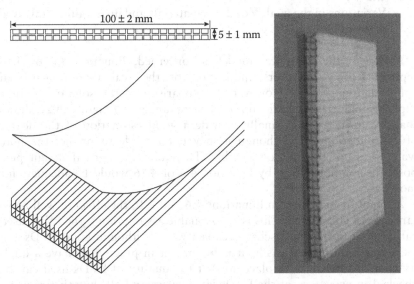

Figure 4.9 Integrated type of PVDs. (Redrawn from Liu, H.L. and Chu, J. (2009). *Geotextiles and Geomembranes*, Vol. 27, No. 2, pp. 152–155.)

Figure 4.10 Circular type of drain used for some vacuum preloading projects.

studies to produce biodegradable PVDs using biodegradable plastics (Park et al., 2010).

4.3.4 Selection of PVDs

The uncertainties involved in the design of PVDs include

1. The bias in the analytical model
2. Uncertainties or errors in the parameters entered to Equations 4.6 and 4.16
3. Variations in the quality of drain materials and uncertainties involved in installation

As far as the analytical model is concerned, Equation 4.6 or 4.16 represents only an oversimplification of the real three-dimensional consolidation process. However, a comparison of the solutions with a sophisticated elasto-plastic finite element analysis (Xiao, 2002) shows that the former can normally provide a good estimation of the degree of consolidation, U_h, although it may not be able to predict the pore water pressure distribution reliably. Therefore, for normal design purpose, the solutions given by Equation 4.6 or 4.16 would be adequate for most cases.

As mentioned before, in Equations 4.6 and 4.16 the effect of well resistance is not considered. This is a reasonable assumption if the PVDs used can provide sufficient discharge capacity, q_w. Although good PVDs can meet this requirement easily, it is important in practice to have a quality-control procedure in place so that the quality of PVDs used can be checked to ensure q_w of the PVDs is adequate and all the other requirements are met.

The quality and suitability of the drains play a key role in the whole soil improvement scheme involving PVDs. Different design situations require different types of PVDs. For example, it is not necessary to use a vertical drain with a high discharge capacity value if the drain is short. The drain filter should also match the soil type. The unit price of vertical drain is another important consideration besides meeting the design requirements. A considerable saving can be achieved without sacrificing the performance of the drain, if the control factors for a vertical drain can be identified and the design requirements are specified accordingly. The factors that control the selection of vertical drain, apart from the cost, are as follows.

4.3.4.1 The discharge capacity

It should be pointed out that the well resistance is controlled not only by the discharge capacity of the drain, q_w, but also by the permeability of soil, k_s, and the longest discharge length, l_m, as can be seen from the last term of Equation 4.13. To evaluate the efficiency of drain in discharging water, a dimensionless parameter, the so-called discharge factor is defined (Mesri and Lo, 1991):

$$D = \frac{q_w}{k_h l_m^2} \tag{4.17}$$

Equation 4.17 reflects the fact that the larger the k_h or the longer the drain, the larger the discharge capacity is required. The effect of well resistance can be evaluated using this discharge factor. Based on a numerical study, Xie (1987) established that in order to meet the assumption that the well resistance can be ignored, the following condition must be met:

$$\frac{\pi}{4} \frac{k_h}{q_w} l_m^2 < 0.1 \tag{4.18}$$

This requires the discharge factor, D, to be:

$$D = \frac{q_w}{k_h l_m^2} \geq 7.85 \tag{4.19}$$

Therefore, the required discharge capacity after applying a factor of safety to consider all the influencing factors on discharge capacity including buckling becomes:

$$q_{req} \geq 7.85 F_s k_h l_m^2 \tag{4.20}$$

where q_{req} = the required discharge capacity, F_s = factor of safety, normally $F_s = 4 \sim 6$.

The condition given in Inequality (4.20) is consistent with the threshold discharge factor of 5 specified by Mesri and Lo (1991). Inequality (4.20) defines the dependence of q_w on k_s and l_m. The relationship among q_w, k_s, and l_m for $F_s = 5$ for negligible well resistance has been plotted in Figure 4.11.

Based on Inequality (4.20), l_m has the most significant influence on the required discharge capacity. If we take $F_s = 5$, $k_s = 10^{-10}$ m/s, and $l_m = 25$ m, then $q_w = 2.45 \times 10^{-6}$ m³/s, or 82 m³/yr. If $l_m = 50$ m instead of 25 m, then $q_w = 9.81 \times 10^{-6}$ m³/s, or 327 m³/yr. Nowadays, most of the drains can provide such a q_w value even under a buckled condition. On the other hand, permeability can also have a great effect when it is not determined accurately. Take the previous case for example, if $k_s = 10^{-9}$ m/s instead of 10^{-10} m/s, then $q_w = 98.1 \times 10^{-6}$ m³/s, or 3,270 m³/yr. In this case, some drains will not be able to meet the requirement.

For the Changi East land reclamation project, the drains used were up to 50 m long. The permeability of the soil normally ranges from 10^{-11} to 10^{-10} m/s. Using Inequality (4.20), q_w required should not be larger than 9.81×10^{-6} m³/s. $q_w = 10 \times 10^{-6}$ m³/s was adopted as the specification for the buckled drain at a pressure of 350 kPa for the Changi East land reclamation project. Back-calculations of q_w were made using the field monitoring data (Bo, 2003). The back-calculated q_w was normally much smaller than the specified value, but some values were as high as 5×10^{-6} m³/s (Bo, 2004). As the field measurements agree reasonably well with q_{req} calculated from

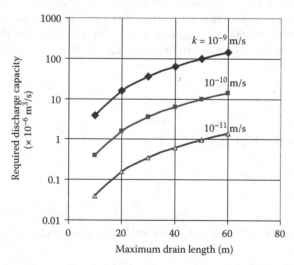

Figure 4.11 Required discharge capacity as a function of PVD length and permeability of soil.

Inequality (4.20), the condition specified in Inequality (4.20) appears to be adequate. It also indicates that the use of a factor of safety of 5 in Inequality (4.20) is reasonable, although a slightly lower value can also be adopted. This is why the factor of safety to be adopted in Inequality (4.20) is suggested to be in the range of 4 to 6.

As the discharge capacity required is controlled by the permeability of the soil and the length of vertical drain, in theory, the discharge capacity required varies from project to project, and even varies from point to point even within the same project. Therefore, the discharge capacity specified for one project may not be applicable to another even when the soil conditions in the two projects are similar.

4.3.4.2 Compatibility of the filter with the soil to be improved

The pore size or the apparent opening size (AOS) of the filter should meet the filter design criteria. On one hand, the AOS has to be small enough to prevent the fine particles of the soil from entering the filter and the drain. On the other hand, the AOS cannot be too small as the filter has to provide sufficient permeability. The two key parameters that indicate the quality of the filter are the AOS and the cross-plane permeability of the filter. Some criteria for AOS have been proposed by several researchers. A commonly used criterion is given by Carroll (1983):

$$O_{95} \leq (2 \sim 3)\, D_{85} \tag{4.21a}$$

and

$$O_{50} \leq (10\ \text{to}\ 12)\, D_{50} \tag{4.21b}$$

where O_{95} is the AOS of filter, O_{50} is the size which is larger than 50% of the fabric pores, and D_{85} and D_{50} refer to the sizes for 85% and 50% of passing of soil particle by weight. $O_{95} \leq 0.075$ mm, or 75 µm, is often specified for PVDs.

The D_{85} of the Singapore marine clay was in the range of 0.01–0.02 mm (Bo et al., 2003). The O_{95} of the PVD filter specified for the Changi land reclamation project was less than 0.075 mm. An AOS of 0.075 mm exceeded the AOS specified by Inequality (4.21a). This was permitted for the following two reasons. First, PVDs with such O_{95} had been used successfully in the previous land reclamation projects in Changi (Choa et al., 1979). Second, there were not many types of PVDs which offer an O_{95} much smaller than 0.075 mm available in the southeast Asian market. To verify whether the filter was adequate, some drains had been pulled out from the site at the end of the project. The inner side of the drain was quite clean. The same was observed from the drains used for some long-term consolidation tests in the Singapore marine clay (Chu and Choa, 1995). Therefore, the filter criterion set by Inequality (4.21a) appears to be too conservative for the Singapore marine clay based on this study. A more relaxed criterion: $O_{95} \leq (4 \sim 7.5) \, D_{85}$, may be applicable to the Changi East land reclamation project. For the Bangkok clay, a more relaxed criterion for O_{95} has also been proposed (e.g., Bergado et al., 1993a).

The permeability of the filter is normally required to be at least one order of magnitude higher than that of the soil. Considering the clogging effect, a much higher permeability should be required for the filter. Nevertheless, even more stringent requirements on the permeability of filter can be met easily as most of the PVDs have a filter permeability higher than 10^{-4} m/s, which is far greater than what is required. For example, the permeability of the Singapore marine clay was in the order of 10^{-10} m/s. The permeability of the filter of the PVDs used for the Changi land reclamation project was higher than 10^{-4} m/s. This is 10^5 times greater. The thickness of the filter is another consideration. Normally the thicker the filter, the better it becomes, given other conditions the same. Based on Wang and Chen (1996), the mass to area ratio should be generally larger than 90 g/m².

4.3.4.3 The tensile strength of drain

PVDs should have adequate tensile strength so that it can sustain the tensile load applied to it during installation. Therefore, the strength of the core, the strength of the filter, the strength of the entire drain, and the strength of the joint need to be specified, normally at both wet and dry conditions. According to Kremer et al. (1983), a drain must be able to withstand at least 0.5 kN of tensile force along the longitudinal direction without exceeding 10% in elongation. It is quite common nowadays to specify the tensile strength of the whole drain at both dry and wet conditions to be larger than 1 kN at a tensile strain of 10%. The same criterion was used for the Changi East land reclamation project. The spliced drain is also required to have strength comparable to that of the original drain. This criterion appeared to be satisfactory for the PVDs used for the project. However, one factor that is often neglected is that some drains can have permanent necking once it is stretched. Such a necking reduces the discharge capacity. Therefore, the amount of elongation and necking should also be observed and reported during a tensile strength test. For the PVDs used for the Changi East land reclamation project, necking only became obvious in the tensile strength test after the tensile strain exceeded 20%, which is unlikely in the field condition.

It should be pointed out that although various methods and equipment have been used for the determination of the properties of PVD, it may not be possible to compare directly the values provided by suppliers. Even within the same method, the values measured can still vary depending on the testing procedure. Furthermore, the discharge capacity is dependent on hydraulic gradient. Thus, one can only compare the discharge capacity when the values are measured using the same method and at the same hydraulic gradient. Unfortunately, this is seldom the case. Under such circumstances, it is necessary to conduct one's own tests to determine the discharge capacity in the same way for all the drains concerned.

4.3.5 Pattern, spacing, and penetration depth for PVDs

PVDs are installed in either square or triangular patterns. In theory, the triangular pattern is slightly more cost effective. However, a square pattern is simpler for layout and installation. The spacing for PVDs can be calculated using the theories presented in Section 4.2. In theory, the closer the drain spacing, the faster the rate of consolidation. However, when the spacing of PVDs is too small, the smear effect becomes more significant as discussed in the earlier section.

Normally PVDs should be installed through the entire depth of the compressible soil layer. However, if the load is applied over a limited area, such as a narrow embankment, the majority of the load may be distributed within a certain depth only. In this case, it is not necessary to install PVDs through the entire depth of the compressible soil layer. One example of such a case is given by Yan et al. (2009).

Prefabricated vertical drains shall be located, numbered, and staked by the surveyor using a baseline and benchmark indicated by the engineer. The as-built location of the PVDs shall not vary by more than 150 mm from the planned locations. During installation the depth and the length of the drain installed at each location should be determined. The drain material shall be cut neatly at its upper end with 100 to 200 mm protruding above the working surface.

4.3.6 Settlement calculations

During the preliminary design stage, the consolidation settlement is normally calculated based on a one-dimensional settlement analysis using data obtained from one-dimensional consolidation tests. One method of using Microsoft Excel spreadsheet for settlement calculation by considering staged loading is shown in Bo et al. (2003). As the one-dimensional settlement calculation has been covered in many papers and textbooks, it will not be elaborated on in this chapter. However, comments related to settlement calculation follow.

First, the settlement prediction using oedometer testing data can only be as reliable as the data. For this reason, it is important to obtain good one-dimensional consolidation test results. This requires good quality samples to be used. It is well known that the sample quality affects the determination of preconsolidation stress and compression index (Holtz and Kovacs, 1981). A small error or uncertainty in the determination of preconsolidation stress can cause a large variation in the settlement. Furthermore, it should be noted that the preconsolidation stress changes with the method of testing and rate of loading when the constant-rate-of-strain method is adopted. Second, the preconsolidation stress and vertical stresses in soil also vary with depth. It is therefore necessary to determine the preconsolidation stresses at different depths and calculate the consolidation settlement using the subdivision method unless the compressible layer is relatively thin. Third, the surcharge applied by fill can change with the settlement of the ground as part of the fill may submerge into water and the effective surcharge load can thus be reduced. One such example is given by Bo et al. (1999). Finally, it should be pointed out that this one-dimensional approach is only reasonable when the extent of the load applied is much larger than the thickness of the compressible soil layer such as in a land reclamation project. Even in this case, settlement prediction has never been an easy task, as elaborated by Duncan (1993). For this reason, settlement prediction using field monitoring data is essential for preloading projects using PVDs.

Sometimes, the settlement due to secondary compression needs to be estimated too. The conventional method of estimating the secondary compression is to use

$$\Delta e = C_\alpha \left(\Delta \log t \right) \tag{4.22}$$

where Δe is the additional compressibility due to secondary compression, C_α is the secondary compression index, and t is time. Mesri and Castro (1987) observed the following relationship for natural medium plasticity index clays:

$$C_\alpha / C_c = 0.04 \sim 0.05 \tag{4.23}$$

where C_c is the compression index of soil.

Equations 4.22 and 4.23 can be used as a preliminary assessment for the order of magnitude of secondary compression. A few advanced methods have also been proposed for the estimation of the secondary compression (e.g., Yin and Graham, 1999). However, the prediction of secondary compression of soil is even more difficult than that for primary consolidation. Part of the reason could be because the secondary compression may not even be separated from the primary consolidation (Leroueil,

1988). If secondary compression has to be considered in the design, the intended preloading should be designed with an additional objective to reduce the secondary settlement. Studies have shown that (Hight, 1999; Mesri et al., 2001) the secondary compression index is much smaller when soil is overconsolidated. Therefore, one effective way to reduce the effect of secondary compression is to preload the soft ground to a consolidation stress higher than the anticipated working load. Then the soil can be in an overconsolidated state during the working stage and thus the settlement due to second compression will be insignificant under the working load.

4.4 CONSTRUCTION

4.4.1 Installation

The PVD installation procedure is illustrated in Figure 4.12. A PVD installation rig is normally used to penetrate a metal mandrel with PVD inside. The PVD installation rigs used on land can be classified into three types: (1) static push-in type for normal ground (see Figure 4.13 for an example); (2) vibratory drive-in type for firm to stiff ground or soft soil with a hard crust; (3) light rigs on trucks for PVD installation on very soft ground, as shown in Figure 4.14. For offshore or above-water PVD installation, a raft or a PVD installation barge has been used. A raft such as the one in Figure 4.15 is only workable in quiet, shallow water. The barge is required in the relatively deep water of an offshore environment. The PVD installation

Figure 4.12 PVD installation procedures. (Redrawn from http://www.americanwick. com/uploads/documents/WICKDrainBrochure1.pdf.)

Figure 4.13 A static push-in type of PVD installation rig.

barge shown in Figure 4.16 can install 12 PVDs at the same time. It has been used for a port project in China (Yan et al., 2009).

The mandrel used for PVD installation is normally much larger than a PVD in terms of cross-section areas as it has to be strong enough to prevent it from bending or buckling. There are four different types of mandrels according to the shape of the cross-sections: rhombic, rectangular,

Figure 4.14 A light-weight truck-supported installation rig. (Courtesy of S.W. Yan.)

Figure 4.15 A floating raft for PVD installation in shallow water (From Chu, J., Bo, M.W. and Arulrajah, A. (2009). Soil improvement works for an offshore land reclamation. *Proceedings of the Institution of Civil Engineers—Geotechnical Engineering*, Vol. 162, No. 1, pp. 21–32.)

square, and circular. The first two are more commonly used. A picture of their cross-sections is shown in Figure 4.17. The typical dimensions are 120–145 mm long, 60–mm wide, and 10–mm thick for rectangular cross-section mandrel and 120–145 mm long, 50–85 mm wide, and 5–15 mm thick for rhombic ones. It should be noted that the smear effect is greatly affected by the cross-section area of the mandrel. Therefore, the cross-section of the mandrel should be as small as possible in order to reduce the

Figure 4.16 A PVD installation barge. (From Yan, S.W., Chu, J., Fan, Q.J., and Yan, Y. (2009). Building a breakwater with prefabricated caissons on soft clay. *Proceedings of ICE, Geotechnical Engineering*, Vol. 162, No. 1, pp. 3–12.)

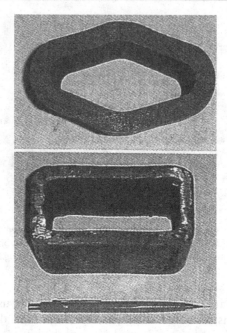

Figure 4.17 Cross-section of mandrels. (From Bo, M.W., Chu, J., Low, B.K. and Choa, V. (2003). *Soil Improvement: Prefabricated Vertical Drain Technique*, Thomson Learning, Singapore.)

Figure 4.18 A bent mandrel coming out of the ground.

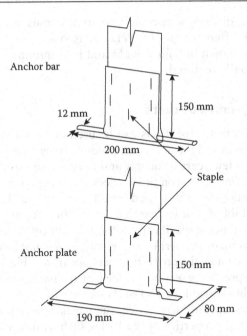

Figure 4.19 Typical designs of anchors. (Redrawn from Bo, M.W., Chu, J., Low, B.K. and Choa, V. (2003). *Soil Improvement: Prefabricated Vertical Drain Technique*, Thomson Learning, Singapore.)

smear effect. Nevertheless, a mandrel must have enough stiffness to ensure the verticality of the drain. A mandrel does bend when it hits a piece of rock or a hard stratum. A mandrel bent inside the soil and coming out behind the installation point is shown in Figure 4.18.

The PVD is pulled down together with the mandrel using an anchor bar or anchor plate as shown in Figure 4.19. The PVD is fixed to an anchor bar or plate using folding and staples. Two pictures illustrating the operation are shown in Figure 4.20.

Figure 4.20 Use of an anchor plate for PVD installation. (Courtesy of J. Han.)

PVDs come in rolls and sometimes have to be connected. Some details of splicing for two different types of PVDs as used for a reclamation project in Singapore are shown in Figure 4.21a and b. A minimum overlapping of 300 mm is normally required.

4.4.2 Quality-control tests

As several million metres of PVDs can be used in even a normal-sized soil improvement project and the drains are produced over a period ranging from a few months to a few years (probably at different factories), it is essential to conduct quality-control tests on site to check the consistency of the products. A list of specifications for PVDs used for different projects in different countries is given in Table 4.4. It is clear that quite different parameters and control values are used in the specifications for different countries. Although it is better to check as many properties as possible, it may not be cost effective or necessary to check on all the properties that appear in Table 4.4. Based on the discussion in the preceding section, we identify the parameters that should be checked for quality control for PVDs: the quality of the filter, the discharge capacity, apparent opening size (AOS), the tensile strength of the drain, as well as the physical properties of the PVDs such as dimensions, weight to area ratio, etc. The next question is how to conduct the quality-control tests. Various methods have been proposed for measuring the various properties of PVDs in the past. However, different tests or even the same test using different control conditions will give different values. It should be pointed out that the specifications given different PVD suppliers were measured using different testing methods. Therefore, one cannot compare the values provided by the PVD suppliers directly. For this reason, the end users should adopt a set of testing methods for conducting not only quality-control tests, but also comparing and selecting PVD products. The set of testing methods should be simple enough that they can be carried out easily in any soil laboratory. One such set of tests has been developed by Chu et al. (2004) and will be described briefly as follows. The basis for the selection of the suitable control values for specifications is also discussed in the following.

4.4.2.1 Determination of the AOS

One of the common methods used to measure the AOS is to conduct sieve analysis using standard beads. This method is applicable to AOS larger than 40 μm. Standard ASTM D4751 (2004) is normally followed in conducting this test, except that the diameters of the silicon beads used in the test range from 40–170 μm, instead of 75–170 μm. The tests are conducted under a relative humidity of 60% and a temperature of 20°C. The percentage of passing of the silicon beads is measured. The AOS of the filter is determined as a correlation with the percentage.

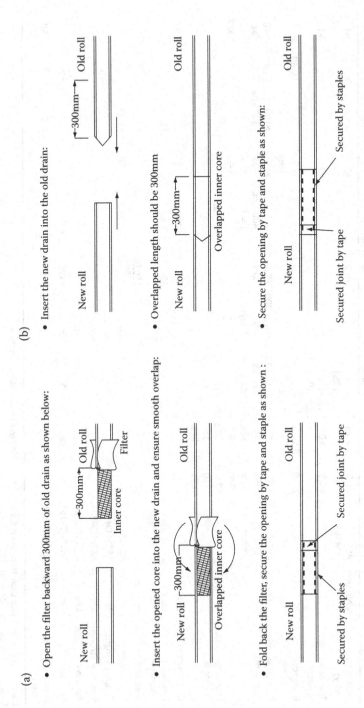

Figure 4.21 Details of splicing: (a) Colbond type of drain. (b) Mebra type of drain. (From Bo, M.W., Chu, J., Low, B.K. and Choa, V. (2003). *Soil Improvement: Prefabricated Vertical Drain Technique*, Thomson Learning, Singapore.)

Table 4.4 Specifications used for different projects in different countries

| Description | Unit | Standard | Netherlands | | Singapore | Thailand | Hong Kong | Malaysia | Taiwan | Australia | Finland | Greece |
			Case I	Case II								
Tensile strength (dry)	kN/10cm	ASTM D4595	>0.5	>0.5	>1 (at 10%)		>0.5		>2		>1	>1
Tensile strength (wet)			>0.5	>0.5	>1 (at 10%)				>2		>1	>1
Elongation	%		2–10 (at 0.5 kN)		<30 (at 1 kN)				<20 (yield)		15–30	
Discharge capacity (straight)	10^{-6} m²/s	ASTM D4716 and others	>10 at 350 kPa 30 days	>50 at 350 kPa 30 days	>25 at 350 kPa 28 days	>16 at 200 kPa, 7 days, i=1.	>5 at 200 kPa	>6.3 at 400 kPa, i=1	>10 at 300 kPa, i=1	>100 at 300 kPa	>10	>10 at 100 kPa
Discharge capacity (folded)	10^{-6} m²/s		>7.5 at 350 kPa	>32.5 at 350 kPa	>10			>6.3 at 400 kPa				
Crushing strength	kN/m³							500				
Equivalent diameter	mm						50			65		
Free surface filter	mm²/m						150,000					
Tear strength	N	ASTM D4533				100		>300	>250	>380		
Graph strength	N	ASTM D4632				>350						
Puncture strength	kN	ASTM D4833				>200						
Bursting strength	kPa	ASTM D3785				>900						
Pore size O_{95}	μm	ASTM D4751	<160	<80	<75	<90	<120	<75			<90	
Permeability of filter	10^{-5} m/s	ASTM D4491			>5		>10	>1	>10	>17		>50

Case I: Stable layer less than 10 m thick; Case II: Unstable layer more than 10 m thick
Source: Modified from Bo et al., 2003.

Figure 4.22 Device to measure the permeability of filter of PVD.

4.4.2.2 Permeability (or permittivity) of filter

The cross-plane (i.e., in a direction perpendicular to the surface of the filter) permeability, or the so-called permittivity, of the filter is measured using a constant head method. A permeameter specially made for this purpose is shown in Figure 4.22. A single layer of filter is used. ASTM Standard D4491 99a (2009) is followed in conducting this test. In this case, the permeability measured under a constant head of 50 mm is reported.

4.4.2.3 Discharge capacity test

The discharge capacity of drain, q_w, is normally measured for both straight and deformed drains. This is because the discharge capacity of PVD reduces after it has buckled in soil. Although the ASTM4716 (2008) is often referred to, it should be pointed out that this standard is set for the determination of the transmissivity of a geosynthetic, not specifically for the discharge capacity of PVDs. Therefore, a standard method for measuring the discharge capacity of vertical drain has not been established yet. As such, various devices and methods have been developed for measuring the discharge capacity of vertical drain (Hansbo, 1983; Kamon et al., 1984; Guido and Ludewig, 1986; Suite et al., 1986; Broms et al., 1994; Bergado et al., 1996; Chu et al., 2004). These methods can be generally classified into two categories. The first is to embed the drain in soil and the second is to warp the drain with membrane or other materials.

The testing devices suggested in the European Standard on Execution of Special Geotechnical Works—Vertical Drainage (BS EN 15237, 2005)

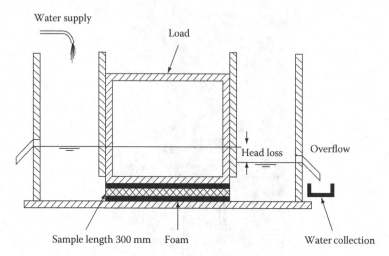

Figure 4.23 Testing apparatus #I for discharge capacity test recommended by BS EN
15237 (2007).

for determining the discharge capacity according to EN ISO 12958
belongs to the first category. The two devices are shown in Figures 4.23
and 4.24, respectively. The specimen in apparatus #1 is covered on both
sides by closed-cell foam rubber with a thickness of 10 mm. The membrane in apparatus #2 is made of latex with a maximum thickness of
0.35 mm.

The studies of Lee and Kang (1996), Chu et al. (2004), and Karunaratne
(2011) indicated that soil should be used in the discharge capacity test as
the results obtained from the tests without soil tend to be substantially
larger than those with soil. This is mainly because of the greater indenta-
tion of filter into the drain groove as shown in Figure 4.25, and the reduction in the water-discharging ability of the filter when soil is in contact with
the filter. For this reason, a method to embed a drain specimen in soil may
be a better alternative. Another problem related to the device suggested
in ASTM4716 or in BS EN 15237 (2007) are shown in Figures 4.23 and
4.24 is that the total head is not measured within the drain specimen. This
is necessary as it is often found out that the hydraulic gradient measured
between two points within the drain is different from that measured using
the head difference of the water reservoirs.

To improve the ASTM D4716 procedure, a new drain tester has been
developed and used for a number of land reclamation projects in Asia
(Broms et al., 1994; Chu and Choa, 1995; Chu et al., 2004). A cross-section
of the new drain tester is shown in Figure 4.26. It consists of a base, a hollow extension plate, and a top cap. In a test, a drain specimen of 100 mm
(or 300 mm) is placed in the base on top of a marine clay layer. The hollow

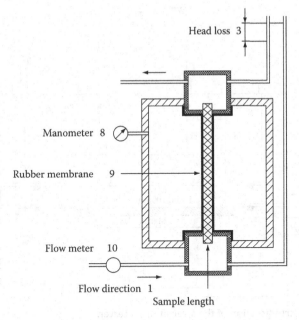

Head loss 3

Manometer 8

Rubber membrane   9

Flow meter   10

Flow direction 1

Sample length

Figure 4.24 Testing apparatus #2 for discharge capacity test recommended by BS EN 15237 (2007).

extension plate is used to anchor the drain and to allow another layer of marine clay to be put on top of the drain. The top cap is then placed and the screws are tightened. Square shaped O-rings are used to seal the tester. The vertical pressure is applied via an oedometer loading frame. This test is simple and can be easily conducted in a site laboratory. The results are

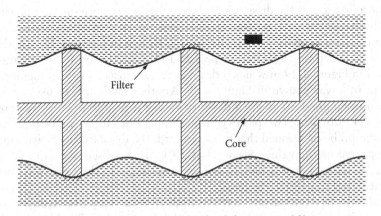

Filter

Core

Figure 4.25 Reduction of flow area caused by the deformation of filter.

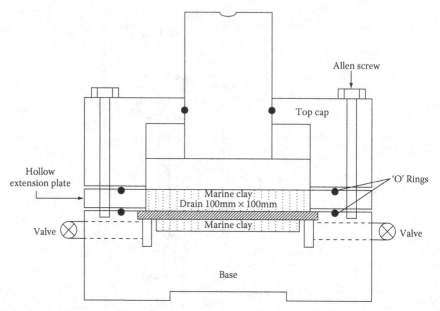

Figure 4.26 A cross-section of the straight drain tester.

easily reproducible and therefore can be counter checked easily. This test also provides a mean to compare and benchmark the discharge capacity values of different drains. The discharge capacity measured using this new drain tester is lower than that measured by the ASTM4716 method. It is safer to use the smaller value measured to compare with the specification.

A buckled drain tester, as shown in Figure 4.27, is also used to measure the discharge capacity of the deformed drain (Chu and Choa, 1995). A buckled sample is shown in Figure 4.28. In conducting this test, it is not necessary to wait until the soil in the drain tester has achieved a high degree of consolidation as the dissipation of excess pore pressure should not affect the discharge capacity measurement. However, it is troublesome to carry out a discharge capacity test using a tester shown in Figure 4.25. An alternative is a method suggested in BS EN 15237 (2007) using an apparatus shown in Figure 4.29 in which a drain specimen inside a rubber membrane is bent in a way shown in Figure 4.29. Another method is to use a device shown in Figure 4.30 where a drain specimen is bent to a 30-degree angle (Figure 4.30b) using two pieces of water impervious foam (Bo et al., 2003).

It should be mentioned that as a standard, the discharge capacity should be reported as the value measured at 20°C. If the discharge capacity test is not conducted at this temperature, a conversion in the same way as for permeability test should be made.

It should be pointed out that the discharge capacity also reduces with time. This might be due to the creep of the drain materials under pressure.

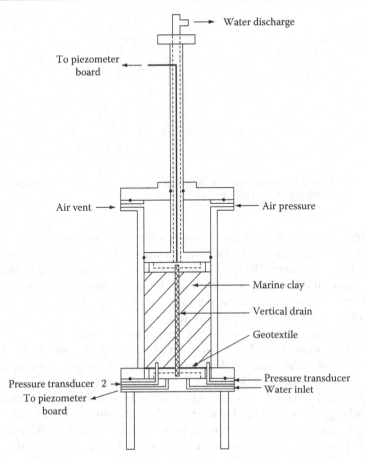

Water discharge

To piezometer
board

Air vent

Air pressure

Marine clay

Vertical drain

Geotextile

Pressure transducer 2

Pressure transducer
Water inlet

To piezometer
board

Figure 4.27 Buckling drain tester.

Figure 4.28 Deformed drain after it is tested in the buckled drain tester.

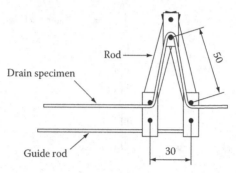

Figure 4.29 Device for measuring the discharge capacity of a bent drain recommended by BS EN 15237 (2007).

The reduction in the discharge capacity over four weeks can be as high as 60% (Chu and Choa, 1995). For this reason, a creep factor is introduced to account for the creep effect. Although it is too time consuming to measure the long-term discharge capacity, some long-time discharge capacity tests should be conducted to assess the rate of reduction in discharge capacity with time or verify the creep factor adopted.

It is generally observed that the discharge capacity reduces with hydraulic gradient (Kamon et al., 1984; Broms et al., 1994; Park and Miura, 1998). As such, when the value of discharge capacity is reported, the hydraulic gradient, i, at which the discharge capacity is measured should be stated. For practical purposes, the discharge capacity should be measured at a hydraulic gradient comparable to the field conditions. However, the in-situ hydraulic gradient is difficult to estimate. Reports on the field hydraulic gradient are also rare, except one case reported by Nakanado et al. (1992) in which the in-situ hydraulic gradient was estimated to be in the range of 0.03–0.8. From the testing point of view, Akagi (1994) pointed out that when the hydraulic gradient is higher than 0.5, the flow inside the vertical drain may not be laminar anymore. He suggested the discharge capacity

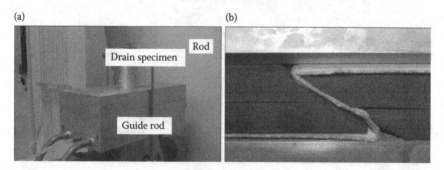

Figure 4.30 Device for measuring the discharge capacity of deformed drain.

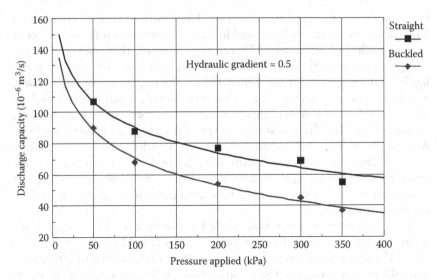

Figure 4.31 Typical PVD discharge capacity test results using the straight and buckled drain testers.

to be measured at a hydraulic value ranging from 0.2–0.5. After analysing the flow behaviour in the drain under different hydraulic gradients, Park and Miura (1998) also suggested that a hydraulic gradient ranging from 0.2–0.5 be used. The data presented in Wang and Chen (1996) also indicate that a steady flow can be difficult to achieve for vertical drain when $i >$ 0.5. Based on these studies, an $i \leq 0.5$ should be used for discharge capacity measurement. BS EN 15237 (2007) suggests the discharge capacity be measured at a hydraulic gradient of $i = 0.1$. However, the testing errors involved in the measurement can be higher when the hydraulic gradient is small. Therefore $i = 0.5$ appears to be the most suitable value. In our method, the discharge capacity of PVD was determined by measuring the discharge capacity at different i values ranging from 0.1 to 1 to establish the relationship between q_w and i under each pressure first, and then using this relationship to determine the discharge capacity at $i = 0.5$. The q_w measured in this way has been used in Inequality (4.20), which has been shown to be compatible with field measurements, as discussed in the early section. Some typical results of discharge capacity tests for a PVD in both straight and buckled conditions are shown in Figure 4.31.

4.4.2.4 Tensile strength tests

ASTM D4632 (2008) is often specified as the method to measure the tensile strength of PVD or the filter of PVD, in which the pull rate is 300 ± 10 mm/min. A tensile strength testing machine that provides such a pull rate is normally too expensive for a site laboratory to equip. On the other hand, a compression machine used for triaxial tests is commonly available in a geotechnical laboratory. Therefore, a method that uses a modified compression machine to conduct tensile strength tests for vertical drains has been developed (Chu and Choa, 1995). The only shortcoming of using a compression machine is that it does not provide a pulling rate as high as 300 mm/min. However, this will not be a problem for the following reasons—first, the drain installation speed is as high as 25,000 mm/min. As even 300 mm/min is far too low, whether a pulling rate is 300 mm/min or lower does not make much difference. Second, the slower the rate, the smaller the tensile strength measured, and the use of a slower pulling rate will result in a more conservative measurement. This is good for quality-control purposes.

When a compression machine is used to conduct tensile strength tests, a pair of clamps designed according to ASTM D4632 need to be used to clamp the drain specimen to the machine. A drain specimen of 200 mm in gauged length is used. The test setup is shown in Figure 4.32. The load was applied under constant-rate-of-extension (CRE). For wet conditions, the specimen, either the entire drain or the filter, was immersed in water for 48 hours before testing. Some typical tensile strength tests for two different types of PVDs are shown in Figure 4.33.

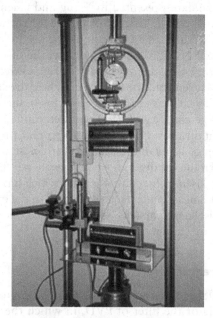

Figure 4.32 Tensile strength test for PVD or the filter.

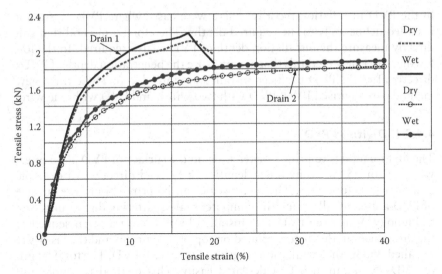

Figure 4.33 Tensile stress versus tensile strain curves measured for 2 typical drains.

4.4.3 Measurement of penetration depth

The effect of soil improvement using PVD is greatly affected by the installation depth. If a full penetration of PVDs is assumed in the design and yet
the PVDs are not installed to the entire depth of soft clay, the predicted rate
of consolidation will be incorrect. Therefore, it is important to measure
the real penetration length of the PVD installed on site. Another reason for
measuring the installation depth of PVD is to gain a more specific knowledge of the depth of soft soil at the PVD installation locations. The installation depth of PVD is normally specified by the designer. However, when
erratic soil profiles are encountered, contractors are allowed to terminate
the PVD only when the stiff or hard formation below the soft soil formation is encountered, which can be gauged based on the efforts required to
penetrate the mandrel. In this case, the thicknesses of the soft soil layer at
different PVD installation points can be known.

At the present, the following three methods have been adopted in measuring the penetration depth of PVD as described by Bo et al. (2003): (1)
using a meter on the mast; (2) using a dial gauge; and (3) using an automatic
digital counter. However, all the three methods measure only the length of
the PVDs that pass through the point where the counter or the dial gauge
is located, not the real length of the PVDs that has been installed into the
soft clay. For this reason, none of the three methods can provide a direct
measurement of the PVD installed in the ground. Therefore, none of the
three methods is suitable to be used for independent checking or auditing
purposes. Without measuring the penetration depth of the PVD directly, it
will be impossible to check whether there is any mistake or even cheating

in the PVD installation records. There were cases where PVDs were not installed deliberately to the required depths. Therefore, a method that can directly measure the installation depth of PVD is also required. In the following, three new methods that can measure the penetration depth of PVDs directly after the PVD has been installed are introduced. These three new methods are digitised PVD, PVD with two wires, and PVD with one wire.

4.4.3.1 Digitised PVD

The first method is to print a meter scale on the surface of PVD at an interval of 20 or 25 cm so its linear length can be read directly. An example is shown in Figure 4.34. The meter scale can be printed onto the surface of PVD automatically using the computer scale-spattering digital metering technology. When the PVDs are installed by following a given sequence, the differences in the meters printed on top of the current and the last PVD installed will be the installation length of the current PVD. The total length of PVDs used can also be calculated easily. This method is simple and incurs almost no extra cost. However, it has the following shortcomings. The numbers printed on the filter of PVD become illegible when the PVDs are stored or exposed outside for too long. Also, the readings are affected by the expansion or contraction of PVDs due to temperature variation or wetting. PVDs may be stretched during installation and the scale printed may not be accurate. Furthermore, it is still not a direct measurement as the length of the PVDs is not measured directly.

Figure 4.34 PVD with scale printed on it for PVD penetration depth measurement. (From Liu, H.L., Chu, J. and Ren, Z.Y. (2009). *Geotextiles and Geomembranes*, Vol. 27, No. 6, pp. 493–496.)

4.4.3.2 PVD with two wires

The second method is to embed two shielded thin copper wires in the PVD, as shown in Figure 4.35. This method has been patented (Ren, 2004). The length of the PVD can be calculated by measuring the resistance of the wires. The two wires are embedded along the overlapping joint of the filter. Before installation, the two wires at the bottom end of the PVD need to be connected together. At the top end of the PVD, the wires are connected to a meter to measure the electrical resistance of the two wires as one loop. A readout unit as shown in Figure 4.36 has been specially designed for this purpose. This readout unit can measure the resistance of the wires, convert it directly into length, and display and store the readings. This is probably the most direct and reliable method available so far. However, this method also has some shortcomings. Firstly, it incurs extra costs to the PVDs to use two wires. Secondly, the connection of two wires before each installation of PVD is troublesome. Furthermore, the connection has to be done properly or the method will not work. Disputes may rise sometimes on whether the PVD is not installed properly or simply because the wires at the end of the PVD are not connected properly. For this reason, a standard procedure should be adopted. It is suggested to use a minimum connection length of 20 mm. The insulation at the connection must be removed by burning or scratching. The connected portion should be put back into the filter.

Figure 4.35 PVD with two copper wires embedded for PVD penetration depth. (From Liu, H.L., Chu, J. and Ren, Z.Y. (2009). *Geotextiles and Geomembranes*, Vol. 27, No. 6, pp. 493–496.)

Figure 4.36 Readout unit for the measurement of PVD penetration depth. (From Liu, H.L., Chu, J. and Ren, Z.Y. (2009). *Geotextiles and Geomembranes*, Vol. 27, No. 6, pp. 493–496.)

4.4.3.3 PVD with one wire

To overcome the problems associated with the two-wire PVD method, a third method has been developed. This method is similar to the two-wire PVD method, but uses only a single wire. This thin copper wire is embedded along the overlapping joint of the filter as shown in Figure 4.37. This method is based on the principle of microwave impedance measurement. When PVD is installed into the ground, the wire in the PVD and another wire connecting to the ground as provided by the readout unit form a two-wire system. Some field verification of this method is given in Liu et al. (2009).

4.4.4 Field instrumentation and evaluation of performance

A preloading plus PVDs soil improvement project is usually carried out until the required degree of consolidation is obtained. Degree of consolidation is an important parameter in evaluating the effectiveness of soil improvement. It is also often used as a design specification in a soil improvement contract. Assessment of the degree of consolidation of the soil therefore becomes one of the most important tasks for construction control. One of the most suitable methods for assessing the degree of consolidation of soil is by means of field instrumentation using settlement or pore water pressure data. For this reason, field instrumentation is normally required to monitor settlements and

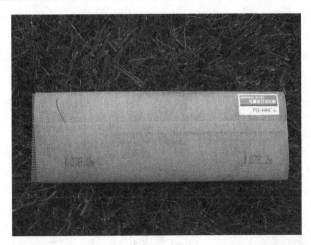

Figure 4.37 PVD with one copper wire embedded for PVD penetration depth measurement. (From Liu, H.L., Chu, J. and Ren, Z.Y. (2009). *Geotextiles and Geomembranes*, Vol. 27, No. 6, pp. 493–496.)

pore water pressures at different elevations as well as groundwater tables, lateral displacement, and earth pressure. One example, for the Changi reclamation project in Singapore, is shown in Figure 4.38. It can be seen that in each soil layer, at least three pore pressure transducers should be used to construct the pore pressure distribution versus depth profile. Some of the commonly used instruments are described in Bo et al. (2003). A case study is presented and some instrumentation issues are discussed in Arulrajah et al. (2009). For projects using PVDs, in particular those that use vacuum preloading, it is highly desirable to measure the pore water pressure inside the PVD. A piezometer can be placed inside a PVD, as shown in Figure 4.39 before PVD installation for this purpose. Some typical settlement and pore water pressure monitoring data are shown in Figure 4.40.

The degree of consolidation is normally calculated as the ratio of the current settlement to the ultimate settlement. However, for a soil improvement project, the ultimate settlement is unknown and has to be predicted. Although consolidation settlement can be estimated based on laboratory oedometer tests, the prediction by this method is normally not very reliable. Methods to estimate the ultimate settlement based on field settlement monitoring data are also proposed. Among them, the Asaoka (1978) and hyperbolic (Sridharan and Rao, 1981) methods are commonly used.

In Asaoka's method, a series of settlement data $(S_1,..., S_{i-1}, S_i, S_{i+1},...S_N)$ which are observed at constant time intervals are plotted in a S_n versus S_{n-1} plot $(n = 1,..., N)$. The ultimate settlement, S_{ult}, is taken as the intersecting point of the curve fitting line with the 45° line (Asaoka, 1978), as illustrated in Figure 4.41. However, S_{ult} obtained from Asaoka's method is affected by

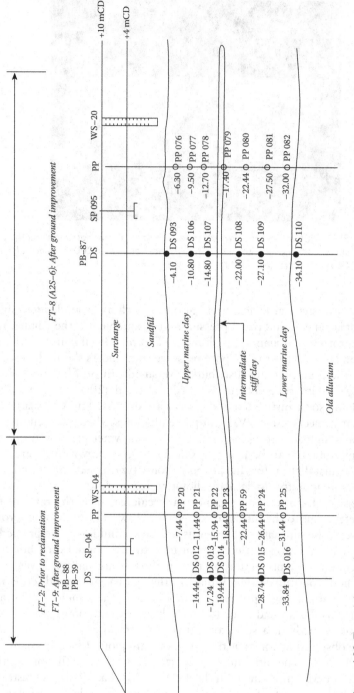

Figure 4.38 Typical details of onshore and adjacent offshore field instrumentation clusters. SP = Settlement plate, DS = Deep settlement gauge, PP = Pneumatic piezometer, WS = Water stand-pipe. (From Arulrajah, A., Bo, M.W. and Chu, J. (2009). Instrumentation at the Changi land reclamation project, Singapore. Geotechnical Engineering, Proceedings of the Institution of Civil Engineers, Vol. 162, No. I, pp. 33–40.)

Figure 4.39 Putting piezometer inside a PVD for pore water pressure measurement.

the time interval used (Matyas and Rothenburg, 1996; Bo et al., 1999). In the hyperbolic method, settlement data are plotted as time/settlement versus time curve (Sridharan and Rao, 1981). The S_{ult} is estimated as the inverse of the linear slope of the plot. However, S_{ult} obtained from this method is affected by the degree of consolidation achieved. The higher the degree of consolidation that the soil has attained, the smaller the S_{ult} obtained as observed by Matyas and Rothenburg (1996), Bo et al. (1999), and Goi (2004). The uncertainties involved in the ultimate settlement calculation will affect the estimation of the degree of consolidation. As a result, different degrees of consolidation are obtained using different methods. As an alternative, pore water pressure data can be used to assess the degree of consolidation.

Once the pore water pressures at different depths are measured during preloading, the initial and final pore water pressure distributions with depth can be plotted (Chu et al., 2000). For generality, a combined fill surcharge and vacuum load case is considered. The typical pore water pressure distribution profiles for a combined vacuum and fill surcharge loading case are shown schematically in Figure 4.42. Using this profile, the average degree of consolidation, U_{avg}, can be calculated as:

$$U_{avg} = 1 - \frac{\int [u_t(z) - u_s(z)]\,dz}{\int [u_0(z) - u_s(z)]\,dz} \qquad (4.24)$$

and

$$u_s(z) = \gamma_w z - \sigma, \ \text{kPa}.$$

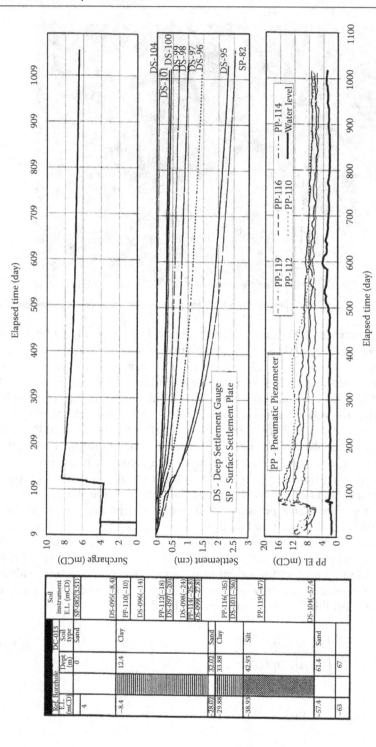

Figure 4.40 Typical settlement and pore water pressure data obtained from the third pilot test. (From Chu, J., Bo, M.W. and Choa, V. (2006). *Geotextiles and Geomembranes*, Vol. 24, No. 6, pp. 339–348.)

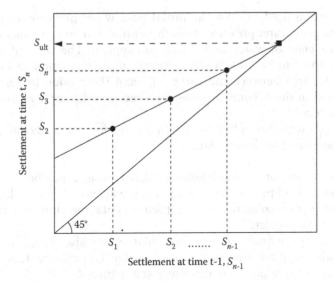

Figure 4.41 Schematic illustration of Asaoka's method.

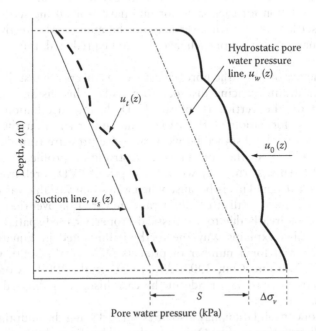

Figure 4.42 Schematic illustration of pore water pressure distributions versus depth under combined surcharge and vacuum load.

In Equation 4.24, $u_0(z)$ = the initial pore water pressure at depth z; $u_t(z)$ = the pore water pressure at depth z at time t; $u_s(z)$ is the suction line, γ_w = unit weight of water, and s = suction applied. The value of s is normally assumed to be 80 kPa. The integral in the numerator in Equation 4.24 is the area between the curve $u_t(z)$ and the suction line $u_s(z)$, and the integral in the denominator the area between the curve $u_0(z)$ and the suction line $u_s(z)$.

The method shown in Equation 4.24 has the following advantages over the method using settlement data:

(1) The degree of consolidation calculated using Equation 4.24 relies only on field pore water pressure data, whereas when calculating the degree of consolidation using settlement data, the ultimate settlement has to be predicted.

(2) Not only the final degree of consolidation, but also the degree of consolidation at any time can be calculated using Equation 4.24, as $u_t(z)$ represents the pore water pressure at any time, t.

(3) For consolidation involving multiple layers, Equation 4.24 can be applied to any single layer to calculate the degree of consolidation achieved in a particular layer. In this case, the upper and lower limits of the integrals in Equation 4.24 are set to be the top and bottom of that soil layer. However, it is not easy to calculate the degree of consolidation for each layer for multilayer soils using settlement, as the settlement of each layer may not be monitored directly and the ultimate settlement of each layer has to be predicted, too.

With the use of PVD, the pore pressure transducers are installed within half of the drains' spacing distance to a PVD. This distance may vary depending on the verticality of the PVD during installation and the subsequent deformation of the PVD. One concern is that the random uncertainties of the distance between the pore pressure transducers and the PVDs will affect the pore pressure distribution profile, as shown in Figure 4.40. This is true only when the depth of PVD is relatively short, say, less than 10 m. This is because when a random variable varies over a long distance, the overall effect of the random variation over the entire distance reduced greatly due to a statistical property called spatial variance reduction. This explains why the method illustrated in Equation 4.21 has worked well for a number of projects (Chu et al., 2000; Chu and Yan, 2005; Yan and Chu, 2003; 2007; Chu et al., 2009). More details and comparisons are also made in the case histories published in these references.

Degree of consolidation may also be estimated using the undrained shear strength distribution profile. One example will be shown in the Case Study in Section 4.7.

It should be mentioned that any method can only be as reliable as the field monitoring data. Any uncertainties involved in the field pore water pressure and settlement measurements will inevitably affect the degree of consolidation estimation. Furthermore, when pore water pressures are measured at only a limited number of points, the spatial pore water pressure distribution cannot be constructed. In this case, the pore water pressure distribution profile established for one section has to be assumed to be the same as that at other sections. This may not be the case, although with the use of PVDs the pore water pressure distributions tend to even out. For the degree of consolidation estimated based on settlement, there is one more source of uncertainty, that is, the uncertainties involved in the ultimate settlement prediction. In view of the various uncertainties involved in the degree of consolidation calculation, it is recommended to estimate the degree of consolidation using both settlement and pore water pressure data. Even if the degree of consolidation is to be calculated using settlement data, the pore water pressure distribution profile provides a way to visualise whether the pore water pressure dissipation is consistent with the degree of consolidation calculated based on settlement. If the differences between the two measurements are too large and the difference cannot be explained, the results should then be examined before they are accepted. For contracting purposes, it will be necessary to specify clearly whether the degree of consolidation should be evaluated based on settlement or pore water pressure or both to avoid future dispute.

4.5 DESIGN CODES, STANDARDISATION, AND SPECIFICATIONS

As discussed in the preceding sections, a number of ASTM D-series standards have been used as the standards for vertical drain testing. However, some of those ASTM standards are not specifically written for vertical drains. Therefore, the testing procedures stipulated in these standards may not be the most suitable methods. Some of the ASTM standards can be compiled using different testing systems. This has been the reason why there are so many different testing methods proposed. As the performance of vertical drains can be affected considerably by the quality of the drains used and the control in the construction procedures, it would be highly desirable to set up some regulations or codes of practice to govern the selection of vertical drain and to regulate the construction activities. Several design codes or standards have been developed. These include the European Standard on Execution of Special Geotechnical Works—Vertical Drainage (BS EN 15237, 2007), the Chinese design code JTJ/T256-96 (1996) that controls the practice for installation of PVDs and JTJ/T257-96 (1996) that stipulates the quality inspection standard for PVDs, and the Australian

Standards for the execution of PVDs, AS8700 (2011). The use of these codes and standards is important in maintaining the quality standards of soil improvement works.

The Chinese Quality Inspection Standard for PVDs (JTJ/T257-96) requires every batch of PVDs and every 200,000 m within the same batch to be sampled and tested for quality-control purposes. What is interesting is that JTJ/T257-96 specifies the thickness of PVDs to vary according to the depth of installation as well as the discharge capacity and tensile strength (see Table 4.5). Although no reasons were given for the specific values used, the use of a thicker and thus larger discharge capacity PVD is in line with the requirement expressed in Inequality (4.20). The other recommended specifications for PVDs are also given in Table 4.5.

The European Standard BS EN 15237 includes the application of PVDs and sand drains and deals with requirements to be placed on design, drain material, and installation methods. For the material properties of PVD, the main properties required and their testing methods are listed in Table 4.6. All the PVD properties used in Table 4.6 have been explained in this chapter except for the velocity index of filter v_{h50} and durability. The 'velocity index of filter' sounds new, but is merely another way to measure filter permeability in Europe. It defines the filtration velocity corresponding

Table 4.5 Minimum thickness and other specifications for PVD as specified by the Chinese Code JTJ/T257-96

(a) Minimum thickness

Type	$L < 15\,m$	$L < 25\,m$	$L < 35\,m$	Stud type
Thickness (mm)	> 3.5	> 4.0	> 4.5	> 6

(b) Other specifications

Description	Unit	$L < 15m$	$L < 25m$	$L < 35m$	Testing conditions
Discharge capacity	cm³/s (m³/yr)	15 (670)	25 (1,115)	40 (1,784)	Under pressure of 300 kPa
Permeability of filter	cm/s		5×10^{-4}		After the sample is immersed in water for 24 h
Pore opening of filter	μm		< 75		O_{95}
Tensile strength of PVD	kN/10 cm	> 1.0	> 1.3	> 1.5	At 10% elongation
Tensile strength of filter (dry)	kN/cm	> 15	> 25	> 30	At 10% elongation
Tensile strength of filter (wet)	kN/cm	> 10	> 20	> 25	At 10% elongation. Sample immersed in water for 24 h.

Table 4.6 Requirements of PVDs properties in BS EN 15237 (2007)

Properties	Requirement	Testing Method
Tensile strength	> 1.5 kN at failure	EN ISO 10319
Elongation at max. tensile force	elongation ≥ 2% at failure; elongation ≤ 10% at a tensile force of 0.5 kN	EN ISO 10319
Tensile strength of filter	> 3 kN/m or 6 kN/m for PVD longer than 25 m.	EN ISO 10319
Tensile strength of seams and joints	> 1 kN/m	EN ISO 10321
Velocity index of filter (v_{h50})	> 1 mm/s	EN ISO 11058
Characteristic opening size of filter (O_{90})	< 80 μm; and < (1.5 to 2.8)d_{50} of soil	EN ISO 12956
Discharge capacity of the drain	See Figure 4.43	EN ISO 12958
Durability	PVDs to be covered on the same day	EN 13252

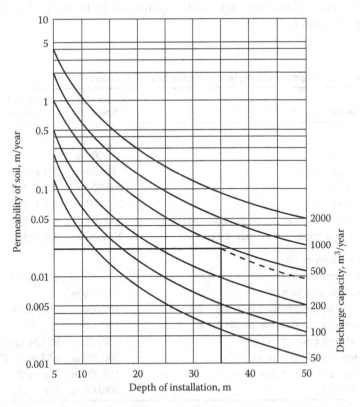

Figure 4.43 Required discharge capacity q_w as a function of permeability of soil and depth of drain installation shown in BS EN 15237 (2007)

to a head loss of 50 mm across the specimen (CEN, 1998). Durability is a general requirement for geotextile- and geomembrane-related products. It is stated in EN 13252 that all the geotextile- and geomembrane-related products shall pass the accelerated weathering test according to EN 12224, unless they are to be covered on the day of installation. Therefore, if we ensure the PVD is to be installed within one day after it is taken out from the store, durability will not be a problem. It is noted that the discharge capacity is not specified as a fixed value, but a function of depth of installation and the permeability of soil as shown from Figure 4.43. It can be seen that the range of values is similar to the Chinese Code JTJ/T257-96. In the past, many PVD specifications used a fixed discharge capacity (some examples are given in Bo et al., 2003). In this respect, both the Chinese Code JTJ/T257-96 (1996) and BS EN 15237 (2007) have set a better design standard of PVDs.

BS EN 15237 has also specified the frequency for conducting quality-control tests as shown in Table 4.7. It is necessary to carry out quality-control tests at a certain frequency, particularly when a huge quantity of PVDs is used. Variations between the quality of PVDs were observed in some past projects.

Table 4.7 Proposed testing frequency for fabrication control

Property	Proposed test frequency	Required standard
Filter:		
Thickness	25,000 m²	EN 9863-1
Mass per unit area	25,000 m²	EN 9864
Pore size	200,000 m²	EN 12956
Velocity index	200,000 m²	EN 11058
Tensile strength in the longitudinal direction	200,000 m²	EN 10319
Tensile strength in the cross direction	200,000 m²	EN 10319
Drain composite:		
Width and thickness	25,000 m	EN 9863-1
Mass per unit length	25,000 m	–
Tensile strength in the longitudinal direction	100,000 m	EN 10319
Elongation at maximum tensile force	100,000 m	EN 10319
Discharge capacity straight	500,000 m	BS EN 15237
Discharge capacity buckled	500,000 m	BS EN 15237
Tensile strength of filter seam	100,000 m	EN 10321
Durability	500,000 m	EN 13252

Source: BS EN 15237. (2007). *European Standard on Execution of Special Geotechnical Works—Vertical Drainage.* European Standard.

4.6 PVD FOR VACUUM PRELOADING

When PVDs are used together with vacuum preloading for soil improvement, some special arrangements may be required. The PVDs in this case will not only discharge water, but also transmit vacuum pressure as shown in Figure 4.44. Therefore, the PVDs used for a vacuum preloading project should possess better quality than normal. Sometimes, prefabricated circular drains (see Figure 4.10) are also used. More detailed description of the vacuum preloading system shown in Figure 4.44 can be found in Chu and Yan (2005b).

The vacuum preloading system shown in Figure 4.44 works well when the low permeability compressible soil extends all the way to the ground surface so that membranes can be used to seal the top surface for vacuum pressure application. However, when there is a relatively thick layer of permeable soil on top, a cut-off wall extending all the way to the bottom of

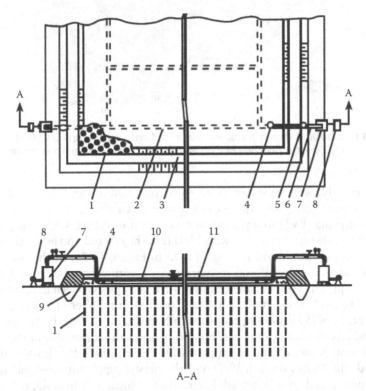

Figure 4.44 Schematic arrangement of vacuum preloading system. 1, drains; 2, filter piping; 3, revetment; 4, water outlet; 5, valve; 6, vacuum gauge; 7, jet pump; 8, centrifugal gauge; 9, trench; 10, horizontal piping; 11, sealing membrane. (From Chu, J., Yan, S.W. and Yang, H. (2000). *Géotechnique*, Vol. 50, No. 6, pp. 625–632.)

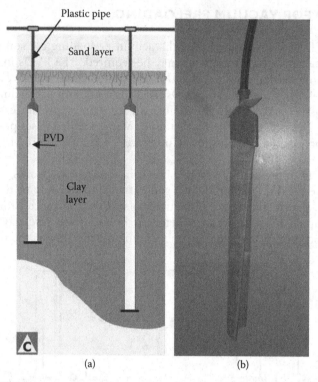

Figure 4.45 BeauDrain vacuum preloading system. (a) Concept. (Courtesy of Cofra, Holland.) (b) Direct connection of PVD with plastic pipe for vacuum application.

the permeable soil layer will have to be installed around the whole soil improvement site. This can be too expensive. An alternative is to connect each individual PVD to a plastic pipe as shown in Figure 4.45. The plastic pipe can be installed together with PVD into the ground and going through the entire permeable soil layer as shown in Figure 4.45a. In this way, the need to use cut-off wall and membrane for creating an airtight seal is no longer required. This so-called BeauDrain technique has been developed by Cofra Holland (Kolff et al., 2004) and used for a number of soil improvement projects (Seah, 2006; Saowapakbiboon et al., 2008). However, this method has its own limitations. First, it requires the soil profile at the PVD installation points to be known precisely. This can be difficult sometimes. Second, the length of each PVD and the plastic pipe connected has to be determined and preconnected beforehand. Third, for installation, each PVD with the plastic pipe has to be pulled through a mandrel with the plastic pipe on top and then installed into the ground using a PVD installation machine, see Figure 4.46. The plastic pipes are then connected to a main vacuum line which is linked to a vacuum pump, see Figure 4.47.

Figure 4.46 Installation of PVDs with plastic pipe connected.

Figure 4.47 Connection of plastic pipes to a vacuum pump.

4.7 CASE STUDIES

4.7.1 Reclamation and soil improvement for a slurry pond in Singapore

PVDs have been used in many soil improvement and land reclamation projects. However, the use of PVDs for the improvement of ultra-soft soil; that is, soil with no or little shear strength are still not common. One such a project was carried out in Singapore to reclaim a slurry pond of about 180 hectares as part of the Changi East Reclamation Projects (Choa et al., 2001, Bo et al., 2005). This slurry pond was created by dredging seabed to an elevation of –22 mCD (Chart Datum) for the storage of silt and clay washings from other sand-quarrying activities. The thickness of the ultra-soft slurry varied from 1–20 m with an average value of 15 m. The grain size distribution curves indicate a fines content in the range of 70%–93%. The upper bound of the mean grain size D_{50} was 0.024 mm, but mostly in the range smaller than 0.001 mm and D_{85} was in the range of 0.004–0.02 mm. The water content of the slurry was mainly in the range of 140%–180%. The bulk unit weight of the slurry ranged mainly from 11–13 kN/m³. As the slurry was deposited recently with little consolidation, it was ultra-soft and highly compressible. Based on the properties of the slurry and the depth of installation, the properties of the PVDs were chosen as shown in Table 4.8. The analysis for the selection of PVD is detailed in Chu et al. (2006).

The procedure adopted for the reclamation of the slurry pond was placing a sand-capping layer on top of the slurry before PVDs could be installed and used for the consolidation of the slurry. As the slurry had essentially no strength, the land reclamation work was carried out by spreading thin layers of sand using a specially designed sand spreader (Chu et al., 2009). To ensure the stability of the fill, small lifts of 20 cm were used in the first phase of the spreading. This phase of sand spreading took about 13 months including the waiting time between the lifts. There was a slurry burst when the fill reached an elevation between –1 and +2 mCD (Chu et al., 2009). Otherwise, the reclamation using sand spreading was workable. After the fill was exposed above the water level, PVDs were installed with 2 × 2 m square spacing. The surcharge was then placed to +6 mCD. The settlement of the fill was monitored. After about 1.5 m of settlement had occurred, a second round of PVDs with the same 2 × 2 m spacing was installed. During the installation of PVDs, slurry was observed to come out through the annulus of the mandrel, as shown in Figure 4.48. This is indicative that the pore water pressure in the ground was still high and the installation process itself helped in the dissipation of pore water pressure and thus the consolidation process. This is another advantage of the two-stage PVD installation method.

Settlement gauges and pore pressure transducers were installed in the ultra-soft clay layer to monitor the consolidation process of the ultra-soft soil. The typical arrangement of instrument layout and profile are shown in Figure 4.49. The surcharge history and settlement and excess pore pressure

Table 4.8 Properties of the PVD used

Permeability of the filter	Discharge capacity of straight drain under 100 kPa pressure	Discharge capacity of buckled drain (under 25% strain) under 100 kPa pressure	Apparent opening Size O_{95}
2.6×10^{-4} m/s	70×10^{-6} m³/s	20×10^{-6} m³/s	Less than 0.07 mm

Figure 4.48 Extrusion of viscous pore water and mud during PVD installation. (From Chu, J., Bo, M.W. and Choa, V. (2006). *Geotextiles and Geomembranes*, Vol. 24, No. 6, pp. 339–348.)

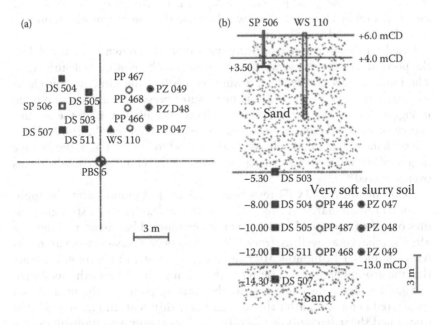

Figure 4.49 Arrangement of instrumentation. (a) Plan view. (b) Elevation view. (From Chu, J., Bo, M.W. and Choa, V. (2006). *Geotextiles and Geomembranes*, Vol. 24, No. 6, pp. 339–348.)

versus time curves are shown in Figure 4.50. As indicated in Figure 4.50a, the second round of PVD installation was carried out about 80 days after the first round of PVD installation, where the top of the slurry had settled nearly 1.5 m. It should be noted that the surcharge load was reducing with time as the fill used for surcharge was gradually submerged into water due to the settlement of the ground. As shown in Figure 4.50b, the slurry had settled for about 2.7 m in 500 days.

The excess pore pressures versus time curves measured by piezometers PZ047, PZ048 and PZ049 (see Figure 4.49) are shown in Figure 4.50c. The piezometers were installed after the first PVD installation. A quick increase in pore pressure was observed after 90 days at all three locations. This was caused by the installation of the 2nd round of PVDs. The excess pore pressure dissipations as measured by PZ047 at −8 mCD and PZ048 at −10 mCD were slow despite of the occurrence of large settlement. The lack of pore pressure dissipation may signify a sedimentation and self-weight consolidation stage prior to consolidation under additional fill. During this stage, slurry was transforming from a liquid to a solid state in which water was dissipating, but the soil particles did not have sufficient contacts to allow the soil skeleton to take up external load. The Mandel–Cryer effect and non-uniform consolidation of soil around the PVD were thought to be the other reasons accounting for the lack of pore water dissipation (Chu et al., 2006). The pore pressure dissipation measured by PZ-49 at −12 mCD was relatively quick. This was because the soil at this elevation was near the silty sand layer below it.

Based on the pore water pressure measurements shown in Figure 4.48c, the pore water pressure versus depth profile can be plotted in Figure 4.51. The initial excess pore water pressure, which had the same magnitude as the surcharge and the hydrostatic pore water pressure line, is also plotted in Figure 4.51. Using the method introduced in Equation 4.24, the average degree of consolidation is estimated to be 42% (Chu et al., 2009). If we use settlement data and apply Asaoka's method to Figure 4.48b, the average degree of consolidation is calculated as 91% (Chu et al., 2009), which was overestimated.

Field vane shear and CPT tests were conducted 14 months after the application of the surcharge. A comparison of the undrained shear strength profiles obtained from field vane shear tests conducted before and 14 months after surcharge as well as from CPT (with pore pressure measurement) tests conducted 14 months after surcharge is shown in Figure 4.52. Note that the ground had settled for more than 2 m within 14 months, as shown in Figure 4.50a. This explains why the starting points of the in-situ tests conducted before and after the surcharge are different. In Figure 4.52, the undrained shear strength profile estimated by assuming a uniform degree of consolidation of 90% is also plotted for comparison. As mentioned, a silty sand was present at an elevation of −12.5 mCD and a sand blanket

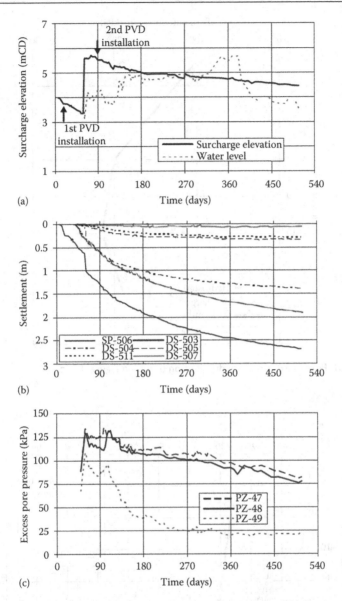

Figure 4.50 Monitoring data during the reclamation of slurry pond. (a) Surcharge variation versus time. (b) Settlement versus time. (c) Excess pore pressure versus time. Initial positions of the instruments: SP-506 at +3.5 mCD; DS-503
at −5.3 mCD; PZ-47, DS-504 and PP-466 at −8 mCD; PZ-48, DS-505 and
PP-467 at −10 mCD; PZ-49, DS-511 and PP-468 at −12 mCD. (From Chu, J.,
Bo, M.W. and Choa, V. (2006). *Geotextiles and Geomembranes*, Vol. 24, No. 6,
pp. 339–348.)

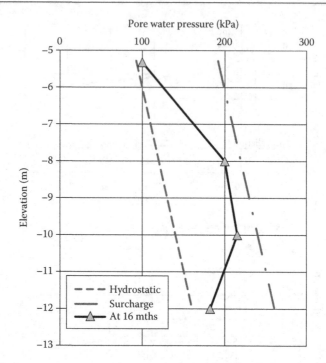

Figure 4.51 Pore water pressure distribution based on Figure 4.50c.

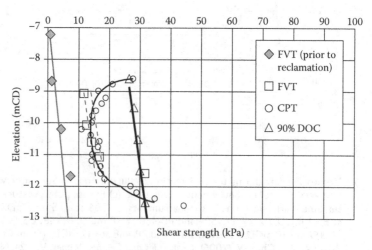

Figure 4.52 Comparison of undrained shear strength profiles measured before and after consolidation for 14 months. (Redrawn from Chu, J., Bo, M.W. and Choa, V. (2006). *Geotextiles and Geomembranes*, Vol. 24, No. 6, pp. 339–348.)

was placed on the top surface. Therefore, there was combined vertical and horizontal drainage near the top and bottom boundaries. This explains why the shear strength increment was the largest at both the top and the bottom. The consolidation in the middle of the clay layer ranging from –9.5 to –12.0 m was contributed mainly by horizontal drainage to the PDVs. Applying a method similar to Equation 4.24, the degree of consolidation can be estimated as 45%, which is similar to that based on pore water pressure.

4.7.2 Combined fill and vacuum preloading case

The second case is for the soil treatment of a storage yard in Tianjin Port, China. The storage yard was located on a 16-m-thick soft clay layer. The top 3–4 m of the clay layer was reclaimed recently using clay slurry dredged from seabed. The remaining 16–19 m was original seabed clay. The soil in both layers was soft and still undergoing consolidation. This soft clay layer needed to be improved before the site could be used as a storage yard. Preloading using fill surcharge alone was not feasible as it was difficult to place a fill embankment several meters high on soft clay. The vacuum preloading method could be used. However, the nominal vacuum load of 80 kPa was not sufficient for this project. Therefore, a combined vacuum and fill surcharge preloading method was adopted. Fill surcharge of a height ranging from 2.53–3.50 m was applied in addition to the vacuum load and 0.3 m of sand blanket. The fill was applied in stages partially for stability consideration and partially due to practical constraints in transporting fill. The fill used was a silty clay with an average unit weight of 17.1 kN/m^3.

The layout of the storage yard is shown in Figure 4.53. It was an L-shape with a total area of 7433 m^2. For the convenience of construction, the site was divided into three sections, I, II and III, as shown in Figure 4.53. The water content of the soil was higher than or as high as the liquid limit at most locations in the soft clays. The field vane shear strength of the soil was generally 20– 40 kPa. For more information on the site conditions and soil properties, see Yan and Chu (2007).

The soil improvement work was carried out as follows. A 0.3-m sand blanket was first placed on the ground surface. PVDs were then installed on a square grid at a spacing of 1.0 m to a depth of 20 m. Corrugated flexible pipes (100 mm diameter) were laid horizontally in the sand blanket to link the PVDs to the main vacuum pressure line. The pipes were perforated and wrapped with a permeable fabric textile to act as a filter layer. Three layers of thin PVC membrane were laid to seal each section. Vacuum pressure was then applied using jet pumps. The schematic arrangement of the vacuum preloading method used is similar to that shown in Figure 4.44.

The soil improvement started from Section I, followed by Section II and then Section III. The loading sequence and the ground settlements induced

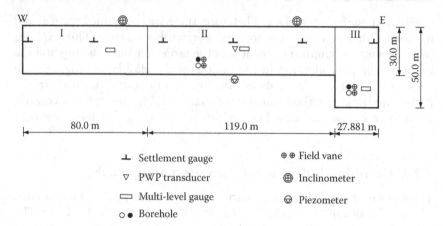

Figure 4.53 Project site layout and plan view of instrumentation. (Redrawn from Yan, S.W. and Chu, J. (2005). *Canadian Geotechnical Journal*, Vol. 42, No. 4, pp. 1094–1104.)

by the vacuum and surcharge loads for Section II are shown in Figure 4.54. The vacuum load was applied for 4–8 weeks before fill surcharge loads were applied in stages. The total fill height applied was 3.5 m for Section II. The maximum surface settlement induced by the vacuum and surcharge loads in this section was 1.614 m.

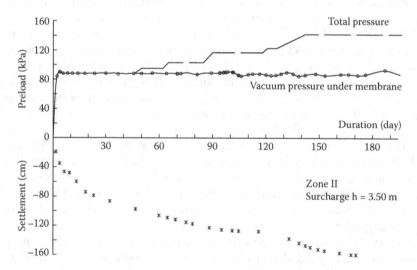

Figure 4.54 Loading sequence and ground settlement measured at Section II. (Redrawn from Yan, S.W. and Chu, J. (2005). *Canadian Geotechnical Journal*, Vol. 42, No. 4, pp. 1094–1104.)

Instruments including surface settlement plates, multi-level settlement gauges, and standpipes were installed in all three sections to monitor the consolidation performance. However, inclinometers were installed in sections I and II only and pore water pressure transducers in Section II only due to budget constraints. The locations of those instruments are shown schematically in Figure 4.53 (plan view) and Figure 4.55 (elevation view). Soil samples were taken from sections II and III both before and after soil improvement for laboratory tests. Field vane shear tests were also conducted at sections II and III both before and after the soil improvement.

Some settlements took place after the vertical drains were installed, but before the vacuum and surcharge loads were applied. The durations between the installation of vertical drains and the application of vacuum loads were 3–4 weeks. The ground settlement measured before the application of vacuum loads was 0.21, 0.31, and 0.25 m for sections I, II, and III respectively. The settlements were induced mainly as a result of the dissipation of the existing excess pore water pressures in the soil as the soil was still under consolidation due to land reclamation. The disturbance to the soil caused by the installation of the vertical drains also contributed to the settlement.

The settlements monitored by the settlement gauges installed at different depths during vacuum and surcharge loadings are plotted versus duration for Section II in Figure 4.56. Settlements were observed in soil up to 14.5 m deep, which indicates that the vacuum preloading was effective for

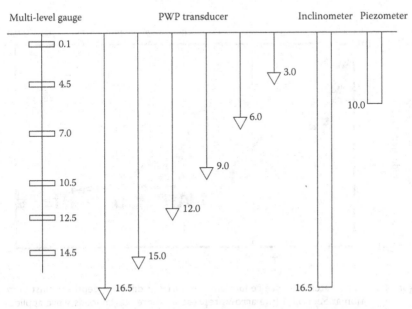

Figure 4.55 Elevation view of instrumentation. (Redrawn from Yan, S.W. and Chu, J. (2005). *Canadian Geotechnical Journal*, Vol. 42, No. 4, pp. 1094–1104.)

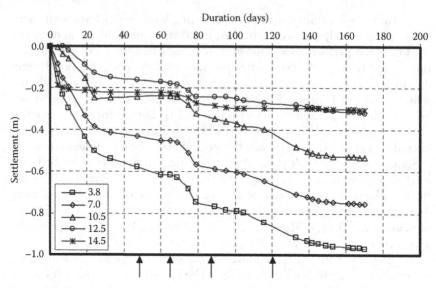

Figure 4.56 Settlement measured at different depths against duration at Section II. (The arrows represent where staged loads were applied)

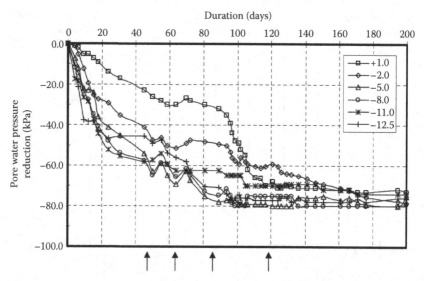

Figure 4.57 Pore water pressure reductions measured at different depths against duration at Section II (the arrows represent where staged loads were applied). (Redrawn from Yan, S.W. and Chu, J. (2005). *Canadian Geotechnical Journal*, Vol. 42, No. 4, pp. 1094–1104.)

the entire 16-m soft clay. The reductions in the pore water pressures measured by the piezometers installed at different depths are plotted versus loading duration in Figure 4.57 for Section II. Under the vacuum load, the pore water pressures reduced quickly with time. However, when the fill surcharge was applied, a localised pore pressure increase occurred. The times at which staged surcharge loads were applied are indicated by arrows in Figure 4.57. It can be seen that the localised increase in pore water pressure coincides with the application of surcharge loads.

Based on the pore water pressure monitoring data shown in Figure 4.57, the pore water pressure distributions with depth at the initial stage, 30 and 60 days, and the final stage are plotted in Figure 4.58. Before the application of vacuum and surcharge loads, the initial pore water

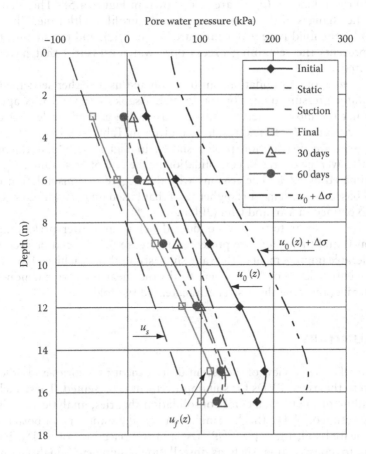

Figure 4.58 Pore water pressure distributions with depth at Section II. (Redrawn from Yan, S.W. and Chu, J. (2005). *Canadian Geotechnical Journal*, Vol. 42, No. 4, pp. 1094–1104.)

Table 4.9 Ultimate settlement and degree of consolidation estimated by different methods

Section	Asaoka's method		Based on pore pressure
	S_∞ (m)	$U_f(\%)$	U_f (%)
II	1.84	87	82

pressures, $u_0(z)$, were greater than the hydrostatic pore water pressure. The total fill surcharge was about 60 kPa for Section II. Using the vertical stress calculated, the initial pore water pressure distribution after the application of the fill surcharge is shown as $u_0(z) + \Delta\sigma$ in Figure 4.58. The suction line for a suction of -80 kPa is also plotted in Figure 4.58 as the line u_s. The pore water pressure distributions at 30 and 60 days and the end of preloading $(u_f(z))$ are also shown in Figure 4.58. These curves show the changes of the pore water pressure profiles with time. The area bound by the final pore water pressure curve, $u_f(z)$, and the suction line, u_s, represents the remaining excess pore water pressures that have not dissipated.

The degree of consolidation can be estimated using either settlement or pore water pressure data. For the former, Asaoka's method was applied to predict the ultimate settlements, S_∞, using the ground settlement data shown in Figure 4.54. The results are given in Table 4.9. Using the pore water pressure distribution profile shown in Figure 4.58 and Equation 4.24, the average degree of consolidation at the end of preloading, U_f, can be estimated as 82%. The reasons why the degree of consolidation estimated based on settlement is higher than that based on pore water pressure were explained in Yan and Chu (2005).

Field vane shear tests were conducted before and after preloading in Section II and the results are presented in Figure 4.59. It can be seen that considerable improvement in the vane shear strength was achieved throughout the entire depth of 16 m where field vane shear tests were conducted. On average, the vane shear strength increased twofold.

4.8 SUMMARY

A comprehensive review of the recent development in theories and practice related to the use of PVDs for soil improvement is presented. These include the outline of the fundamental consolidation theories, analytical methods for the design of PVDs, the determination of soil parameters for both intact and smeared soil, the types of PVDs, and the selections of PVDs. Issues related to construction such as installation equipment, quality-control tests, measurement of penetration depth, field instrumentation, and methods for the performance evaluation of PVDs, design codes, standards, and

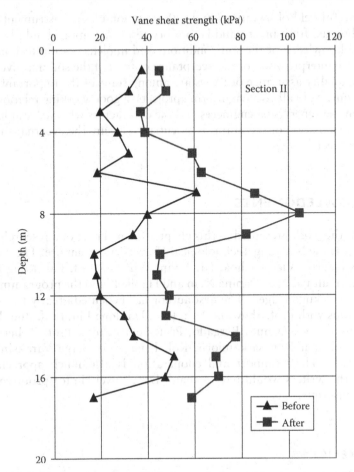

Figure 4.59 Field vane strength profile before and after soil improvement at Section II. (Redrawn from Yan, S.W. and Chu, J. (2005). *Canadian Geotechnical Journal*, Vol. 42, No. 4, pp. 1094–1104.)

specifications are also discussed. Some specific methods for using PVDs for vacuum preloading are also introduced. Finally, two case studies, one for the use of PVDs for soil improvement of slurry type of soil and another for the use of PVDs for a combined fill and vacuum surcharge for the improvement of the soft soil below a storage yard, are presented to illustrate the practical applications of PVDs for different soil improvement projects. In conclusion, sufficient research development has been made and practical experience been gained over the years on the use of PVDs for soil improvement. However, it is still more of an art than science as far as for the prediction of the outcome of a soil improvement scheme is concerned. A holistic approach is thus advocated for the implementation of PVD techniques for soil improvement. The design methods proposed should be used together with careful field observations and interpretation of soil instruments and data obtained from in-situ and laboratory tests. Attention should be paid to local knowledge of the soil conditions and implementation techniques. A sensible interpretation of the acceptance criteria of the soil improvement and the quality assurance of PVDs are among some of the important considerations. As in all civil engineering projects, a good working relationship between the supervising engineers and the contractors who are responsible for the installation of the drains is essential to ensure the eventual success of the project.

ACKNOWLEDGMENTS

Some of the work presented in this chapter formed part of a research programme at the Nanyang Technological University, Singapore. The contributions of Prof. Victor Choa, Dr. Myint Win Bo, Prof. Bak Kong Low, Prof. A. Arulrajah, Dr. Daping Xiao and Haojie Liu to the programme are gratefully acknowledged. The first author has benefited greatly from collaborations with Prof. Shuwang Yan, Prof. Hanlong Liu, Prof. Buddhima Indraratna, Prof. Dennes Bergado, Prof. Pedro Pinto, Prof. Robert Lo, Prof. Cholachat Rujikiatkamjorn, Kok Pang Lam, Serge Varaksin and Kenny Yee. Their support and cooperation is also much appreciated. Finally the writers would like to thank Dr. Zhiwei He for proofreading the manuscripts.

REFERENCES

Akagi, T. (1994). Hydraulic applications of geosynthetics to filtration and drainage problems – with special reference to prefabricated band-shaped drains, *Proc. 5th Int. Conf. on Geotextiles, Geomembranes and Related Products,* Singapore, preprint of keynote lecture, pp. 99–119.

Arulrajah, A., Bo, M.W. and Chu, J. (2009). Instrumentation at the Changi land reclamation project, Singapore. *Geotechnical Engineering, Proceedings of the Institution of Civil Engineers,* Vol. 162, No. 1, pp. 33–40.

Arulrajah, A., Nikraz, H., and Bo, M.W. (2004). Factors affecting field instrumentation assessment of marine clay treated with prefabricated vertical drains, *Geotextiles and Geomembranes,* Vol. 22, No. 5, pp. 415–437.

Asaoka, A. (1978). Observational procedure of settlement prediction. *Soils and Foundations,* Vol. 18, No. 4, pp. 87–101.

ASTM D4491-99a. (2009). *Standard Test Methods for Water Permeability of Geotextiles by Permittivity.* ASTM Standards.

ASTM D4632. (2008). *Standard Test Method for Grab Breaking Load and Elongation of Geotextile.* ASTM Standards.

ASTM D4716. (2008). *Standard Test Method for Determining the (In-plane) Flow Rate per unit Width and Hydraulic Transmissivity of a Geosynthetic Using a Constant Head*, ASTM Standards.

ASTM D4751. (2004). *Standard Test Method for Determining Apparent Opening Size of a Geotextile*, ASTM Standards.

AS8700 (2011) *Execution of Prefabricated Vertical Drains*, Standards Australia.

Balasubramaniam, A.S., Bergado, D.I., Long, P.V. and Thayalan. (1995). Experiences with sand drains and prefabricated vertical drains in ground improvement of soft clays, *Seminar on Engineering for Coastal Development*, Singapore.

Barron, R.A. (1948). Consolidation of fine-grained soils by drain wells, *Trans ASCE*, Vol. 113, Paper 2346, pp. 718–54.

Bergado, D.T., Singh, N., Sim, S.H., Panichayatum, B., Sampaco, C.L. and Balasubramaniam, A.S. (1990). Improvement of soft Bangkok clay using vertical geotextile band drains compared with granular piles, *Geotextiles and Geomembranes*, Vol. 9, No. 3, pp. 203–231.

Bergado, D.T., Sampaco, C.L., Shivashankar, R., Alfaro, M.C., Anderson, L.R. and Balasubramaniam, A.S. (1991). Performance of a welded wire wall with poor quality backfills on soft clay, *Proc. ASCE Geotechnical Congress*, Boulder, Colorado, pp. 909–922.

Bergado, D.T., Alfaro, M.C., and Balasubramaniam, A.S. (1993a). Improvement of soft Bangkok clay using vertical drains, *Geotextiles and Geomembranes*, Vol. 12, No. 7, pp. 615–663.

Bergado, D.T., Mukherjee, K., Alfaro, M.C. and Balasubramaniam, A.S. (1993b). Prediction of vertical-band-drain performance by the finite-element method, *Geotextiles and Geomembranes*, Vol. 12, No. 6, pp. 567–586.

Bergado, D.T., Long, P.V., Lee, C.H., Loke, K.H. and Werner, G. (1994). Performance of reinforcement embankment on soft Bangkok clay with high strength geotextile reinforcement, *Geotextiles and Geomembranes*, Vol. 13, pp. 403–420.

Bergado, D.T., Manivannan, R., Balasubramaniam, A.S. (1996a). Proposed criteria for discharge capacity of prefabricated vertical drains, *Geotextiles and Geomembranes*, Vol. 14, pp. 481–505.

Bergado, D.T., Anderson, L.R., Miura, N. and Balasubramaniam, A.S. (1996b). *Soft Ground Improvement in Lowland and Other Environments*, ASCE Press, ASCE, New York.

Bergado, D.T., Balasubramaniam, A.S., Fannin, R.J. and Holtz, R.D. (2002). Prefabricated vertical drains (PVDs) in soft Bangkok clay: a case study of the New Bangkok International Airport project, *Canadian Geotechnical Journal*, Vol. 39, pp. 304–315.

Bo M.W., Chu, J. and Choa, V. (1999). Factors affecting the assessment of degree of consolidation, *Proc. International Conference on Field Measurements in Geomechanics*, Balkema, Rotterdam, pp. 481–486.

Bo, M.W., Chu, J., Low, B.K. and Choa, V. (2003). *Soil Improvement: Prefabricated Vertical Drain Technique*, Thomson Learning, Singapore.

Bo, M.W. (2004). Discharge capacity of prefabricated vertical drain and their field measurements, *Geotextiles and Geomembranes*, Vol. 22, No. 1–2, pp. 37–48.

Bo, M.W., Chu, J., Choa, V. (2005). Changi East Reclamation and Soil Improvement Project. In: Chapter 9, *Ground Improvement—Case Histories*. Indraratna, B., Chu, J. (Eds.), Elsevier, Amsterdam, pp. 247–276.

Broms, B.B., Chu, J., and Choa, V. (1994). Measuring the discharge capacity of band drains by a new drain tester, *Proc. 5th Int. Conf. on Geotextiles, Geomembranes and Related Products*, Vol. 3, Singapore, pp. 803–806.

BS EN 15237. (2007). *European Standard on Execution of Special Geotechnical Works—Vertical Drainage*. European Standard.

Carillo, N. (1942). Simple two- and three-dimensional cases in the theory of consolidation of soils. *Journal of Mathematics and Physics*, Vol. 21, No. 1, 1e5.

Carroll, R.G., (1983). Geotextile filter criteria. Transportation Research Record, No. 916, pp. 46–53.

Chai, J.C., Miura, N., Sakajo, S. and Bergado, D.T. (1995). Behaviour of vertical drain improved subsoil under embankment loading, *Soils and Foundations*, Vol. 35, No. 4, pp. 49–61.

Chai, J.C., Shen, S.L., Miura, N. and Bergado, D.T. (2001). Simple method of modelling PVD improved subsoil, *Journal of Geotechnical and Geoenvironmental Engineering*, ASCE, Vol. 127, No.11, pp. 965–972.

Choa, V., Bo, M.W., and Chu, J., (2001). Soil improvement works for Changi East reclamation project. *Ground Improvement*, Vol. 5, No. 4, pp. 141–153.

Choa, V., Vijiaratnam, A. Karunaratne, G.P., Ramaswamy, S.D. and Lee, S.L. (1979). Pilot test for soil stabilisation at Changi Airport, *Proceedings 7th European Regional Conference on Soil Mechanics and Foundation Engineering*, Brighton, Vol. 3, pp. 29–36.

Chu, J. and Choa, V. (1995). Quality control tests of vertical drains for a land reclamation project, Compression and Consolidation of Clayey Soils, In: Yoshikuni, H., Kusakabe, O. (Eds.), *Proceedings of an International Symposium*, Hiroshima, Japan, 10–12 May, A.A. Balkema, Rotterdam, pp. 43–48.

Chu, J., Yan, S.W. and Yang, H. (2000). Soil improvement by the vacuum preloading method for an oil storage station. *Géotechnique*, Vol. 50, No. 6, pp. 625–632.

Chu, J., Bo, M.W., Chang, M.F., and Choa, V. (2002). Consolidation and permeability properties of Singapore marine clay, *Journal Geotechnical and Geoenvironmental Engineering*, ASCE, Vol. 128, No. 9, pp. 724–732.

Chu, J., Bo, M.W. and Choa, V. (2004). Practical consideration for using vertical drains in soil improvement projects, *Geotextiles and Geomembranes*, Vol. 22, pp. 101–117.

Chu, J. and Yan, S.W. (2005a). Estimation of degree of consolidation for vacuum preloading projects, *International Journal of Geomechanics*, ASCE, Vol. 5, No. 2, pp. 158–165.

Chu, J. and Yan, S.W. (2005b) Application of the vacuum preloading method in land reclamation and soil improvement projects. Chapter 3, In *Ground Improvement—Case Histories*, B. Indraratna and J. Chu, (Eds.), Elsevier, pp. 91–118.

Chu, J., Goi, M.H. and Lim, T.T. (2005). Consolidation of cement treated sewage sludge using vertical drains. *Canadian Geotechnical Journal*, Vol. 42, No. 2, pp. 528–540.

Chu, J., Bo, M.W. and Choa, V. (2006). Improvement of ultra-soft soil using prefabricated vertical drains, *Geotextiles and Geomembranes*, Vol. 24, No. 6, pp. 339–348.

Chu, J., Varaksin, S., Klotz, U. and Mengé, P. (2009a). Construction processes. *State-of-the-Art-Report, Proc. 17th International Conf on Soil Mechanics and Geotechnical Engineering,* Alexandria, Egypt, 5–9 Oct., Vol. 4, pp. 3006–3135.

Chu, J., Bo, M.W., and Arulrajah, A. (2009b). Reclamation of a slurry pond in Singapore, *Geotechnical Engineering, Proceedings of the Institution of Civil Engineers,* Vol. 162, No. GE1, pp. 13–20.

Chu, J., Bo, M.W. and Arulrajah, A. (2009c). Soil improvement works for an off-shore land reclamation. *Proceedings of the Institution of Civil Engineers— Geotechnical Engineering,* Vol. 162, No. 1, pp. 21–32.

Chu, J., Indraratana, B., Yan, S. and Rukikiatkamjorn, C. (2012). Soft soil improvement through consolidation: an overview. State-of-the-art Report, *Proc. International Conference on Ground Improvement and Ground Control,* Vol. 1 (in print).

Collazos, O.M., Bowders, J.J. and Bouazza, A. (2002). Prefabricated vertical drains for use in soil vapor extraction applications. *Transportation Research Record,* No. 1786, pp. 104–111.

Duncan, J.M. (1993). Limitations of conventional analysis of consolidation settlement, 27th Karl Terzaghi Lecture, *Journal Geotechnical and Geoenvironmental Engineering,* ASCE, Vol. 119, No. 9, pp. 1333–1359.

Fang, Z. and Yin, J.H. (2006). Physical modelling of consolidation of Hong Kong marine clay with prefabricated vertical drains. *Canadian Geotechnical Journal,* Vol. 43, pp. 638–652.

Goi, M.H. (2004). Use of stabilised sewage sludge for land reclamation, MEng thesis, Nanyang Technological University, Singapore.

Guido, V. A. and Ludewig, N. M. (1986). 'A compative laborotry evaluation of band-shaped prefabricated drains'. *Consolidation of Soils: Testing and Evaluation.* ASTM STP 892, Yong, R. N. and Townsend, F. C. Eds. 642–662.

Hansbo, S. (1960). Consolidation of clay, with special reference to influence of vertical sand drains, *Proc. Swedish Geotechnical Institute 18,* Linkoping.

Hansbo, S. (1979). Consolidation of clay by band-shaped prefabricated drains, *Ground Engineering,* Vol. 12, No. 5, pp. 16–25.

Hansbo, S. (1981). Consolidation of fine-grained soils by prefabricated drains, *Proc. 10th International Conference Soil Mechanics and Foundation Engineering,* Stockholm, Vol. 3, pp. 677–682.

Hansbo, S. (1983). How to evaluate the properties of prefabricated drains, *Proc. 8th European Conf. on Soil Mechanics and Foundation Engineering,* Helsinki, Vol. 2, pp. 621–626.

Hansbo, S. (1997). Aspects of vertical drain design: Darcian or non-Darcian flow, *Géotechnique,* Vol. 47, pp. 983–992.

Hansbo, S. (2001). Consolidation equation valid for both Darcian or non-Darcian flow, *Géotechnique,* No. 1, pp. 51–54.

Hansbo, S. (2004). Band drains. In *Ground Improvement* (2nd edition), Spon Press, pp. 4–56.

Hansbo, S. (2005). Experience of consolidation process from test areas with and without vertical drains. In B. Indraratna and J. Chu (Eds.), Chapter 1, *Ground Improvement—Case Histories,* Elsevier, Oxford, pp. 3–50.

Hight, D.W., Bond, A.J. and Legge, J.D. (1992). Characterisation of Bothkannar clay: an overview, *Géotechnique,* Vol. 42, No. 2, pp. 303–347.

Hird, C.C., Pyrah, I.C., and Russell, D. (1992). Finite element modelling of vertical drains beneath embankments on soft ground, *Géotechnique*, Vol. 42, No. 3, pp. 499–511.

Hird, C.C. and Moseley, V.J. (2000). Model study of seepage in smear zones around vertical drains in layered soil, *Géotechnique*, Vol. 50, No. 1, pp. 89–97.

Holtz, R.D. and Holm, B.G. (1972). Excavation and sampling around some sand drains at Ska-Edeby, Sweden, *Proc. 6th Scandinavian Geotechnical Meeting*, Trondheim, Norwegian Geotechnical Institute, pp. 75–89.

Holtz, R.D. and Kovacs, W.D. (1981). *An Introduction to Geotechnical Engineering*, Prentice-Hall, Englewood Cliffs, New Jersey.

Holtz, R.D. (1987). Preloading with prefabricated vertical strip drains, *Geotextiles and Geomembranes*, Vol. 6, No. 1–3, pp. 109–131.

Holtz, R.D., Jamiolkowski, M., Lancellotta, R. and Pedroni, R. (1991). *Prefabricated Vertical Drains: Design & Performance*, CIRIA Ground Engineering Report, Butterworth-Heinemann Ltd., London.

Holtz, R.D., Shang, J.Q. and Bergado, D.T. (2001). Soil improvement, in *Geotechnical and Geoenvironmental Engineering Handbook*, R. K. Rowe (Ed.), pp. 429–462.

ICE. (1982). *Vertical Drains*, Thomas Telford Ltd., London.

Indraratna, B., Balasubramaniam, A.S. and Ratnayake, P. (1994). Performance of embankment stabilized with vertical drains on soft clay, *Journal of Geotechnical Engineering*, ASCE, Vol. 120, No. 2, pp. 257–273.

Indraratna, B. and Redana, I.W. (1997). Plane strain modelling of smear effects associated with vertical drains, *Journal of Geotechnical and Geoenvironmental Engineering*, ASCE, Vol. 123, No. 5, pp. 474–478.

Indraratna, B. and Redana, I.W. (1998). Laboratory determination of smear zone due to vertical drain installation, *Journal Geotechnical Engineering*, ASCE, Vol. 124, No. 2, pp. 180–184.

Indraratna, B. and Redana, I.W. (2000). Numerical modelling of vertical drains with smear and well resistance installed in soft clay, *Canadian Geotechnical Journal*, Vol. 37, pp. 132–145.

Indraratna, B., Chu, J. (Eds.) (2005). *Ground Improvement Case Histories*, Elsevier, Oxford, 1115 pp.

Indraratna, B., Rujikiatkamjorn, C., Balasubramaniam, A.S., and Wijeyakulasuriya, V. (2005). Predictions and observations of soft clay foundations stabilised with geosynthetic drains and vacuum surcharge. In: Indraratna, B. and Chu, J. (Eds.), *Ground Improvement—Case Histories*. Elsevier, Amsterdam, pp. 199–229 (Chapter 7).

Jamiolkowski, M., Lancellotta, R. and Wolski, W. (1983). Precompression and speeding up consolidation, *General Report—Specialty Session 6, Proc. 8th European Soil Mechanics and Foundation Engineering*, Helsinki, Vol. 3, pp. 1201–1226.

JTJ/T256-96. (1996). *Code of Practice for Installation of Prefabricated Drains*. Ministry of Communications, China.

JTJ/T257-96. (1996). *Quality Inspection Standard for Prefabricated Drains*. Ministry of Communications, China.

Kamon, M., Pradhan, B.S., and Suwa, S. (1984). Laboratory evaluation of the prefabricated band-shaped drains, soil improvement, *Current Japanese Materials Research*, Vol. 9, Cambridge University Press, Cambridge, UK.

Karunaratne, G.P. (2011). Prefabricated and electrical vertical drains for consolidation of soft clay, *Geotextiles and Geomembranes*, Vol. 29, No. 4, pp. 391–401.

Kitazume, M. (2007). Design, execution and quality control of ground improvement in land reclamation, *Proc. of the 13th Asian Regional Conference on Soil Mechanics and Foundation Engineering*, keynote lecture.

Kjellman, W. (1948). Accelerating consolidation of fine grained soils by means of cardboard wicks. *Proc. 2nd International Conference on Soil Mechanics and Foundation Engineering*. Vol. 2, Rotterdam, pp. 302–305.

Kolff, A.H.N., Spierenburg, S.E.J. and Mathijssen, F.A.J.M. (2004). BeauDrain: a new consolidation system based on the old concept of vacuum consolidation, *Proc. 5th International Conference on Ground Improvement Techniques*, Kuala Lumpur, Malaysia.

Kremer, R.R.H.J., Oostven, J.P., Van Weele, A.F., Dejager, W.F.J., and Meyvogel, I.J. (1983). The quality of vertical drainage, *Proc. 8th European Conf. on Soil Mechanics and Foundation Engineering*, Vol. 2, Helsinki, pp. 721–726.

Lee, S.L., Karunaratne, G.P. and Aziz, M.A. (2003). Design and performance of Fibredrain in soil improvement projects. *Ground Engineering*, Vol. 10, No. 4, pp. 149–156.

Lee, C.H. and Kang, S.T. (1996). Discharge capacity of prefabricated vertical band drains, *Final Year Report*, Nanyang Tech. Univ., Singapore.

Leroueil, S. (1996). Compressibility of clays: fundamental and practical aspects. *Journal Geotech. Engrg. Div.*, ASCE, Vol. 122, No. 7, pp. 534–543.

Li, A.L. and Rowe, R.K. (2001). Combined effects of reinforcement and prefabricated vertical drains on embankment performance, *Canadian Geotechnical Journal*, Vol. 38, No. 6, pp. 1266–1282.

Liu, H.L. and Chu, J. (2009). A new type of prefabricated vertical drain with improved properties, *Geotextiles and Geomembranes*, Vol. 27, No. 2, pp. 152–155.

Liu, H.L., Chu, J. and Ren, Z.Y. (2009). New methods for measuring the installation depth of prefabricated vertical, *Geotextiles and Geomembranes*, Vol. 27, No. 6, pp. 493–496.

Matyas, E.L. and Rothenburg, L. (1996). Estimation of total settlement of embankments by field measurements, *Canadian Geotechnical Journal*, Vol. 33, pp. 834–841.

Mesri, G and Castro, A. (1987). The C_α/C_c concept and K_0 during secondary compression. *Geotech. Engrg. Div.*, ASCE, Vol. 112, No. 3, pp. 230–247.

Mesri, G. and Lo, D.O.K. (1991). Field performance of prefabricated vertical drains, *Proc. Int. Conf. on Geotechnical Engineering for Coastal Development*, Yokohama, Vol. 1, pp. 231–236.

Mesri, G., Kjlouni, M.A., Feng, T.W. and Lo, D.O.K. (2001). Surcharging of soft ground to reduce secondary settlement. *Soft Soil Engineering*, Lee, C.F. et al. (Eds.), pp. 55–65.

Mitchell, J.K. and Katti R.K. (1981). Soil improvement—state of the art report, *Proceedings 10th International Conference Soil Mechanics and Foundation Engineering*, Stockholm, Vol. 4, pp. 509–565.

Moseley, M.P. and Kirsch, K. (2004). *Ground Improvement* (2nd ed.), Spon Press.

Nakanado, H., Hanai, H. and Imai, W. (1992). Permeability of plastic board drains. *Technical Report*, Vol. 17, Hukken, Co. Ltd., pp. 139–151.

Onoue, A., Ting, N.H., Germaine, J.T. and Whitman, R.V. (1991). Permeability of disturbed zone around vertical drains, *Proc. of ASCE Geotechnical Engineering Congress*, Colorado, pp. 879–890.

Park, Y.M. and Miura, N. (1998). Soft ground improvement using prefabricated vertical drains (PVD). In: Miura, N., Bergado, D. (Eds.), *Improvement of Soft Ground, Design, Analysis, and Current Researches*, pp. 35–48.

Park, J.H., Yuu, J. and Jeon, H.Y. (2010). Green geosynthetics applications to sustainable environmental fields from the viewpoint of degradability, *Proceedings International Symposium and Exhibition on Geotechnical and Geosynthetics Engineering: Challenges and Opportunities on Climate Change*, 7–8 Dec Bangkok, pp. 43–50.

Raison, C. (2004). *Ground and Soil Improvement*: Thomas Telford, London.

Ren, Z.Y. (2004). *A prefabricated vertical drain with capability for penetration depth measurement*, China patent No. 0119469.0.

Rujikiatkamjorn, C., Indraratna, B. and Chu, J. (2008). 2D and 3D Numerical Modeling of Combined Surcharge and Vacuum Preloading with Vertical Drains, *International Journal of Geomechanics*, Vol. 8, No. 2, 144–156.

Saowapakbiboon, J., Bergado, D.T., Chai, J.C., Kovittayanon, N., and de Zwart, T.P. (2008). Vacuum-PVD combination with embankment loading consolidation in soft Bangkok clay: A case study of Suvarnabhumi Airport project, *Proc. 4th Asian Regional Geosynthetics Conference*, Shanghai, China, pp. 440–449.

Schaefer, V.R., Abramson, L.W., Drumheller, J.C., Hussion, J.D., & Sharp, K.D. (1997). Ground improvement, *Ground Reinforcement and Ground Treatment: Developments 1987–1997*, ASCE GSP 69.

Seah, T.H. (2006). Design and construction of ground improvement works at Suvarnabhumi Airport. *Geotechnical Engineering Journal of Southeast Asian Geotechnical Society*, Vol. 37, pp. 171–188.

Shang, J.Q. (1998). Electroosmosis-enhanced preloading consolidation via vertical drains. *Canadian Geotechnical Journal*, Vol. 35, pp. 491–499.

Sivaram, B. and Swamee, P. (1977). A computational method for consolidation coefficient. *Soils and Foundations*, Vol. 17, No. 2, pp. 48–52.

Sridharan, A. and Sreepada Rao, A. (1981). Rectangular hyperbola fitting method for one-dimensional consolidation, *Geotechnical Testing Journal*, Vol. 4, No. 4, pp. 161–168.

Suits, L.D., Gemme, R.L. and Masi, J.J. (1986). Effectiveness of prefabricated drains on laboratory consolidation of remolded soils. In: Yong, R.N., Townsend, F.C. (Eds.), *Consolidation of Soils: Testing and Evaluation*, ASTM STP 892, pp. 663–683.

Towhata, I. (2008). *Geotechnical Earthquake Engineering*, Springer.

Varaksin, S. & Yee, K. (2007). Challenges in ground improvement techniques for extreme conditions: Concept and Performance, *Proc. 16th Southeast Asian Geotechnical Conference*, 8–11 May: Kuala Lumpur, pp. 101–115.

Wang, T-R. and Chen, W-H. (1996). Development in application of prefabricated drains in treatment of soft soils, *General Report, Proc. 3rd Symp. On Weak ground Improvement using PVDs*, Oct., Lianyuangang, China, pp. 13–40.

Xiao, D.P. (2002). Consolidation of soft clay using vertical drains, PhD thesis, Nanyang Tech. Univ., Singapore.

Xie, K.H. (1987). Consolidation theories and optimisation design for vertical drains, PhD Thesis, Zhejiang University, China.

Yan, S.W. and Chu, J. (2003). Soil improvement for a road using the vacuum pre-loading method, *Ground Improvement*, Vol. 7, No. 4, pp. 165–172.

Yan, S.W. and Chu, J. (2005). Soil improvement for a storage year using the combined vacuum and fill preloading method. *Canadian Geotechnical Journal*, Vol. 42, No. 4, pp. 1094–1104.

Yan, S.W., Chu, J., Fan, Q.J., and Yan, Y. (2009). Building a breakwater with prefabricated caissons on soft clay. *Proceedings of ICE, Geotechnical Engineering*, Vol. 162, No. 1, pp. 3–12.

Yin, J.-H. and Graham, J. (1999). Elastic visco-plastic modeling of the time-dependent stress-strain behavior of soils. *Canadian Geotechnical Journal*, Vol. 36, No. 4, pp. 736–745.

Chapter 5

Permeation grouting

Gert Stadler and Harald Krenn

CONTENTS

5.1 INTRODUCTION

Grouting techniques in general are intended to fill voids in the ground (fissures in rock and porosity in sediments) with the following aims:

- Increase resistance against deformation
- Supply cohesion, shear, and uniaxial compressive strength
- Reduce conductivity/transmissivity via interconnected porosities in an aquifer (this is the most common goal)

Grouting uses fluids like thin mortars, particulate suspensions, aqueous solutions, and chemical products like polyurethane, acrylates, or epoxy injected into the ground under pressure, via boreholes and packers, providing the geometrical layout of 'points of attack' in the underground space. By displacing gas or groundwater, these fluids fill pores and fissures in the ground and thus—after setting and hardening within a predetermined lapse of time—attribute new properties to the subsoil. The degree of saturation and the properties of the hardened grout do define the degree of achieved improvement.

The first applications of grouting were in the fields of mining (shaft-sinking) and hydro-engineering (grouting under dams). These go as far back as 1802, when Berigny repaired the foundation of a sluice at Dieppe (France), followed by similar applications at Rochefort, where leaks into a dock were stopped by mortar injections. Ground stabilisation around city excavations for high-rise structures and subways (Metro) have been prominently added to these examples, as well as immobilisation of waste, grouting behind tunnel linings, and rehabilitation of concrete structures of dams. A prominent example for the latter is shown in Figure 5.1.*

Commercial considerations and costs are, of course, at all times a matter of the market, and therefore difficult to generalise. In general, grouting is only viable if other more economical and 'designable' ground engineering techniques would be physically impossible, and if the process may be accomplished within acceptable construction time using drilling techniques and grout material both available and economical. Grouting pressures applied must stay below the pressure level causing ground fracturing, and the technical result (for instance, increase in strength or reduction of permeability) must be reasonably anticipated during design considerations.

Costs of grouting works are governed by technical and operative 'boundaries'. Typical performance rates are

- Average grouting rate per pump: 5–20 l/min
- Average man-hour [H] per operative pump-hour [h]: 1.1–3.5 H/h
- Average man-hour per ton of cement of a neat Ordinary Portland Cement (OPC) grout: 5–10 H/ton
- Average minimum borehole spacing equal to the thickness of the treatment, or <3 m
- Average percentage of voids on which to base the grout consumption: in sediments, 22%–35%; in rocks, only 0.5%–1.5%
- Average metre of borehole per m³ of soil/rock grouted: 0.15–0.8 m/m³
- Average cost for depreciation (plus interest), and for maintenance and repair of equipment and machinery: 3.6%–4.1% (of replacement value) per month

---

* Structural repair of cracked concrete in a double curvature arch dam (the Koelnbrein Dam, Figure 5.1) was accomplished by a specialized application of grouting with epoxy resins of high viscosity and strength (applied under a considerable head of water!). Lombardi provided the design for this repair work and took the occasion to apply his concept of a Grouting Intensity Number (GIN) at this major repair project. Another typical example of grouting applications is grouting of horizontal barriers (blankets) in sand below city excavations. To reduce seepage during excavation of construction pits at gradients of as much as 10, it is possible to reduce permeabilities to around 1×10^{-7} m/s, which corresponds to seepage values of 1.5 l/s per 1000 m².

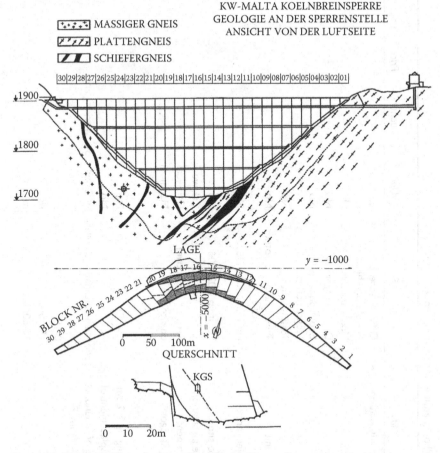

KW-MALTA KOELNBREINSPERRE
GEOLOGIE AN DER SPERRENSTELLE
ANSICHT VON DER LUFTSEITE

Figure 5.1 Kölnbrein Dam, Austrian Drau Hydro Power Company, typical sections.

For a standard grouting application using these performance rates, suitable estimates of total time and costs can be derived.

On top of these costs, approximately 25% of the cost of the operative personnel should be foreseen for supervision and infrastructural services. Mobilisation and demobilisation costs are to be added to such approximate overall budget considerations. These costs specifically depend on suitable access to the construction site, conditions for transport, possibly confined working space, climate and time of the year, location of the drilling points, and other factors.

Measurement of and payment against grouting works should be based on respective Bill of Quantities/Schedule of Rates, sufficiently detailed to address various operations and services related to grouting. For reference, see the suggested standard schedule in Table 5.1, which could be adapted to suit individual project situations.

Table 5.1 Proposed standard schedule of rates/ bill of quantities for grouting works

Item No	Group	Main	Sub	Unit	Quantity	Price	Item price
1.01	Mobilization, site installation, demobilization	General site installation	Offices, store, personnel W/Shop, vehicles, etc	LS			
1.02		Rigs & equipment	Mixing/batching plant	No			
1.03			Drill rig (Type)	No			
1.04			Grout pump (single, containerised)	No			
1.05			Testing unit	No			
1.06		Additional units	Type as above or other specified	No			
1.091	Relocating rigs & equipment (Type)	Within project area	Per item 1.02-1.06	No			
1.092		From site to site	Per item 1.02-1.06	No			
2.01	Rental of equipment	General site installation		Cal. week			
2.02		Rigs & equipment	Per item (as above)	Cal. week			
3.0	Idle/downtime (as specified)	Personnel (Category)	on site	Man hour			
4.00	Setup & rigging drill over hole						
4.01		incl displacing rig > 2.5m		No			
4.02		Without displacement of rig		No			
5.00	Drilling for coring or grouting, in all types of ground, all directions and inclinations, collaring < 2.0m above working area						
5.01	Coring 75-115mm						
5.011	0-15m			m			
5.012	15-30m			m			
5.013	30-60m			m			

5.014	Extra over for casing		m
5.02	Roto-Percussion drilling 56 bis 76 mm		
5.021		0-6m	m
5.022		6-12m	m
5.023		12-20m	m
5.03	Over burden drilling 115 to 133mm (in loose ground)		
5.031		0-15m	m
5.032		15-30m	m
6.01	Supply, install and sheath grout sleeve pipes (tubes a manschettes),	MS 2" dia, port distance 66cm	m
6.02	ditto in HDPE		
6.0	Supply, maintain, position and remove.	Packers: all dia, all depths	
6.01		Single packer	No
6.02		Double Packer	No
6.03		MPSP-inflatable packer	No
6.04		Circulation-packer incl return line	No
6.05		Inflatable double-packer for tubes a manschettes	No
7.0	Operating grout pump, incl weighing, batching, storing, ducting of grout mixes of all kind, operating grout pump under pressure, electronic data aquisition	Documentation of pressure, rate and quantity as per EN 12715 requirement	
7.01		Hour operation of one only (first) pump	H

Continued

Table 5.1 Proposed standard schedule of rates/ bill of quantities for grouting works (Continued)

Item No	Group	Main	Sub	Unit	Quantity	Price	Item price
7.02		Hour operation of a second pump at same site location, simultaneously with operating the first pump		H			
7.03		Hour operation of a third pump at same site location, simultaneously with operating the first and second pump		H			
8.0	Material for grout mix						
8.01		OPC (Blaine > 3.900 cm^2/g)		to			
8.02		UFC (D$_{80}$ < 12μm)		to			
8.03		Sodium-Bentonit		to			
8.04		Calcium-Bentonit		to			
8.05		Sodiumsilicat (liquid, 38°Bé)		kg			
8.06		PU (single shot)		kg			
8.07		PU (two-component mix)		kg			
8.09		Acrylat-(as per tenderers proposal)		kg			
9.0	Borehole test	Water pressure test in rock (Lugeon), permeability test in loose ground (Lefranc), incl all pumps, ducts, packers and data recording/ documentation, all depth		No			
Total, net				Currency			

5.2 PREPARATORY WORKS AND DESIGN

It is important that the designer is aware of both the possibilities and the limitations of grouting. The principles of fluid mechanics might on first glance seem to govern grouting in the same way as they do the propagation of fluids in other media like pipes and ducts. The lack of knowledge of the intricate rheology of the grout and of the complex geometry of flow paths in the ground, however, poses serious problems to arriving at mathematically 'exact solutions'. Data on the rheology of the fluid do lack information on interstitial adhesion, surface tension relative to wetted surface, and whether capillary forces would tend to support or prevent penetration of the grout. It is particularly the small sections of interconnected porosities for which it is difficult to develop a good flow model. This is because the very narrow parts of flow channels will – particularly when its diameter is getting close to the grain size of the suspension – govern the penetration of grout more by actions of surface tension and affinity of the fluid relative to the wetted surface of the ground than by filtration, viscosity and yield.

Cementitious grouts are the most common type of particulate grouts; that is to say they contain particles (i.e., the grains of cement in a water suspension). Cement as a material requires a water-cement ratio by weight of about 0.38 to achieve complete hydration. However, in this form it would be an extremely stiff paste, so for injection purposes additional water is be added to the mix for the purpose of transporting the cement grains within the fissure (or pores of a sediment). The addition of water has the combined effects of reducing the strength of the grout, increasing its shrinkage, and increasing its setting time. The higher the water-cement ratio employed, the weaker the grout, the greater the shrinkage, and the longer the setting time that will result.

The question of setting time is important. Cements are manufactured so that they have a setting time of about 4–5 hours. This period is standardised to provide a suitable period of workability for normal structural applications. If we greatly dilute cements the setting time is delayed; 10–16 hours may result for water-cement ratios of 2:1 and 3:1, respectively. The addition of clays or bentonite into the mix will further delay the setting of the cement. Accelerator admixtures may be employed to reduce setting times, but these work best on low water-cement ratio mixes and have the disadvantage that they tend to increase the viscosity of the mix.

The penetrability of a cement-based grout into fissures depends on two main factors: the grain size of the cement used and the rheological (and dispersive, particle-separating) properties of the suspension. As is well known, the success of the grout is characterised by the size of the solid particles of the grout in relation to those of the fissures to be grouted. However, to study the penetrability of a mix by merely studying the size of a single dry grain is misleading: single dry grains have a tendency to grow in size during hydration and to agglomerate, thereby producing 'flocs' larger than the single dry particle. Therefore, to improve the penetrability of a particulate grout, it is necessary to both keep the grain size low and reduce or prevent the tendency for single grains to flocculate in the mix.

5.2.1 Aspects of rheological laws for particulate grout mixes

If we discuss penetrability we have to therefore discuss rheological properties of the mix. These are normally characterised by three parameters: plastic viscosity, cohesion, and internal friction (surface tension which strongly governs the penetration of fine fissures, however, still remains unresearched).

Figure 5.2 shows two laws of rheologic behaviour. Curve (1) is typical of a purely viscous (Newtonian) fluid. Water and many chemical grouts such as silicates and acrylamides follow this law. Curve (2) represents the behaviour of a so-called Bingham fluid, which is characterised not only by viscosity but also by cohesion. As discussed, cement grouts are not solutions but particulate suspensions in water. If these suspensions are stable (i.e., during grouting only a minor portion of water does become separated from the cement) they do behave as a Bingham fluid.

Assuming a stable, perfectly viscoplastic mix, Lombardi (1989) analysed the flow conditions of a mix through a smooth rock fissure. He concluded

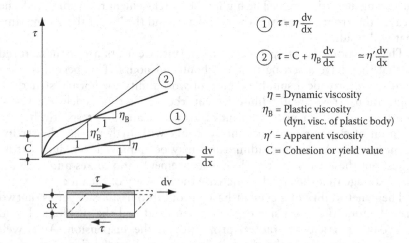

$$\text{(1)}\quad \tau = \eta \frac{dv}{dx}$$

$$\text{(2)}\quad \tau = C + \eta_B \frac{dv}{dx} \quad = \eta' \frac{dv}{dx}$$

η = Dynamic viscosity
η_B = Plastic viscosity (dyn. visc. of plastic body)
η' = Apparent viscosity
C = Cohesion or yield value

Figure 5.2 Rheogram of fluids. (From Kasumeter, International Society for Rock Mechanics, Widmann, R. (1996), Int. J. Rock Mech. Min. Sci. & Geomech., 33(8):803–847.

that the cohesion determines the maximum distance the grout can reach, and the viscosity determines the flow rate, and therefore the time necessary to complete the injection at given pressures.

Simple theoretical considerations and elementary experimental evidence show that, as soon as internal friction appears in a particulate mix, grouting is no longer possible. If the cement grains are not transported by the fluid and come into contact, they will develop friction between the particles and effectively stop grouting. This phenomenon is particularly important since during the grouting process the initial water-cement ratio may decrease due to loss of water under the applied pressure (i.e., pressure filtration) or simply due to gravity (bleeding).

As we have seen from the previous discussion on grout theory, a low viscosity grout mix is required to achieve suitable flow rates in fine aperture fissures. The viscosity of the grout is normally measured by means of the cone efflux test whereby the time required for a measured quantity of grout to drain from a conical funnel of fixed dimensions is measured. Various funnel geometries are used for different types of grout. For testing common grouts used for injection purposes, the Marsh cone is normally employed.

The ability of particulate grouts to penetrate into fine aperture fissures is controlled by the ability of the particles themselves to enter the fissure and by the degree of elastic widening of the fissure that occurs during the grouting process. Various authors have attempted to correlate these parameters. The following comments indicate the range of results obtained.

The D_{95} particle size of a cement means that 95% of the particles within the cement are smaller than this size. A Type I Ordinary Portland Cement typically has a D_{95} particle size of about 60 microns, and a Type III rapid hardening OPC (of the type commonly used for grouting) may typically have a D_{95} particle size of about 40 microns. Microfine cements are available with D_{95} particle sizes as small as 12 microns. The D_{95} particle size of Bentonite clay is typically also about 60 microns.*

Following common filtration criteria, Karol (1990) like other authors proposed for rock grouting that for cement-based grouts to penetrate a fissure within a rock mass, the aperture of the fissure must be at least three times the particle size of the cement grains. We consider that in practice the joint width must be wider than this rule and that in addition to this initial aperture, further elastic widening of the joint during injection is necessary.

The ACEL research programme,† like other research (e.g., by Baban 1992) revealed (in carefully instrumented strain measurement tests using microfine cements and epoxy resin components) elastic opening and clos-

---

* A micron being a millionth part of a metre or one thousandth part of a millimetre.
† By Keil et al. of 1989 (quoted in Weaver and Bruce, 2007)

ing of the fracture zone by as much as 100 microns during the grouting operation. This dimension being some eight times the D_{95} particle size of the microfine cement decidedly proves the importance of the elastic deformation of the rock mass for a successful treatment.

In attempted grouting of water bearing rock at a depth of 830 metres using microfine cement grouts, Naudts (1990) reported that it was not possible to inject a stable microfine cement grout into a strata observed to have a permeability of over 100 Lugeons. This value would be consistent with very heavily fractured rock (of >>10 fracs/meter and corresponding frac widths of 50 to 100 microns), which would normally be expected to be readily treatable with such materials. In this case it could be suspected that insufficient pressure might have contributed to the phenomenon.This is because grouts, when entering a fissure, have to initially overcome resisting forces of friction, being frequently in the range of specified maximum grouting-pressure limits! (Feder, see FN, ISRM 1996).

Suppliers' catalogue information for microfine cements indicates that for the repair of concrete structures, microfine cements with a D_{95} particle size of 12 microns will penetrate cracks as fine as 0.25–0.3 mm. In this application overpressurisation and elastic opening of the crack is not appropriate, and this rule of thumb represents a ratio between crack aperture and particle size of some 20 times.

In order for cement grouts to be successfully injected it is necessary for the cement particles to remain in suspension during injection. When injected under pressure the mix may lose water into the fissure. This loss of water will cause a thickening of the mix and the generation of internal friction, increased viscosity, and rigidity of the grout with (in the end) the formation of a dense dry cake. These phenomena may eventually block any further flow of the grout into the fissure.

If high grouting pressures are to be employed, the grout mix must have a pressure filtration characteristic that is stable at the required pressure. Pressure filtration testing requires either laboratory equipment like an API filter press or (for lesser requirements) may commonly be replaced by a simple bleeding test, which measures the tendency for the cement particles to settle out of suspension. In this test a fresh sample of grout is placed in a glass measuring cylinder and covered to prevent evaporation. The amount of free water left on top of the cylinder after two hours is generally termed the bleeding characteristic of the grout. A grout is considered to be 'stable' if there is less than 5% of free water of the total volume of the grout sample after two hours.

In order to produce stable grout (albeit with a relatively high viscosity) without further additives or admixtures a water-cement ratio of about 0.6 to 0.8 is required for OPC cements and about 0.9 to 1 for microfine cements. The difference between the two would mostly arise from the finer particle size of the latter. However the rheological properties of these grouts are unsuitable for normal injection applications, being too viscous and cohesive.

Two courses of action are possible. It is possible to add additional water to the grout mix to give suitably low values of viscosity and cohesion and to stabilise the mix by the use of a colloidal additive. The most common of these colloidal additives is bentonite clay, which is normally added to the mix water in the ratio of 1%–6% by weight. When added to water and hydrated, bentonite gives the mix water thixotropic gel properties, and this acts to inhibit the settlement of the cement grains. However, the addition of bentonite strongly increases the cohesion, and to a lesser extent the viscosity, of the mix. The bentonite particles are of a similar size to OPC particles and do not affect the penetrability of these materials. However, the bentonite particles are considerably larger than microfine cement particles and stabilising microfine cement grouts in this manner would limit their performance. Other, more powerful, colloidal admixtures in the form of long chain polymers such as methyl cellulose are available, but at high concentrations these materials will significantly increase the cohesion (reducing the reach of grouts) and viscosity of the mix. Catalogue information indicates that a 2% solution of methyl cellulose in water at 20°C has a viscosity of 4,000 centipoises. Note that the viscosity of water at 20°C is approximately 1 centipoise.

The stability of grouts may also be improved to some degree by the addition of very fine-grained pozzolanic materials such as silica fume. This material is the by-product of the ferro-silica industry and generally has to be imported in bulk from Norway. It therefore attracts substantial transport costs for those outside Europe. The material is extremely fine, having a grain size with a mean diameter of 0.1–0.15 microns, about 100 times finer than Ordinary Portland Cement. The material reacts with the lime of the cement liberated during the hydration to form an amorphous gel. Vipulanandan et al. (1992) reported tests on grout mixtures comprising an Ordinary Portland Cement that had been stabilised by either the addition of 5% bentonite, or 5% silica fume material. Both mixes had a water/binder (cement + silica fume where used) of 1:1. He observed that the measured bleeding of both mixes were similar, about 8% after two hours. Penetrability, however (of the silica fume mix), would in comparison be expected to be much better.

The alternative method is to use a stable grout mix with a lower water-cement ratio and to achieve the required rheological properties by adding of a super-plasticiser such as a Na-metacrylate. These admixtures are surface active agents that negatively charge the cement grains and act to reduce the grain agglomeration, apparent cohesion, and viscosity of the grout. Grout mixes formulated with this type of admixture have the added advantages that they are stronger, set faster, and suffer less shrinkage than conventional cement bentonite grouts. Set and workability are also easier to control when using accelerator admixtures.

Stadler and Hornich (2008), for the purpose of design and preparatory works, classify grouting into five different categories. It should be mentioned that all techniques associated with intended deformation while grouting will not be considered in this section of the chapter. However, it still is important to realise that every grouting application under pressure is hydraulically introducing energy into the ground. Making hydraulic forces act onto surfaces of grains in sediments, or onto the surface of fissures in rock, causes displacements even of a minor order be it intended for deformation or not. So let us assume and accept that even permeation grouting is a process where such (largely elastic, and only to a minor extent plastic) deformations do occur, and in fact do support the penetration of grout, the saturation of voids, and thus the success of the treatment.

Design specifications for permeation grouting should carefully view this aspect, particularly when limitations on grouting pressures are stipulated. The difference between a useful and unavoidable (intrinsic) deformation by the pressure in a propulsed grout flow, and the avoidance of a frac pressure which overpowers the structural resistance in the ground (causing undesirable deformation and heave) is not necessarily reconcilable with the weight of the overburden! Today, this still remains the governing concept for the stipulation of such a pressure limit. Viscous Bingham-fluid grouts disperse considerable energy in the first decimetres after entering the ground, and exhibit a strongly digressive pressure distribution in the porosities penetrated. Thus, the corresponding uplift forces remain limited. The ultimate proof, however, still is a field test to verify the genuine ground reaction (instrumented with proper deformation gauges) under different pumping velocities and grout types.

An overview on grouting techniques (principles and methods) is given in Figure 5.3 and in EN 12715, Execution of Special Geotechnical Work, Grouting (2001), under Pt 7.3.1.1. For the case of rocks, the *Report on Grouting* (ISRM, 2000) is the literature of common reference. *Ground Improvement* (Second Edition, 2005) contains a reference where a practical crossover between virgin permeabilities in soils and rock is combined with types of grout mixes to choose, placement techniques, and grouting protocols (for operative parameters like pressure, quantity, energy, etc.).

Adequate knowledge of the relevant properties of the subsoil is of prime importance for designing grouting works. It cannot be overemphasised that with grouting being a predominantly hydraulic process, site investigation does primarily require reconnaissance of hydraulic properties of the ground. In particular, the stratification of sediments and the type, frequency, and orientation of discontinuities in rock are important features.

Type	Soil		Rock		
	Fine	Coarse	Diffuse fissurisation, Kakinites	Stable rock	Discrete joints (fracs)
			Collapsble/Unstable		Unstable
Grout ways, ports	Open-ended pipes	Drillrods & lances	Multiple packer Sleeve Pipe single/double packer		
	Perforated pipes				
	Sleeved manchette pipes				
System	Displacement (frac. compaction)	Penetration (pore grouting, permeation)	Stage grouting	Bottom up grouting	Stage
			Frac (Fissure) grouting	Permeation grouting	
Grout mixes	Silicate/Acrylate		Acrylate/Epoxy		
	Microfine binder		Microfine binder		
	Bentonite/Cement		Ordinary Portland Cement		
		Mortar		Mortar	
Virgin Kf / Virgin Lug.	10.E-6	10.E-5	10.E-4	10.E-3	10.E-2	>>10.E-1
	1	5	10	25	50	>>100
Grouting parameters	Energy and displacement criteria	Grout limited quantities below frac pressure until resurgance or interconnections do occur	Limitation of quantity and pressure		
	Energy and displacement criteria	Energy and saturation criteria	Pressurelimitation/Energy criteria		

split-spacing/from inside outwards//from outside inwards

Figure 5.3 Overview of grouting techniques. (From Stadler, G. (2004). 'Cement grouting,' in Moseley, M.P. and Kirsch, K., *Ground Improvement*, 2nd ed Spon Press: Abington.)

For grouting of soils, the following information is required:

- Stratification, typical grain-size distributions, conductivity profile, K_h/K_v
- Porosity, saturation, specific surface [m^2/m^3]
- Density of packing (CPT, SPT), grain shape, deformation modulus
- Mineralogical composition of the soil layers
- Groundwater table, gradient, GW-chemistry
- Position of wells, rivers, sewers, gullies, lines and ducts relative to the intended grouting area; building foundations, basements, underground structures and their respective conditions and properties adjacent to any intended treatment zone
- Soil pollution

The dimension of the measured unit of conductivity (permeability coefficient) is m/s: This 'velocity', however, is related to the cross-section of the ground as a whole. The 'true velocity' of a fluid in the ground may therefore be established only by relating the respective flow rate (m^3/sec) to the available porosity.

In rocks, the following information is required:

- Lithological stratification, stereo plot of discontinuities, transmissivity profile
- Frequency of discontinuities, modulus of deformation, porosity
- Anisotropy of transmissivity, RQD, mineralogical composition, weathering
- Groundwater table, gradient, sources, barriers and wells, groundwater chemistry
- Position and conditions of any underground structures

Hydraulic testing in rock aims at quantifying the capacity of absorption of water in litre per minute and per metre of borehole at 10 bar (excess) pressure. The respective unit value is 1 Lugeon, which thus corresponds to the volume-flow of water at 1 l/min per linear meter of hole (irrespective of diameter, but 76 mm as a standard) at 10 bar pressure into rock. The corresponding term is transmissivity (T) and the respective dimension consequently is m^2/s. Elastic or permanent deformations of rock, turbulent flow conditions, and so on may be identified when interpreting test data based on multiple pressure steps. Lugeon testing gives valuable information on geotechnical and hydraulic conditions of the underground, but does not necessarily relate to grout takes during later grout treatment because penetration into fissures and corresponding deformations do differ between using water or grout.

Since its early application at the Aswan project (built in the years 1960 to 1971, Figure 5.4), grouting of alluvial ground is accomplished by using the sleeve pipe method (tube à manchette). Pipes inserted into boreholes usually are of 1½ to 2 inches in diameter (single-port 1/2-inch pipes are now also used in uniform sands). The annular space between sleeve pipe and borehole is sealed by a 'plastic' sheathing grout of cement—bentonites. This sheath grout is intended to prevent grout escaping to the surface instead of penetrating into the ground. However, this technique is not suitable in rock.

Grouting through drill rods or driven pipes (lances) are techniques for grouting applications of lesser requirement in coarse-grained ground of high porosity and at low pressures. Grout mix in these cases will be placed through perforated pipes which are driven or inserted in predrilled boreholes; via borehole casings during withdrawal from the borehole; and through the drill bit itself when drilling the grout hole.

The range of penetration and the degree of filling of voids using drill rods or lances is limited. Fine-grained and cohesive soils are less apt to treatment with particulate grouts or chemicals for reasons of filtration. Schulze (1993) did some research on pore size distribution and penetrability of sediments by relating grain size distribution of OPC and microfine binders to the sieve analysis of soil samples. Comparable efforts to define the application of

Figure 5.4 Recent view of a cross-section of Aswan High Dam (on-site tourist poster).

different grouts in soils lead to 'groutability ratios' commonly used in the United States. There the D_{15} (diameter at which 15% of the soil sample is passing) is related to a D_{85} (diameter at which 85% of the grouting material is passing); according to this, groutability may be expected at ratios >24.

Also, success is not likely if grouting fine-grained sediments with a silt content >5% when using particulate grout based on Portland Cement (PC) which at 85% passing contains material of diameter >40 microns. For such cases the use of microfine binders is recommended instead. For application where these microfine suspensions cannot successfully penetrate, the only remaining solution is to use chemical grouts.

Compaction grouting and frac grouting may be resorted to as a means to consolidate or tighten the ground with soil-displacing methods of grouting. Such systems make use of quite intensive pressures of 40 bar and higher.

Grouting in rock formations in the majority of cases aims at tightening of fissures against percolation of water. Groundwater that migrates and flows under varying gradients in fissures, joints, and tectonic discontinuities under dams or in the form of seepage into deep tunnels will be reduced or stopped by grout from such migration. Below 3 Lugeon, only microfine binders, acrylates or silicates (with organic hardeners) and sometimes resins may be efficient.

To arrive at an assessment of likely average fissure widths prevailing, Cambefort (1964) explored relations between transmissivity (Lugeon values), fissure frequencies, and opening widths. In ISRM's *Report on Grouting* (2000) this approach is updated with more recent comparable research and field experience (see Figure 5.5).

Discontinuities in rock are dominantly two-dimensional in shape and, moreover, are frequently intersected by other sets of fissures and joints. All of which neither appears plane and parallel nor is the opening width constant. Consequently, the flow regime in fissures varies from 'channel flow' at low transmissivity (of <5 Lugeon), to concentrically 'planar flow' starting off a singular intersection of a grout hole with a fissure plane, and finally to 'spherical flow,' which activates a multitude of criss-cross fissurisations, suggesting quasi isotropic conditions. Flow equations are proposed by Hässler and Gaisbauer (Widmann, 1993).

Grouting pressures generally drop exponentially with increasing distance from the point of injection. Bingham fluids tend to accentuate this pressure drop compared to Newtonian fluids, and because of this the problem of fissure widening or the danger of frac propagation is linked more to the use of the latter.

Grouting open holes in stable rock is carried out in sections of 1.5 to 6.0m and from bottom up. Mixes for grouting in rock originally did make use of unstable suspensions (separating >5% or even less free water in 2 h under gravity) relying on the phenomena of pressure filtration for the success of grouting. Nowadays, stable suspensions are preferred—and may

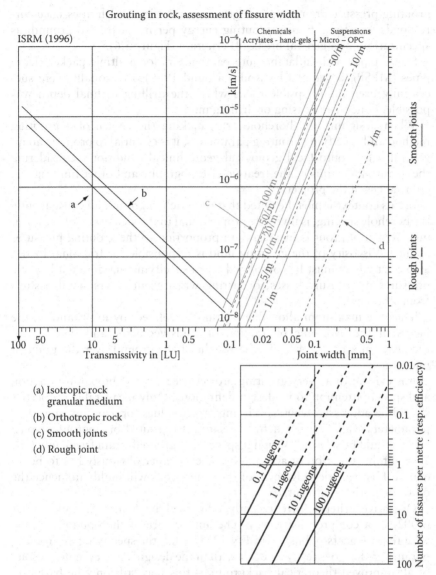

Figure 5.5 Grouter's diagram (From ISRM, modified by Stadler, G., Howes and Chow (1998). '100 years of engineering geology,' symposium, TU-Wien).

also be easier produced than before. This is achieved by adding cement additives or by using higher quality cement. However, a proper laboratory testing procedure needs to be established and carried out.

Grouting to a predetermined refusal pressure has been abandoned as well, in favour of a dual criterion in which grouted quantity and effective grouting pressure are multiplied (grouting intensity, which in essence corresponds to a limitation of grouting energy per m^3 of treated ground) as specification for a general break-off criterion (Figure 5.6).

Sleeve pipes with inflatable jute bag packers (or multiple packer sleeve pipes (MPSP), developed by Rodio around 1980) occasionally were successfully used in collapsible rocks where the drilling to final depth was possible, either using casing or drilling muds.

When designing the borehole grid, making the choice of a grouting method and specifying grouting parameters, it is essential to properly adapt general 'rules' onto local geological geotechnical conditions, considering the quantifiable aims of the treatment, topography and other limiting circumstances of the project.

In this context it must be noted that the reach of grout, and consequently the borehole spacing, is inversely proportional to the yield value of the grout mix. Reach of grout is increasing in proportion to the grouting pressures applied. Viscosity of the grouting fluid is responsible for frictional losses, an aspect which must be considered when specifying maximum allowable pressures or pressure losses, occurring when grout enters small fissures from a borehole.

Thus the maximum allowable pressure is defined 'by itself' and on the basics of the hydraulic interaction of voids-geometry and rheology of the grout mix rather than on the weight of the ground over the point of injection.

It is evident that every grouting project needs expert preparatory action and special attention to its design definitions. Only measurable properties should be addressed when specifying target values for grouting, and only parameters which have a defined relation to grouted or ungrouted soils/rocks should be selected. Visual inspection of grouted strata frequently fails as a suitable criterion of acceptance. The designer obviously has to be an expert. His realistic judgement of these questions will highly influence the outcome of the treatment.

Operative criteria were recently addressed by Semprich and Stadler (2002) in a comprehensive way. The authors are of the opinion (in line with requirements published in EN 12715) that the specifying of grouting parameters has to already be dealt with in the design itself, even if—despite of all improved theoretical background—this may still only be based on experience and empirical data from past project realisations.

Whereas formerly the grout consumption alone (grouting rate and amount at which a passé can be injected) determined the next steps (Weaver, 1993), today it is a dual strategy which facilitates a more global assessment of the proceedings. The maximum allowable grouting pressure frequently will be fixed at around 80% of prevailing so-called frac pressure. This frac pressure (at which the ground is separating and/or is losing its cohesive state) may only be established by testing the ground at the individual project site using systematically stepped-up pumping rates. This methodical approach makes the (guesswork) specification of maximum pressures on the basis of depth (in relation to the surcharge weight of the ground) obsolete. In alluvial soils, the respective allowable grouting pressures range between 5 and 35 bar with pumping rates varying between 5 and 15 l/min, respectively.

For fissures in rock exhibiting <0.15mm width and using highly viscous epoxies, these pressures might rise to as much as >120 bar without causing any damage. The reason for this is that the pressure drop at the entry of the fissure is already consuming most of the destructive energy.

The specification of a maximum quantity of grout to be injected per passé or per unit volume of ground is based on the plausible estimate of accessible porosity. Accordingly, for sediments these estimates vary between approximately 25% and 40%. For rock these quantitative limits are specified—frequently for economical than technical reasons—to prevent uncontrolled loss of grout. Porosities in rock generally vary between 0.5% and 5%.

The grouting rate results from interactions between hydraulic frictions in cross sections of porosities exposed to flow depending on the rheology of the fluid (grout mix). Common applications of particulate suspensions are operated at rates (as mentioned above) between 3 and 20 l/min. In karstic rock this value might rise to even 100 l/min, or the limiting capacity of the pump. Highly viscous epoxies, on the other hand, might have to be grouted into fissures of <0.15mm at rates of as low as <1 l/min.

In an effort to optimise extent and result of grouting works, a careful monitoring of grouting data is recommended in EN 12715. There (among others) the interpretation of Transient Pressure Data (TPA) and the limitation of applied grouting energy (GIN, as the product of quantity of mix grouted times grouting pressure, per linear metre of hole) provide new diagnostic tools to the grouting process, which make it possible to quantitatively discuss the applied grouting parameters against the original design. Regarding adjustments to the grouting procedure and the recommended steps that lead to the final halt of the grouting operation, Weaver (1991) formulated respective criteria that have been successfully applied in grouting under dams and may be adapted to similar applications. His flow charts do supply the respective logic, indicating when to change rate, mix, or pressures of grouting. Based on such or comparable considerations (TPA and GIN), it becomes possible to formulate the design of modern grouting practice, particularly for grouting in rock.

An indication regarding relative costs for grout material may be drawn from Table 5.2.

The definition of operative parameters for penetration grouting of sediments is more dependent on the relation between geometry of pore sizes, composition of particulate grouts, and rheology of the mix. Diagnostic interpretation of the process during the grouting operation itself at present remains limited. The interpretation of success or failure of grouting

Table 5.2 Relative cost of grout material

Types of grout material			Relative cost of diff. type grout material, per kg (provided but not injected)
Ordinary Portland Cement (OPC)			1
Binder			1–3
Microfine Binder (MFC)	Blaine value	8,000 cm²/g	5
	Blaine value	>12,000 cm²/g	10
Silicate gel (hardener: aluminate/acetate)			215
Resin products (e.g., polyurethane, specialised epoxies)			>30–150

in distinctively orthotropic situations, as is the case with the stratigraphy in most alluvial sediments, will therefore be even more dependent on the relation between a K_f horizontal and K_f vertical than on an observation of the development of the grouting pressures or rates, even an intricate one. One of the methods practised is to either observe or interpret rates at constant pressures, or pressures while keeping the rate of grouting at a constant value.

The reality in tunnel grouting (as a modern application and revival of grouting techniques), however, is that it is not possible to 'design' the work with comparable precision in advance, which in many ways prohibits its comparability to this 'design' process. The design of tunnel grouting operations is limited to the best estimates of the permeability and geometry of fissures in the rock through which the tunnel is to be driven, frequently based on the average values only. Therefore, the basic design for the grouting operation for tunnelling has to be reduced to an empirical, observational basis.

5.3 EXECUTION OF WORKS

In general, grouting works should always be carried out by trained and skilled personnel under competent and experienced supervision. Drilling should make use of systems that least disturb the access for the grout into subsoil porosities.

In spite of some drawbacks in terms of influencing the size of pores and fissure intersections near the hole, roto-percussive systems are favoured—mainly for economic reasons—and make use of external or down the hole hammers, with or without casing, in rock as in alluvium. Rod size is normally 1¼ inches in diameter, and casings are up to 139 mm in diameter.

Direction and inclination of holes must follow the intentions of the design. Two-percent deviation is normally an acceptable limit up to a depth of 20 m. However, it has to be kept in mind that horizontal holes and holes drilled by percussion tend to deflect more than others. Flushing of holes with the aim to wash out fines or clayey materials from the ground has limited effect and should in any case not be carried out at length.

The most effective way to fully provide grout into all underground voids would be to address individually each and every fissure, and each and every individual stratum of sediment. Each of the porosities' hydraulic properties could then be matched by the application of a rheological corresponding mix, applied at optimum pressures, and supplied at optimum pumping rates. However, this is neither technically feasible nor economically viable. Therefore, an 'averaging' process is chosen as an economical compromise, having the grouting ports installed at predetermined intervals (*tube à manchette*, TAMs)—irrespective of details in sedimentary stratification—or (in rocks) by separating individual borehole sections by packers at regular, uniform intervals, for example, at 1–6 m.

Different layouts and designs of grout-pipes and packers should therefore be considered at the time when deciding the drilling method.

- Manchette pipes (TAM, *tube à manchette*); their undisputed advantage is the reuse of the individual ports when grouting successive phases using differing grouts.
- Single-port outlet mounted as a nonreturn valve at the bottom end of a ½-inch pipe. This grout pipe may also be installed in bundles of several individual supply lines, connecting to ports at different elevations in the same hole, the advantage being that no manoeuvring of packers is required when grouting at different depths.
- Multiple packer sleeve pipes (MPSP) do consist of a combination of manchette pipes activated between jute bags inflated by cement grout. Thus, even collapsible ground may be systematically treated in well-defined sections.
- Open-ended or perforated lances driven into the ground by hammer or hydraulics. These grouting devices provide access for grout in situations of lesser requirement, or lose ground exhibiting high conductivities. They are also limited in depth and installation accuracies.
- Single or double packers are used when grouting in rock or, the latter, when grout is pumped into TAMs. Single packers set at the collar of a hole in rock are frequently screw type expandable rubber packers whereas, gas-inflatable single or double packers (between 0.3 and 1.5 m in length) may be lowered into holes as deep as 50 m. At greater depth the risk increases of packers getting stuck and lost.
- Self-inflating rubber packers using the back-pressure of the grout (being pumped through a nozzle in the packer or breaking through a metal-membrane of defined bursting pressure) to inflate the sealing element. This packer type may not be retrieved.

Mixing of suspensions sounds like a trivial task; however, it is an art if performed well to predetermined requirements. The requirements depend on the task the grout has to fulfil. For compaction grouting, the strength is of minor interest, but the volume stability and expansion is important. For grouting jobs in the tunnel environment, very often the strength criteria combined with bleeding less than 1% and expansion in the range of 0.2%–1%. Uniaxial compressive strength of up to 35 N/mm^2 is nowadays standard. Depending on the water-cement ratios, bleeding and volume stability become important issues, especially with ratios above 0.7.

Stationary plants using silos not only for cements but also for premixed bentonites (for full hydration) and fine sands (in the case of using mortars and pastes) do provide for sufficient automatic functions to limit manpower and increase capacity and accuracy. This helps to reduce mistakes and keep to tolerances. In the last decade. major developments have been achieved in terms of software developments and automation of supply plants. Fully automated systems are not yet standard for all grouting jobs except large operations, but this will probably change.

Stable mixes are now preferred and standard. This means that under gravity no more than 5% free water should appear in a settlement test after 2 hours. However, for w/c-ratios below 1 using quality standard cement or ready mixed binders bleeding in the range of 0.1% can easily be achieved. Pressure filtration according to ASTM should not give more than 100 ml of filtration water. Insufficient stability of the mix not only affects the final volume (lost by filtration), but also increases viscosity and yield, and reduces setting time which reduces penetrability. To achieve a high quality grout, conforming to design requirements, a proper mixing unit has to be used. Fully automated mixers that do not only mix mechanically but also circulate the grout are preferred. Modern mixer can rotate the grout up to 2,000 times per minute. In case a large volume of grout needs to be prepared, agitators shall be used to keep the ready mixed grout in motion until it is pumped to the injection point.

Grout pumps are mainly of the double-acting piston or reciprocative plunger type. They are hydraulically driven and regulate any flow rates within the range of the capacity of the pump (usually 3 to 20 l/min). Pressures may range up to 250 bar (for highly viscous epoxies); usually pumps should be able to handle up to 100 bar at the corresponding minimum rate (i.e., 50 bar at 6 l/min, or 20 l/min at around 15 bar). The introduction of digital hydraulic control management in the grouting pumps technique enables to preset specific values and injection parameters such as 'pressure switch off limits', 'delivery flow rates', 'pressure prognostic curves', and 'GIN curves'. Other pumps in use (for minor applications and standards) are sometimes of the screw-feed type. Every pump is connected to a single grouting port (packer position). Manifolds connecting more than one hole to a pump are reducing the quality of the treatment.

Table 5.3 Grouting strategies according to EN 12715 (CEN)

| | Rock | | | | Soil | |
| | Stable | Collapsible | | | | |
	Open borehole	TAM	Drillrod	TAM	Lance, casing
Single phase	×		×	×		×
Multiple phase		×			×	
Bottom up	×	×	×	×	×	×
Top down		×	×		×	×

Pumps are connected to recording systems, which do help to follow pre-determined quality assurance measures. In Table 5.3, some of the more frequent strategies of grouting are presented, as published in European Standards EN 12715.

In stable rock, it is common to drill the grout hole to the designed/required depth and to start grouting in passes from bottom to top. Single or double packer may be used. The use of single packer might result in reactivating flow of grout in the preceding pass. A new hole has to be drilled if the same grouting area wants to be taken up a second time. Open boreholes in collapsible rock are either treated top down (stage grouting, Table 5.3), through TAMs or drillrods, or with multiple packer sleeve pipes. 'Top down' stage grouting means that in a first step the hole is drilled to a depth, where the borehole walls still remain reasonably stable (but less than 6 m to assure decent spread of grout penetration), a single packer is set at the collar of the hole, and grout pumped into this first section. TAMs will only work if rocks permit sufficient deformation for the sleeves to open; therefore, MPS pipes are used, where the section between the jute packers remains unsheathed by sealing grout, and is open for the cement grout to spread and flow into existing fissures.

Another important consideration must concern not trapping water in the pores or fissures between already grouted areas, or to prevent grout from escaping into areas where the treatment is not foreseen or to avoid grout being lost outside the intended zone of treatment.

Prominent and typical examples for these kinds of problems is grouting behind tunnel linings (Figure 5.7). Particularly if precompression of the concrete lining is aimed at, it becomes of structural importance to avoid anisotropic hydraulic loading.

A proper reporting system is recommended and, indeed, required to keep track of operations and take adequate and timely decisions on the changes to the procedures during the process. Electronic data acquisition is the standard today for the reporting of grouting parameters such as rate, quantity, and pressures. Online transfer of these data may occasionally be arranged

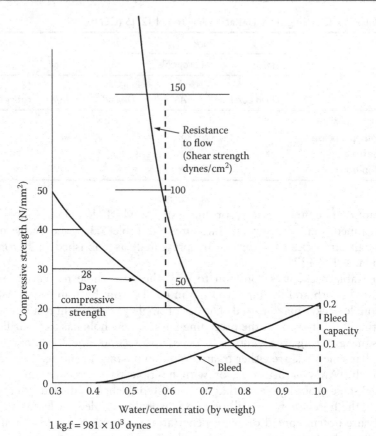

$$1 \text{ kg.f} = 981 \times 10^3 \text{ dynes}$$

Figure 5.6 Viscosity, shear strength, and bleeding of OPC suspensions. (From Littlejohn, G. and Stadler, G. (1976). Joint lectures on anchoring and grouting at SAICE, South Africa; and EUROCK 2004 & 53rd Geomechanics Colloquium, Salzburg, Austria.)

even for remote control and interpretation. Storage and handover to the engineer on discs for documentation purpose is standard. Interference in the daily routines of a grouting operation from remote interpretation of data is not recommended.

Quality assurance of the grout mix and its consistency are of prime importance:

- Density is to be checked for correct content of solids (aerometer) or on site using a scale
- Viscosity (at least Marsh flow cone time in seconds, if not by a shearometer). Be aware that flow cone diameter are different in most of the countries and not all flow cones are suitable

Figure 5.7 Drakensberg Pumped Storage Scheme, RSA, Tailrace Tunnel, Precompression grouting, ESCOM, Rodio SA, 1970s.

- Yield (fluid cohesion; Kasumeter, 2003; Heinz et al., 2003).
- Setting time (not at ambient but at ground temperature) of dehydrated grout (remainder of an ASTM pressure filter metre test), regular 250 ccm samples (including 28 day uniaxial strength), or of a film (adhesive layer shed over a suitable base)
- Dispersion test (drop of mix squeezed between two 10 by 10 cm glass plates and viewed under scaled magnifying-glass: 50×, or microscope, against light).

5.4 MONITORING, CONTROLS, AND ACCEPTANCE TESTS

The acquisition of grouting data makes an interpretation of the grouting process possible, and conclusions may be drawn from these data as to the success of the treatment.

- Development of pressure and rate against time
- Hydraulic fracs or other respective movement of ground
- Interconnecting holes or so-called resurgences (spurt of grout to the surface) and *renards* (French for larger 'foxholes' through which grout may escape to surface) of grout.

Permeability tests before and after grouting may be helpful to assess the degree of saturation achieved by grouting. The higher the virgin permeabilities of the ground, the greater the chance of a considerable improvement. Wherever possible, upstream/downstream piezometers or seepage rates should form the acceptance criteria for a successful impermeabilisation scheme and reduction of take of subsequent passes would indicate the progress of filling.

Drilling energy measured when drilling (roller or fish-tail bit, borehole supported by drilling mud) test holes before and after grouting indicate the gain of strength. Core sampling for laboratory testing is more suitable in rocks than in soils. The system is inevitably failing in soils, even using diamond core drills with uniaxial strengths of the cored material dropping below 5 MPa, since obtaining intact samples, which would satisfy laboratory requirements, is almost impossible. However, recovered sample can still be used for visual inspection of the grouting success. Open pit inspection sometimes is sufficient to ascertain an improved cohesion of grains or the visible presence of grout in the ground but limited in depth.

5.5 RESOURCES AND EQUIPMENT

Drill rigs for the production of grout holes are of diesel/electrohydraulic design with different mast configurations and kinematics. The power heads might provide (hollow stem if rods are passing through the power head) spring-loaded or hydraulic jaws, or swivel-type connections (rods only connected below a drill head) between driven rotating parts of the head and the drill rods. The length of free travel on the mast is essential for productivity; every breaking and connecting manoeuvre of rod couplings reduces production time for drilling. In Europe, percussion and core drills by Atlas Copco, Casagrande, Huette, Wirth, and Klemm produced respective machinery, which has satisfactorily performed over decades.

Mixers and pumps are electrically driven and (as far as pumps are concerned) have secondary hydraulic systems installed. Mixing may be categorised into mixing by agitation (paddle mixers, unstable mixes, batch mixing), mixing by generating high shear forces (comparable to centrifugal pumps), and mixing time, which should be limited to 30–120 s. Overlong mixing heats the grout and triggers the hardening process of hydration at too early a stage. The mix should be kept (if at all) in agitating tanks where low energy paddles keep the grout in motion and prevent particles from sedimenting.

Many experts are of two minds about the requirements on the 'evenness' (continuity) of the grout flow, though a slightly pulsating regime finds many supporters. But what is even more essential is the possibility to regulate flow and pressure of the pump in a way that makes either a constant energy concept possible or a constant rate or constant pressure scheme.

Pressure limiters with ON/OFF function are not suitable and hence not recommended; pumps with an uncontrollable direct drive need a bypass system which is prone to early wear; abrasive grouts should be handled by low wear plunger pumps; ease of cleaning and maintenance is of great importance (downtime!) especially when using chemicals.

5.6 GROUT MATERIAL

The correct choice of grouting material for the grouting works is of major importance. In recent years, new chemical products have been introduced on the market by several suppliers. Experience shows that the new materials are often only tested under off-site ambient conditions and rarely under conditions corresponding to the ones prevailing on site. Studying datasheets may not be considered sufficient. Depending on the grouting application, laboratory tests need to be set up in accordance between designer and contractor in an endeavour to appropriately test envisaged grouting materials. The user needs to be aware that laboratory standards and experience may be different in each country and even more different from continent to continent.

Portland cement as a grouting material is well known and suitable for most standard grouting applications. Very often in a two-stage grouting process it can be found that in a primary grouting phase OPC is injected with a Blaine value of around 3900 cm^2/g or higher, and for the second stage (to fill voids of a smaller cross-section) an ultra-fine cement (UPC) may be applied. In Europe, the use of UPC is well established but on other continents scarcely available. Also, a combined application of OPC followed by a chemical grout such as a gel is common. Experience using foams show that foams are not always suitable for permanent applications. Foams should be used for tasks such as stoppage of water inflow into a tunnel or into an excavation pit, and applying a second run using cement-based grouts for 'filling up' or as a more rigid supplement.

In 'modern grouting methods' the requirement for controlled (low) viscosity and yield (cohesion) may be achieved by adding of a 'super-plasticiser' as an admixture, making a much lower water/cement ratio possible.

The latter is recommended for making the grout more stable, less prone to washout, producing less excess water under grouting gradients, and avoiding pocketing of filtration water in the ground. The yield value (or 'cohesion' of the fluid at zero flow, measured by ball harp or Kasumeter) should be carefully monitored to avoid producing 'sticky' grout, which prevents free travel of the fluid beyond a certain range.* Typically, when using an OPC grout, the water/cement ratio would range from 0.8 to 1.5. For microfine cements (due to the greater fineness of these materials) the water-cement ratio would be in the range of 1.1 to 2.0. A lower water-cement ratio of any particulate grout makes it stable by its own constituents, and there is no requirement to add bentonites or other clay material. The combination of low water/cement ratios and the absence of clay in the mix has the advantage of

- The strength of the grout and its durability being high
- The shrinkage and permeability of the grout being reduced
- The normal setting times of the grout remaining maintained and, if required, be controlled by the admixture of accelerators to the grout

Chemical grouts[†] typically are used as supplementary materials for special situations and purposes, examples being where there are very strict requirements in terms of permissible water or transgress, or where there is need for the spontaneous local stoppage of percolating water flow (Stephen and Gert, 1999).

Where the indicated fissure apertures (in rock) are assumed to be larger than 0.2 to 0.3 mm the use of an OPC grout would still be appropriate. In such cases the use of chemical grouts (such as silicate grouts) would not only show poor economics but, may also be ineffective as a single-stage grouting material. They could be ineffective because of shrinkage due to syneresis (a chemical reaction depending on individual grout volumes), or inability, due to low gel strength, to resist the higher water pressures at depths. To be effective, chemical grouts should be used in a secondary or tertiary grouting phase after the major fissure structures have been filled by stronger (and cheaper) cement or other particulate grouts.

A large number of chemical grouts are in fact available on the market (Figure 5.8). Of the chemical grouts most commonly in use, silicate gels are primarily mentioned (Hornich and Stadler, Grundbautaschenbuch, 2009) of which the two main types are the so-called 'hard gels' and 'soft gels', mainly differing in strength. With hard gels, strength of 1 to 10 N/mm^2 can

---

* Remember that the formula (for the 'range' of grout travelling from a borehole) follows principally: R=P$_{grout}$*A$_{width}$ of void/$\tau_{f\ yield\ of\ fluid\ (grout)}$ (where the dimensions would be for R [m], P[bar], A [m], τ [bar]).

† Stephen and Gert (1999)

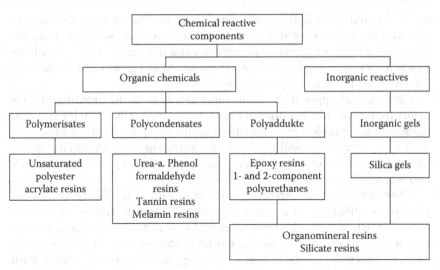

Figure 5.8 Chemical Grouts, ISRM Commission on Rock Grouting, 1996.

be achieved. Soft gels are mainly used for sealing applications (for example, in excavations) and do attain strength of 0.1 to 0.5 N/mm^2. Gels consist of approximately 50%–70% of water, 30%–45% of sodium silicate plus hardener or flocking agent. The major advantage of a hard gel compared to cement mixes is the respective setting/hardening time. Hard gels may develop their final strength within hours. For a more detailed summary on chemical grouts and special chemical grouts, the reader is referred to Hornich and Stadler (2009) and subsequent paragraphs on silicate and acrylate grouts.

5.7 CHARACTERISTICS AND APPLICABILITY OF CHEMICAL GROUTS[*]

'[Chemical grouts] is a generic term that can be applied to all forms of grout that contain chemicals in solution either in water or with each other. The family includes silicates, phenolic resins, lignosulphates, acrylamide, acrylates, acrylic, soluble lignates, sodium carbomethylcellulosis, amino resins, polyurethane, polyester, epoxies, etc. Commonly these chemicals are dissolved in water to form aqueous solutions and rely on a chemical reaction to cause a change of state from a fluid to either a foam gel or a solid. In this type of grout there are no suspended particles, hence by definition chemical grouts are stable and the application of a bleeding test to this class of grout

---

[*] Stephen and Gert (1999).

is not required. Chemical grouts generally act as Newtonian fluids; displaying viscosity but not (or only very low) values of cohesion. Consequently there are no absolute limits on penetration and this factor is limited only by practical or economical considerations of acceptable grouting/gel times and (to a marginal extent) on pressure limitations.

The principal application of chemical grouts is in the grouting of soils rather than rock. The majority of products commercially available and economically acceptable are formulated for this market and these materials, with some exceptions, will generally lack sufficient strength and durability for applications involving stemming the flow through open fissures in rock and when subjected to high hydrostatic pressures.

When reviewing the literature, one may find that silicate grouts are used in conjunction with a pre-injection of cements stabilised by bentonites. Misleadingly, the term 'chemical grout' has become to a certain extent (and in certain areas of the world) synonymous with the use of silicate (and acrylamide) grouts only. These grouts therefore are dealt with in some detail.

5.7.1 Silicate grouts

Sodium silicate grouts have been extensively used in urban areas for more than 15 metro schemes, including those in London, Paris, Vienna, Cairo, Caracas, and Hong Kong. They consist of liquid silicate plus water and hardener, are generally considered to be nontoxic, and there have been no incidents of significant pollution or environmental damage recorded, although unpleasant smells have been reported associated with the use of particular hardeners during the construction of Auber station on the French RER project. In that very case, the reagent used was ethyl acetate (which is now out of use).

Silicates used for grouting are usually manufactured by fusing a mixture of silica sand and sodium carbonate at 1400°C in a furnace. The vitreous silicate obtained is subsequently dissolved in water under pressure at a temperature of 150°C to give a syrupy liquid: Liquid silicate or 'water glass'. The liquid silicate in its concentrated form (38° Beaumé) is too viscous for direct injection and may have a viscosity in the range of 40 to 180 centipoises at 20°C depending upon balance between the silica and sodium molecules in the formulation. The viscosity of water at similar temperature is approximately 1 centipoise.

For grouting purposes, it is necessary to dilute the liquid silicate with water. The greater the dilution, the lower the viscosity of the grout and the greater the flow rate that may be achieved. However, the greater the dilution, the weaker the gel strength of the set grout and the greater the tendency to syneresis problems. To achieve a viscosity of five centipoises (as specified) implies a high dilution, low strength gel of the type typically employed for injection into completely weathered rocks and soils after having been pretreated with particular suspensions.

Liquid sodium silicates are highly alkaline, with a pH in the order of 10.5 to 11.5. They will react with an acid or acid salt to form a gel. Amongst the gelling agents commonly used were sodium bicarbonate, sodium aluminate, and various other inorganic or organic acids. In the late 1950s a new generation of gelling agents was developed consisting of methyl and/ or ethyl diesters formed from the action of aliphatic diacide mixtures on methanol and/or ethanol. These gelling agents when dispersed in a sodium silicate solution in the correct proportions go through a slow saponification, which after a predetermined time provokes the liquid to gel in the form of a white mass: SILICA GEL. These reagents are proprietary chemical systems, such as Hardener 600, and are marketed by specialist companies such as Rhone Poulenc of France. The silicate grouts in common usage today are a combination of a sodium silicate resin with a proprietary chemical agent (hardener).

In application for grouting, the gelling agent or hardener is mixed with the diluted sodium silicate shortly before injection into the soil. This grout penetrates into the interstitial voids between the soil particles, conferring on the formation the required cohesion and impermeability when solidifying during setting.

Silica gels suffer from the phenomena of syneresis, which refers to the progressive extrusion by the gel of a significant quantity of water. The phenomenon is particularly problematic in dilute low viscosity gels, which have been formulated for a long setting time. When injected into fine sands syneresis is not normally a problem when the interstitial pore dimensions within the soil are small. However, the phenomenon may become problematic if injection takes place into coarse sands or open fissures in rock where the pore aperture dimensions are larger. In these materials the strength of the gel itself becomes of importance and syneresis can lead to the final failure of the gel.

Silica gel can also be subject to washing out. When grouted sand samples are immersed in flowing water it is observed that the gel will progressively break down, leading eventually to a complete disintegration of the sample for certain gel types. The water solubility of the silica gel is due to the presence of non-neutralised soda, which attacks the silica. Laboratory tests have shown that the higher the non-neutralised soda concentration, the greater this solubility becomes. Shirlaw (1987) described the piping failure in coarse beach and alluvial sands treated with silicate grout and attributed the failure to these mechanisms. The same formulations had previously been successfully when injected into fine-grained sands and weathered rocks that had a smaller pore structure.

It should be noted that the suppliers of gelling agents for sodium silicate grouts do not recommend these materials for grouting of fissures in rock, preferring to recommend acrylamide-based materials which do produce a stronger and more stable gel for these applications.

5.7.2 Acrylamide grouts

This type of grout has been used successfully in various rock grouting projects where very low permeability results were required to be achieved. Acrylamide-based grouts consist of a mixture of two organic acrylamide monomers, which forms between 90%–97% of the mixture, and a cross-linking agent such as methylene-bis-acrylamide that forms the balance. The higher the percentage of the cross-linking agent, the stronger the resultant gel. Grout solutions up to 20% solid have viscosities less than 2 centipoise and are readily injectable into very fine fissure structures—but only at fairly low pressures! (Remember: 'injectability' of chemicals depends more on structural properties of the fluid (chemical-molecular chain-length, size and complexity, surface tension and cohesion), whereas viscosity mainly governs friction of flow and thus, grouting pressures applied—but not so much penetrability itself.) Such solutions when properly catalysed will, after a length of time dependent upon the catalyst concentration, change almost instantly into a solid, irreversible gel.

The principal difficulties with these materials are their very high cost and the potential toxicity of the components. Acrylamide grouts are (depending on make and specific concentration) a potentially neurotoxic poison and as such may represent a considerable hazard to the operatives employed in their use. In practical applications it is not always possible to ensure complete neutralisation of the grout and it may not always be possible to prevent grout from leaching into local watercourses, leading to acrylamide poisoning. Acrylamide grouts were first introduced in the United States in 1953 but due to the toxicity problems were withdrawn in 1978. A similar product, Nitto SS, was withdrawn in Japan following a careless application near a well, which led to several cases of acrylamide poisoning. A similar French product, Rocagil BT, was used with success on the Hallendsas project in Sweden but had to be withdrawn after the material leached into a local stream and poisoned livestock. Neurological problems associated with loss of sensation and motor control in limbs were also reported from this project.'

5.7.3 Grouts having to fulfil environmental standards*

Chemical grouts have to be formulated and applied in a way so as not to create undue hazards or transgression of environmental standards. In this respect two types of product descriptions of grouting materials may be relevant.

---

* Stadler (2001).

First, the new Eurocode on grouting EN 12715 of 2000, which states:

6.2.5 Chemical products and additives:
6.2.5.1 Chemical products such as silicates and their reagents, lignin based materials, acrylic or epoxy resins, polyurethanes or others can be used in grouting work subject to compliance with environmental legislation.
6.2.5.2 The effects of all products and by-products resulting from reaction of the chemical products with other components of the grout or with the surrounding ground shall be considered.
6.2.5.3 Admixtures are organic or inorganic products added in small quantities during the mixing process in order to modify the properties of the grout and to control the grout parameters such as viscosity, setting time, stability, and strength, resistance, cohesion and permeability after placement.
6.2.5.4 Admixtures to grout such as super plasticisers, water retaining agents, air entrainers and others are subject of parts 1, 3, 4 and 6 of prEN 934 and prEN 480-1 to 480-12.

Second, refer to legislation in relation to the term 'toxicity'. In fact, many international standards do differ considerably on this subject.
Relevant standards do require that any

...environmental impact, particularly the toxicity of the grout and the grout components and their effect on the ground and drinking water should be considered before grouting. When testing the grouting material for environmental impact, the following aspects should be considered: (a) whether during processing, transport or grouting, substances can be generated or released which could be hazardous to the environment or the grouting crew; (b) whether noxious substances can spread upon mixing with groundwater; (c) whether reaction products can be produced or released which influence the water quality; (d) the type of particles eroded from the hardened grout; (e) chemical reactions between hardened grout and groundwater.

It seems, however, that no official/reliable definition of the term 'toxic' does exist, and frequently codes like the Canadian Environmental Protection Compendium may have to be consulted, which states, 'The term "toxic" refers to the ability of a physical, biological or chemical agent to provoke an adverse effect or deleterious response in an organism.' The compendium further notes that: '... some jurisdiction extends the term organism to the environment as a whole.'
Unfortunately it is these words: '...*harmful, adverse, deleterious, noxious, hazardous, dangerous, negative, lethal...*' which may leave engineers

with a semantic confusion when e.g. preparing the wording of a construction specification, be it for temporary or permanent grouting measures.

The toxicity of a substance is commonly measured in terms of the effects caused by oral ingestion. However, the wider definition of the term also includes adverse effects such as dermal irritation through skin contact or damage to eyes and the respiratory system. Such reactions are of no less concern, although toxicity levels for such reactions are poorly defined. Where such reactions are of concern, standards such as the EEC directives concerning labelling conservatively enforce safety procedures based simply on the classification of a substance but irrespective of its concentration.

The oral toxicity of substances is tested on laboratory animals. The lethal single oral dose of the material that will kill 50% of the sample population is termed the LD_{50} and is quoted in terms of milligrams of the substance per kilogram of body weight of the animal.

Various classification systems existing for toxicity and the principal systems are shown here:.

UK—EEC

Very toxic	$LD_{50} = 0$–25 mg/kg
Toxic	$LD_{50} = 25$–200 mg/kg
Harmful	$LD_{50} = 200$–2000mg/kg

USA

Very toxic	$LD_{50} = 5$–50 mg/kg
Moderately toxic	$LD_{50} = 50$–500 mg/kg
Very slightly toxic	$LD_{50} = 500$–5000mg/kg

Canadian Environmental Protection Compendium, in an inverse order to the above:

Practically non toxic	$LD_{50} = > -15,000$ mg/kg
Slightly toxic	$LD_{50} = 5000$–15,000 mg/kg
Moderately toxic	$LD_{50} = 500$–5000 mg/kg
Very toxic	$LD_{50} = 50$–500 mg/kg
Extremely toxic	$LD_{50} = 5$–50 mg/kg
Super toxic	$LD_{50} = <5$mg/kg

Karol describes the LD_{50} for acrylamides (used in acrylamide grout products such as AM9, Nitto SS, AV100, etc.) to be 200 mg/kg and the LD_{50} for the methylene-bis-acrylamide commonly used as the cross-linking agent as 390 mg/kg. Karol quotes the methanol acrylamide used in ROCAGIL BT as having a toxicity of 50% of Nitto SS.

Obviously and in consequence, when working with chemicals which to any principal extent might be toxic, it is necessary to establish safe exposure levels significantly below the fatal dose and to ascertain any degree of cumulative toxicity.

Many and most of the 'adverse effects' (from skin irritation to neurotoxicity) are not fatal, however, but obviously do already occur at considerably lower levels of exposure than the LD_{50} dosage. In summary, the safe application of chemical grouts has to be meticulously established by proper planning before any use on site. Grouts which require extensive quality assurance programmes (as a consequence of differentiated dosing and mixing programmes on site) should be used last. Grouts which are particularly reaction-sensitive to even marginal dosing errors and temperatures should be avoided. The fact that ever-demanding specifications are aiming at permeability coefficients below 10^{-7} m/s in sandy gravels or below 0.1 Lugeon in rock will increasingly make the use of chemicals beyond the range of application of microfine binders inevitable in future.

The more questions of material interrelate with aspects of engineering application, the more it becomes the duty of engineers to perceive both disciplines when realising demanding grouting works. They also have to be prepared and educated enough to accept responsibility not only for the engineering aspects, but for questions of safety and the proper environmentally responsible use of such materials.'

For special grouting applications where high strength is a criterion, cement-based grouts need to be modified with additives to fulfil design criteria such as volume stability, degree of expansion, and degree of sedimentation. Quite a number of additives are available on the market. The user needs to be aware that the majority of the additives have been developed for the concrete industry and just been modified to be used in grouts. It is important to undertake proper laboratory testing to verify the suitability and successful functioning of the additive. The average volume of additives used in cement-based grouts is in the range of 1%–2%. The laboratory tests shall also consider the way grout will be mixed on site, considering the rotation per minute, the time of mixing, and the order of adding the materials into the mixer. The latter is particularly important when mixing chemical grouts. Chemical grouts such as resin or gels do often consist of two to four components. For small grouting jobs this is suitable; however, for larger jobs with large quantities the appropriate equipment for preparing a grout out of four components may often not be available. As mentioned before, cement is the most common material for producing grout, especially Portland cement. Blended hydraulic cements, blast furnace cements, and other special cements such as microfine cement are less used for grouting. When designing a grouting application, it is important to be aware that cement standards and definitions differ between countries. For example, in Europe cement is defined by its strength followed by its mineralogical composition and original prime material. In North America, cement is basically defined by its application, originating materials, and its characteristics of strength development. In recent years, cement manufacturers have specialised and developed ready-mixes of cement-based material (so-called binders) for standard grouting applications and also for filling the annular space between the manchette tubes and the ground (sheath grout). One of the popular standard additives to grouting mixes is clay containing high proportions of montmorillonite. In the construction industry these clays are better known as bentonites. Bentonite is usually added to reduce the sedimentation of aggregates (providing for 'stability' of the mix), and does change the flow characteristic (rheology) and viscosity of the mixed grout. This can be achieved by adding just 1 to 2% of bentonite (by weight of cement).

5.8 TECHNICAL SUMMARY

Literature over the last few decades has seen quite a number of reports on successful grouting applications, with detailed reports about their design, execution, and performance, including many hydro and irrigation dams (ICOLD). Several useful conclusions may be drawn from these experiences, but it is not always wise to compare project situations without detailed knowledge on ground conditions and drilling-grouting technologies applied, or targets set and (measurably) achieved. In order for cement grouts to be successfully injected, it is necessary for the cement particles to remain in suspension during injection. Equally important are effective grouting pressures, sufficiently high to overcome substantial pressure losses when entering fine voids and to enlarge fissures elastically in order to facilitate the entry of the grout particles.

Simple theoretical considerations and elementary experimental evidence show that, as soon as internal friction in a particulate mix occurs, grouting is no longer possible. The penetrability of a cement-based grout into fissures depends on two main factors: the grain size of the cement used and the rheological properties of the suspension.

However, merely studying the size of a single dry grain is misleading: single dry grains have a tendency to grow in size during hydration and agglomerate, producing 'flocs' larger than the single dry particle. To improve the penetrability of a particulate grout (suspensions are most popular because of being cheap), it is necessary to both keep the grain size low, stay within or delay the start of hydration, and reduce/prevent the tendency for single grains to flocculate in the mix.

The question of setting time (irrespective of the type of grout) is important for the management of the grouting process against time, and the choice of a correct treatment system altogether. Cements are manufactured so that they have a setting time for industrial applications of about 4–5 h. If we greatly dilute cements the setting time is first delayed (10–16 h may result for water-cement ratios of 2:1 and 3:1, respectively), and then accelerated again during filtration. The addition of clays, bentonites, or accelerator admixtures reduces setting times (simultaneously increasing the viscosity of the mix). It is quite clear that the rheological behaviour of the suspension follows delicate relationships which have to be monitored and engineered on a continuous basis.

In conclusion, the essential ingredients for a successful grouting project are

- To go about any grouting project as open (educated) and engineering-minded as possible
- To perform under continuous questioning/reaffirming of the geotechnical model of the ground (in partnership with the designer - preferably an experienced Engineer or an Engineering geologist)
- Under permanent perception of the phenomena observed and interpretations derived from these (possibly cross-checking these with an experienced grouting foreman)

REFERENCES

Akjinrogunde, A. (1999). 'Propagation of cement grout in rock discontinuities under injection conditions,' Institute for Geotechnics, University of Stuttgart, Vol. 46.

Baban, O.R. (1992). 'Crack-injection in to massive concrete with synthetic resin,' PhD thesis, Institute of Materials Sciences, Faculty of Civil Engineering, Graz Technical University.

Baker, W. Hayward (ed.) (1982). 'Grouting in Geotechnical Engineering,' *Conference Proceedings, New Orleans*.

BS EN 12715. (2000). 'Execution of special geotechnical work. Grouting.' July.

Cambefort, I. (1964). *Injections des sols*, Eyrolles, Paris.

Doran S. and Stadler G., unpublished, generalised expertise on chemical grouting, Hong Kong, Nov. 1999.

Ewert, F.-K. (1985). *Rock Grouting with Emphasis on Dam Sites*, Springer Verlag: Berlin–Heidelberg.

Heinz, A., Hermanns Stengele, R., and Plötze, M. (2003). 'How to measure rheological properties of bentonite suspensions on construction sites,' *Annual Transactions of the Nordic Rheology Society*, Vol. 11.

Houlsby, D. (1989). 'Cement Grouting,' Dep. of Water Affairs, Sydney, Australia.

Hornich, W. and Stadler, G. (2009). *Grundbau-Taschenbuch*, 7. Auflage, Teil 2, Kapitel 2.3 'Injektionen,' Ernst & Sohn: Berlin.

Karol, Reuben, H. (1990). *Chemical Grouting*, 2nd ed., Marcel Dekker: New York.

Kasumeter, International Society for Rock Mechanics, Widmann, R. (1996), Commission on Rock Grouting, *Int. J. Rock Mech. Min. Sci & Geomech.*, 33(8):803–847.

Kasumeter, (2003), 'How to measure rheological properties of Bentonite suspensions on construction sites', Heinz, Hermanns Stengele, *Annual Transactions of the Nordic Rheology Society*, 11.

Keil et al. (1989). Quoted in Weaver & Bruce, 'Dam foundation grouting,' ASCE Publications, 2007.

Littlejohn, G. and Stadler, G. Joint lectures on anchoring and grouting at SAICE, 1976, South Africa; and EUROCK 2004 & 53rd Geomechanics Colloquium, Salzburg, Austria.

Lombardi, G. (1989). 'The role of cohesion in grouting,' *ICOLD Proceedings*, Lausanne.

Naudts, A. (1990). 'Revolutionary changes in the grouting industry resulting from polyurethane injection technology,' Second Canadian International Grouting Conference, Toronto.

Nonveiller, E. (1989). *Grouting, Theory und Practice*, Elsevier: Amsterdam.

Schulze, B. (1993). Neuere Untersuchungen über die Injizierbarkeit von Feinstbindemittel-Suspensionen. *Berichte der Int. Konf. betr. Injektionen in Fels- und Beton*, A.A. Balkema: Rotterdam, 107–116.

Semprich, S. and Stadler, G. (2002). 'Grouting.' In: U. Smoltczyk (ed.), *Geotechnical Engineering, Geotechnical Handbook*, Ernst & Sohn: Berlin.

Shroff, A.V. and Shah, D.L. (1993). *Grouting Technology in Tunnelling and Dam Construction*, A.A. Balkema: Rotterdam.

Stadler, G. (1992). Transient Pressure Analysis of RODUR Epoxy Grouting at Koelnbrein Dam, Austria, Diss. Thesis, MUL, Leoben.

Stadler, G. (2001). Seminar on waterproofing of tunnels, CUC-Seminars, F. Amberg, Sargans, Switzerland.

Stadler, G. (2001). *Permeation grouting*, ASCE Seminar Publications: New York.

Stadler, G. (2002). 'Was hat die internationale Normung der Injektionstechnik gebracht,' Injektionen in Boden und Fels, Ch. Veder Colloquium, Graz University of Technology, pp. 1–20, April.

Stadler, G. et al. (1989). 'Pressure sensitive grouting,' Technical University, Vienna,

Stadler, G., Howes and Chow (1998). '100 years of engineering geology,' symposium, TU-Wien.

Stadler, G. (2004). 'Cement grouting,' in Moseley, M.P. and Kirsch, K., *Ground Improvement*, 2nd ed. Spon Press: Abington

Stephen, D. and Stadler, G., unpublished, generalized expertise on Chemical Grouting, Hong Kong, Nov. 1999.

Verfel, J. (1989). *Rock Grouting and Diaphragm Wall Construction*, Elsevier: Amsterdam.

Vipulanandan, C. and Shenoy, S. (1992). 'Properties of cement grouts and grouted sands with additives,' *Proceedings, ASCE Specialty Conference on Grouting, Soil Improvement and Geosynthetics*, pp. 500–511.

Weaver, K. (1991, 2007). *Dam Foundation Grouting*. American Society of Civil Engineers: New York.

Weaver, K. (1993). 'Selecting of grout mixes – some examples from US practice,' *Proceedings of Grouting in Rock and Concrete*, A.A. Balkema: Rotterdam, pp. 211–218.

Widmann, R. (ed.) (1993). *Grouting in Rock and Concrete*, A.A. Balkema: Rotterdam.

Chapter 6

Jet grouting

George Burke and Hiroshi Yoshida

CONTENTS

6.1 INTRODUCTION

Of all forms of ground improvement systems, jet grouting must be regarded as one of the most versatile. With this technique it is possible to strengthen in-situ soils, cut off groundwater, and provide structural rigidity with a single application. In particular, jet grouting can create the highest-strength treated ground (soilcrete) of all the ground improvement systems. It can also be regarded as one of the most technically demanding of ground improvement systems requiring both technical excellence in design and construction because failure of either component will result in failure of the product.

Figure 6.1 shows the principal method of application whereby either high-pressure water or grout is used to physically disrupt the ground, in the process modifying it and thereby improving it. In normal operation the drill string is advanced to the required depth and then high-pressure water or grout is introduced while withdrawing the rods.

As discussed in Section 6.2, jet grouting has a long history of development from its initial use to current practice. In the field of jet grouting, the most notable advancements have been in Japan where the technique has been refined to its present-day capability by careful attention to detail in all aspects of the system. Through the years, careful research and execution has resulted in increasing column diameter and range of applicable soils. This development is also set out in Section 6.2.

After reading this chapter, it is hoped that the practicing engineer will understand how jet grouting came into existence, the technical complexity and design requirements needed for a successful application, and the range of applications for which jet grouting can be used.

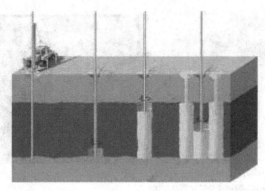

Figure 6.1 Jet grout column construction.

6.2 HISTORY

The scouring power of water has probably been employed as a soil excavation method since early times, especially in the mining industry, where use of it is documented in the Middle Ages.

The earliest patent regarding jet grouting was applied for in England in the 1950s; however, the real practical development of jet grouting took place for the first time in Japan. This technology was initially aimed at improving the effectiveness of water tightness in chemical grouting by eroding the untreated or partially treated soil, which was then ejected to the surface for disposal, being replaced with cement-based slurry for imperviousness. Subsequently, jet grouting was first applied to create thin cut-off walls, as shown in Figure 6.2.

For preventing water ingress, a derivative of panel jet grouting was evolved which sealed the gap between declutched sheet piles, for example.

Figure 6.2 Exposed jet grout panels.

Figure 6.3 Jet grout sealing between piles.

This derivative allowed the formation of part columns (shown in Figure 6.3) by causing a twin-angled jetting motion or a windscreen wiper motion of the monitor during lifting.

In the early 1970s, rotating jet grouting emerged in Japan because of the fact that panel jet grouting could hardly create satisfactory products due to varying thickness and somewhat fragile strength.

In the mid-1970s, jet grouting was exported to Europe and since then has become popular worldwide. According to required geometry, three main variants of jet grouting have emerged in the same period, of which conceptual schematics are illustrated in Figure 6.4. One of the variants is called the single system (S), which is the simplest form of jet grouting, ejecting a fluid grout to erode and mix with the soil.

In certain soil types spoil can be more viscous and, without the aid of an airlift, cannot easily travel up to the surface and heave may consequently occur. In cases where heave would cause serious damage (e.g., when underpinning a building), 100% relief through the annulus of the borehole has to be guaranteed. When drilling significantly below the groundwater level, eroding effectiveness can be considerably reduced on account of the absence of the shrouded air, which increases cutting energy.

The double system (D) adds compressed air, which surrounds (shrouds) the grout jet to enhance the erosive effect, especially below the water table.

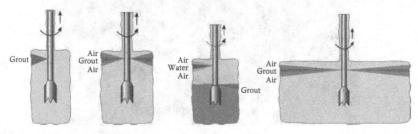

Figure 6.4 Single, double, triple, and SuperJet grouting.

However, a considerable percentage of the grout is lost to the surface due to the airlift. The double system has proven to be most effective in noncohesive coarse-grained soils like gravel and sand.

The third method, which is called the triple system (T), utilises three fluids—grout, jetting water, and compressed air shrouding the water. This system normally consists of a grouting nozzle below a water jetting nozzle, added in order to convey as many excavated soil particles as possible to the surface while limiting the grout ejected. While with the double system the ratio between water and cement is fixed or can only be adjusted through the water-cement content, the triple system achieves erosion and grout injection independently and can thus be optimised for the required performance. This has potential advantages in fine-grained and cohesive soils like silt and clay, allowing adjustment of the cement content independent from the energy used to cut the soil and where the double system would require a separate pre-cutting step.

In the 1980s, experience and confidence with jet grouting spanned a very wide range of application. Since the early 1990s, newer methods of jet grouting capable of a considerably larger treatment range or column diameter have been developed on grounds of cost and programme. This enabled jet grouting to obtain a column with a diameter in excess of 5 m, or even 9 m in softer ground (Figures 6.5 and 6.6 show examples of such an oversized body). This method could improve volumes of soil 20 times as large as the previous conventional systems, due to equipment development providing significantly higher flow rates at higher pressures.

The successful construction of a large column requires the use of focused jets, of which an example is shown in Figure 6.7, maintained in pristine condition; otherwise a large proportion of the jetting energy is lost within the system itself. Thus, jet grouting emerged capable of spanning a very wide range of applications.

The results of jet grouting can vary according to both equipment and soil types. Given these constraints, many measurements have been taken by varying the values of key parameters as a basis of theoretical solutions; however, even these trials cannot provide exact solutions because of the

Figure 6.5 Exposed SuperJet columns during technology development.

Figure 6.6 Trial SuperJet columns.

limited investigation into the soil and a lack of understanding of the real elementary process that occurs when the jet meets the soil.

In the late 1980s, a new concept provided an innovative progress for jet grouting systems, namely, dual jets colliding with each other to limit their eroding capability, thus achieving an exact intended diameter regardless of soil type. The arrangement of these jets is shown in Figure 6.8a while an exposed column is shown in Figure 6.8b.

The conceptual comparison of conventional and colliding methods is shown on Figure 6.9, noncolliding jets producing columns of variable diameter in variable ground. Colliding jet grouting has raised the required design quality since its appearance under the name of 'crossjet grouting'. In the early 1990s, colliding jetting was further evolved to include the deep mixing method to substantially increase the range of application. Conventional *in-situ* soil mixing suffers from a serious drawback of imperfect continuity when executed adjacent to walls; however, attaching an assembly of colliding jetting equipment at the tip of a drilling bit or blade as in Figure 6.10 has enabled the construction of optimal interlocking, as shown in Figure 6.11.

Furthermore, the enhancement in this in-situ mixing system results in more than four times the treated volume using the same equipment. This is shown in Figure 6.12, the conceptual schematic of the jet and churning system management (JACSMAN) system.

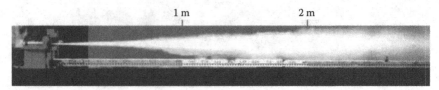

Figure 6.7 Focused jets.

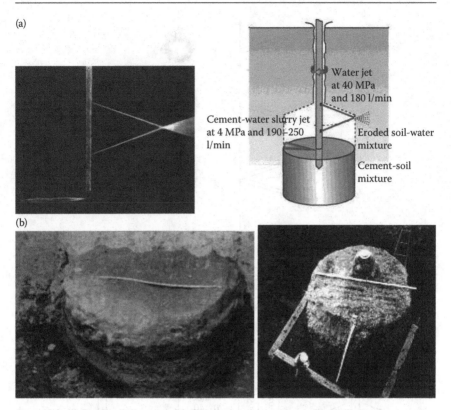

Figure 6.8 (a) Dual colliding jets. (b) Columns produced by dual colliding jets.

6.3 THEORY OF JET GROUTING

Many factors influence the efficiency and effectiveness of the jet grouting process and require consideration when designing and constructing jet grout columns.

6.3.1 Effect of dynamic pressure

When eroding soil with a high-pressure jet, the eroding distance radically increases after the pressure exceeds a certain level called the limit break pressure (Figure 6.13). The erosion distance radically increases after the jet pressure exceeds the unconfined compressive strength of both cohesive and sandy soil. Figure 6.14 illustrates the distribution of dynamic pressure versus the distance from nozzle, in which the soil is eroded to the distance at which the jet impact pressure attenuates to the level equal to the unconfined compressive strength of the soil (scaled laboratory conditions).

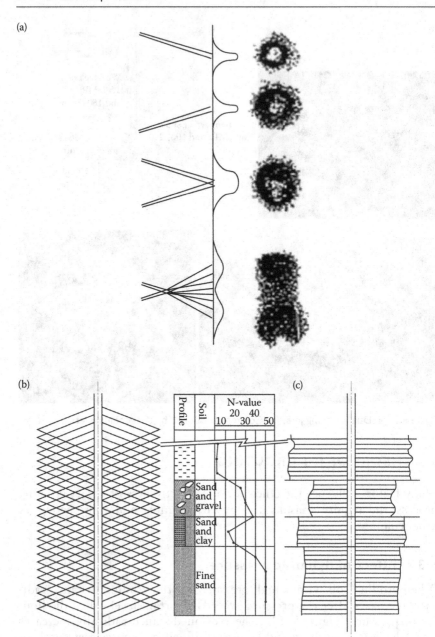

Figure 6.9 Principles of cross jetting. *(a)* Strike marks of dual colliding jets on pressure-sensitive film. *(b)* Soil cutting by a dual colliding jet. *(c)* Soil cutting by a single jet.

Figure 6.10 JACSMAN tool details.

Figure 6.11 JACSMAN column abutting sheet piling.

It is possible, with a lower impact pressure, to erode the same distance over a longer time; however, the high pressure saves time for most practical applications. Typically, jet pressures between 30 and 60 MPa for an overburden soil such as silt, sand, etc., and more than 200 MPa for rock formation are employed.

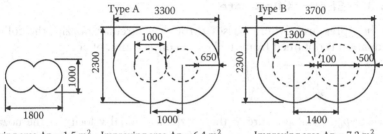

Figure 6.12 Treated area enhanced by colliding jets in JACSMAN (all dimensions are in mm).

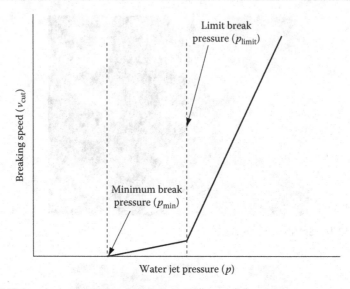

Figure 6.13 Jet pressure regulated for soil erosion.

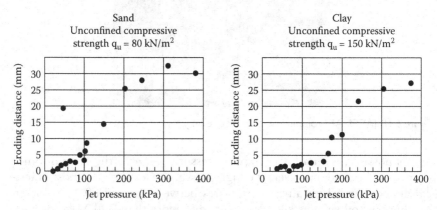

Figure 6.14 Eroding distances vs. jet pressure for sandy and clayey soil (results of small-scale laboratory tests).

6.3.2 Effect of flow rate

When pressurised fluid passes through a circular nozzle, the following equation is obtained from the law of conservation of energy:

$$v_0 = m\sqrt{2g\frac{p_0}{\gamma}} = m\sqrt{2\frac{p_0}{\rho}} \tag{6.1}$$

where p_0: initial pressure at the nozzle, v_0: initial velocity at the nozzle, g: acceleration of gravity, m: nozzle efficiency, γ: fluid unit weight, and ρ: fluid mass density.

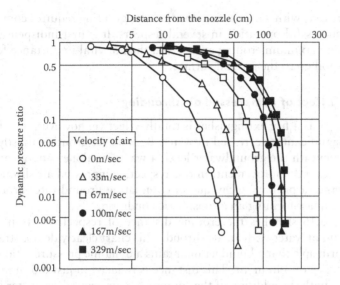

Figure 6.15 Dynamic pressure ratio (p/p_o) along jet centre axis with various air velocities.

A practical example of a calculation for pressure effect and flow effect is given here. If a water jet is discharged at 40 MPa through a fine nozzle of 2 mm in diameter such that the velocity of shrouded air is 100 m/sec, we can obtain an eroding distance of 1 m (2 m in diameter) at the point of 4 MPa from Figure 6.15 (dynamic pressure 0.1 times nozzle pressure p_o). This may be regarded as the effective limit of the column for most practical purposes.

Since an excellent nozzle has m = 0.92 as an efficiency coefficient, Equation 6.1 results in a flow rate Q:

$$Q = vA = m\sqrt{2gp_0}\,\frac{\pi}{4}d^2 \qquad (6.2)$$

= 49 *l/min*, where A: nozzle area, d: nozzle diameter.

If a 5-mm nozzle of the same efficiency is used instead, then in order to achieve the same required column diameter, the flow rate must be altered in accordance with the square of the nozzle diameters:

$$\frac{Q_1}{Q_2} = \left(\frac{d_1}{d_2}\right)^2$$

Hence the flow rate is 306 l/min.

6.3.3 Effect of compressed air

An increased air velocity with even low pressure can extend the eroding potential considerably, as illustrated in Figure 6.15 (dynamic pressure along jet centre axis with various air velocities). Jet grouting requires compressed air for successful operation in several respects. It is first indispensable for obtaining maximum eroding energy and then of vital importance for conveying spoil up to the ground surface.

6.3.3.1 Effect of compressed air shrouding

A water jet as a fire extinguisher is totally effective; however, its effectiveness is significantly decreased in water. Because jet grouting mostly treats the soil beneath the groundwater level, a water jet alone cannot cause significant ground improvement. In this respect, compressed air shrouding of liquid jets is a primary technique in eliminating groundwater around the jets, thus quasi-forming an atmospheric condition.

Figure 6.16 sketches the eroding distance of respective jets in air, in water, and in water with an air shroud. This chart clearly demonstrates the jetting principle that a liquid jet maintains a dynamic pressure ratio of 0.01 at a distance of 3 m in air. This distance is reduced to just 0.5 m in water; however, with the addition of the compressed air around the water jet, it is extended again to 1.1 ~ 1.2m.

6.3.3.2 The velocity and volume of compressed air

As stated previously, the mere presence of the air shroud does not always prove successful, but it should also maintain a higher velocity than half

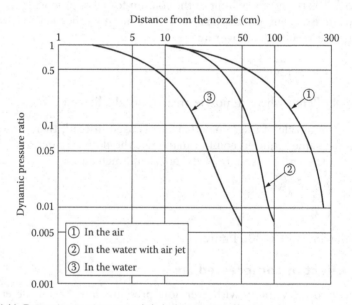

Figure 6.16 Dynamic pressure ratio (p/p_o) along jet centre axis.

the sonic velocity to ensure the formation of an atmospheric condition, as is clearly outlined in Figure 6.15. Additionally, an air nozzle has to be ring shaped or annular surrounding the nozzle, which preferably includes a minimum straight length before the air discharge point.

The width of this annulus must be approximately 1mm thick as standard which should provide sufficient air flow and yet does not allow any foreign particles like sand to flow upstream. Compressed air may be generated by a low-pressure compressor rated at 0.7 MPa for work up to 20 m deep; however, a high-pressure compressor is required to withstand the groundwater pressure for deeper works.

6.3.4 Effect of the soil

Soil type and stratigraphy influence the quality of soilcrete (the soil-cement product of jet grouting) and geometry of erosion. Figure 6.17 presents a qualitative scale for soil 'erodibilty'. In the local region of fluid injection, the turbulence created alone is enough to disaggregate cohesionless soil types. As plasticity and stiffness increase, erodibility decreases to a point where jet grouting may not effectively erode stiff cohesive soils.

Effective diameters of single fluid jet grouting typically range from 300–500 mm, and for double fluid typically range from 800–1,300 mm.

The Japanese Jet Grouting Association proposes standard jet grout (soilcrete) diameters for the triple fluid (Table 6.1) and the SuperJet methods (Table 6.2) to be used in different soil types.

In Table 6.2, the standard diameters listed represent special optimised tooling. Conventional tooling may yield considerably smaller geometries. Soil stratification is also a consideration, as variable soil conditions lead to variable soilcrete quality. Also, the jet grouting parameters may need to change versus depth to create uniform geometry, or variable geometry may result.

Gravels, cobbles, and boulders, although considered cohesionless, may range from highly erodible to very difficult to erode depending on *in-situ*

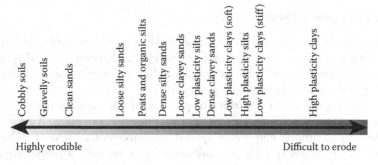

Figure 6.17 Soil erodibility scale.

Table 6.1 The standard soilcrete diameters (m) for sandy and clayey soil using triple fluid jet grouting

N value[a]						
Gravel[b]						
Sandy soil	N ≤ 30	30 < N ≤ 50	50 < N ≤ 100	100 < N ≤ 150	150 < N ≤ 175	175 < N ≤ 200
Clayey soil	–	N ≤ 3	3< N ≤5	5< N ≤7	–	7 < N ≤ 9
Organic soil[c]						
Effective diameter vs. depth[d]						
0 < Z ≤ 30m	2.0	2.0	1.8	1.6	1.4	1.2
30 < Z ≤ 40m	1.8	1.8	1.6	1.4	1.2	1.0
Lifting rate (min/m)	16	20	20	25	25	25
Pumping rate (m³/min)	0.18	0.18	0.18	0.14	0.14	0.14

Note: For cohesion c around 50 kN/m², the standard diameter may be difficult to obtain. For sandy soil with N > 150, and clayey soil with N > 7, the grouting specifications must be determined with considerable examination such as a field verification trial.

a   The largest SPT-N value of the soil to be treated should be used.
b   For gravelly soils, the soilcrete diameter expected is tabulated diameter less 10%. A field trial to verify the diameter is recommended prior to production.
c   For organic soil, considerable examination is recommended to determine the grouting specifications.
d   For the depth Z > 40, considerable examination is recommended to determine the grouting specification.

density, soil matrix, and other conditions. When soil contains more than 30% gravel, or the size of gravels exceeds 10 cm in diameter (cobbles), resulting soilcrete diameter may be smaller than expected. Reducing tool rotation or increasing slurry pump rate may solve this type of problem. Boulders will block the jet stream and a 'shadow' of untreated soil will exist beyond. Buried obstructions can also include trees, utilities, or cemented soil.

6.3.5 Other effects

The quality of the material and internal finish of the nozzle is of vital importance as well as its dimensions and geometry. Furthermore, in reality, care must be taken that even a perfect nozzle before use may be easily damaged owing to anomalies in the jetting stream.

In order to account for this, inspection of the condition of nozzles before and after each jet grouting operation has to take place. An optimal inspection technique employs a special measurement system of dynamic testing in association with pressure-sensitive films with a predetermined range.

Table 6.2 The standard soilcrete diameters (m) for sandy and clayey soil using superjet grouting

N value				
Sandy soil	N ≤ 50	50 < N ≤ 100	100 < N ≤ 150	150 < N
Clayey soil	N ≤ 3	3 < N ≤ 5	5 < N ≤ 7	7 < N ≤ 9
Organic soil				
Effective diameter vs. depth				
0 < Z ≤ 30m	5.0	4.5	4.0	3.5
30 < Z	4.5	4.0	3.5	3.0

Notes:
1. A field trial prior to production works is recommended to verify the diameter. The tabulated diameter is sometimes difficult to obtain in soil with certain characteristics.
2. For gravelly soil, a field trial to verify the diameter must be performed prior to production works. Soilcrete diameter with the tabulated diameter less 10% may be used for preliminary designing.
3. For soft soil (sandy soil N < 10, clayey soil N < 1), the diameter sometimes exceeds the tabulated diameter, which leads to shortage of solidifying material and results in lower compressive strength than expected. A field trial to verify the diameter is recommended prior to production.
4. Sandy soil with N < 150 in the table applies only to nonsolidified sandy soil. For clayey soil with N > 9, you may consult the SuperJet Association for some specifications to achieve the construction objective. A field trial to verify the diameter is recommended prior to production.
5. For cohesion c around 50 kN/m2, the standard diameter may be difficult to obtain.

If the jet is sound, the pressure-sensitive film reveals an annulus, with the centre destroyed, which is the so-called core of the jet still maintaining sufficient eroding energy to penetrate the film, as sketched in Figure 6.18, left (a focused flow). For a defective jet, the film reflects a totally coloured spot, with no central penetration as sketched in Figure 6.18, right (a turbulent flow).

Apart from dynamic pressure and flow rate, there are other parameters that have an influence on the eroding power of a liquid jet. An experimental equation explains this:

$$R = \left(4.95 K p_0^{-1.4} \ Q_w^{-1.6} \ N_t^{-0.2} v_n^{-0.3}\right)^{-1/1.4} \tag{6.3}$$

where:
 R = Eroding distance (m)
 K = Improvement factor (experience based) (m/sec)
 p_0 = Pumping pressure (tonnes/m^2)
 Q_w = Flow rate (m^3/sec)
 N_t = Repetition frequency (number of times a jet nozzle passes the same point)
 v_n = Rotational velocity of nozzle (m/sec) = $D_m \times \pi \times$ Rs/60 (D_m: Diameter of the monitor, R_s: rpm of the monitor).

Figure 6.19 provides experimental results for the optimal repetition frequency of the eroding jet, indicating that frequencies in excess of 5 only

Figure 6.18 Focused flow (left) and turbulent flow (right).

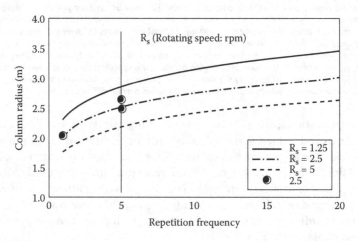

Figure 6.19 Experimental results for optimal repeating frequency of eroding jet.

marginally increase the column diameter. Lifting up the jetting rods in steps provides the necessary rotation using an integral number, which is not possible with a steady lift as shown in Figure 6.20 (lifting methods). Each step corresponds to an intended diameter; however, practical experience gives a maximum 5 cm lift for up to 2 m in diameter, and a maximum 10 cm lift for more than 4 m in diameter, as optimal increments, but is soil-type dependent.

6.3.6 Practical considerations

In order to successfully design a jet grout project, both theoretical and practical considerations need to be taken into account. For a successful project,

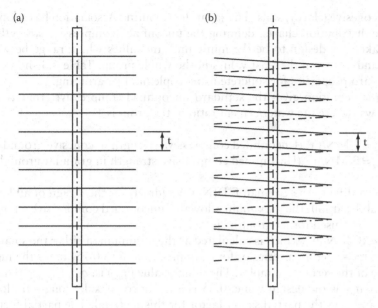

Figure 6.20 Lifting methods. (a) Intermittent lift. (b) Steady lift.

both columns must be installed correctly and the achieved properties must be in accordance with those values required by the design.

6.3.6.1 Design parameters for jet grout material

Strength of treated ground is usually assessed on the basis of unconfined compressive strength tests on samples obtained by coring and/or *in-situ* grab samples cast into moulds. The histograms shown in Figure 6.21 demonstrate experiential unconfined compressive strengths in granular (sandy)

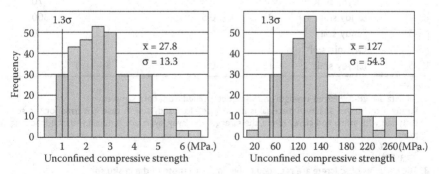

Figure 6.21 Histograms of unconfined compressive strength using triple fluid jet grouting in clayey soil (left) and sandy soil (right).

and cohesive (clayey) soils. The Japan Jet Grouting Association has adopted these distribution charts, defining the unconfined compressive strength to be taken for design to be the minimum safe values which range between 1% and 3% from the least values in the whole group. Table 6.3 shows the standard properties for soilcrete using triple fluid jet grouting.

This definition gives the standard unconfined compressive strengths as follows (where the water/cement ratio of the grout is 1):

$q_u = 1$ MN/m^2 (Unconfined compressive strength in cohesive ground)
$q_u = 3$ MN/m^2 (Unconfined compressive strength in granular ground)

According to the German E DIN 4093 (draft) for the design of all kinds of stabilised soil, the maximum allowable unconfined compressive strength $(q_{u,k})$ to be used for jet grouting is:
$q_{u,k} < 10$ MN/m^2, which is calculated as the minimum of either the smallest value measured in a series of four samples or (0.6 to 0.75) times the mean value of the series of samples. The design value $(q_{u,d})$ is: $q_{u,d} = q_{u,k} \times 0.85/\gamma_m$ where $q_{u,d}$ is the design value, 0.85 is a factor to consider long-term loading, and γ_m is the partial safety factor for this material. The partial factors for the loads (load case 1) vary between 1.35 (for dead weight) and 1.5 (live loads). The resulting global safety factor between mean value and design

Table 6.3 The standard properties for soilcrete using triple fluid jet grouting

Grout material	Soil type	Unconfined compressive strength (MN/m^2)	Cohesion (MN/m^2)	Bonding strength (MN/m^2)	Tensile strength (MN/m^2)	Modulus (E_{50}) of deformation (MN/m^2)
JG-1(H)	Sandy soil	3	0.5	1/3C	2/3C	300
	Clayey soil	1	0.3			100
JG-1(L)	Sandy soil	2	0.4			200
	Clayey soil	0.7	0.2			70
JG-2	Sandy soil	3	0.5			300
JG-3	Sandy soil	1	0.3			100
JG-4	Organic soil	3	0.3			30
JG-5	Clayey soil	1	0.1			

Notes:
1. All data are 28-day cured strength and were determined from core samples.
2. Strength-controlled soilcrete material is usually used for sandy soil. In case it is used for the soil stratified with sandy and clayey layers, the strength of clayey layers are reduced with the following rates:
   JG-2: 70% of JG-1
   JG-3: 50% of JG-1
3. The densities of soilcrete are regarded to be similar to those of the in-situ soil.
4. For gravelly soil, sandy soil data are to be used.
5. Seven-day strength of soilcrete is regarded to be 30%–40% of the four-week strength.

strength from this is between 4.4 and 3.2. For this purpose, samples are tested after a curing time of 28 days.

The directory from Japan Road Association states that permeability is in a range of 1×10^{-6} to 1×10^{-7} cm/s. The uses of the grout material are as follows:

- JG-1: High-strength soilcrete (standard material)
- JG-2: Strength-controlled soilcrete (medium strength)
- JG-3: Strength-controlled soilcrete (low strength)
- JG-4: For organic soil
- JG-5: For clayey soil

The design standard strengths of cohesion, bond and tension in bending are then determined with reference to the values shown in the Table 6.3. If it is desired to use alternative values it is recommended that laboratory mix design testing with representative soil precede production work.

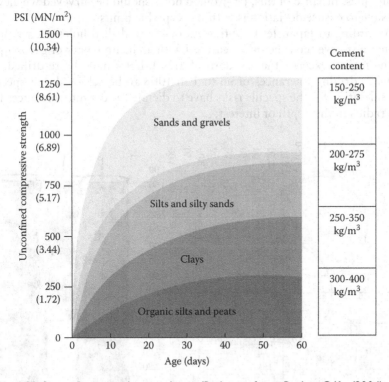

Figure 6.22 Strength–material type chart. (Redrawn from Burke, G.K. (2004). Jet grouting system: advantages and disadvantages, *Proceedings of Sessions of the GeoSupport Conference: Innovation and Cooperation in the Geo-Industry*, Jan. 29–31, 2004, Orlando, Florida, United States; sponsored by International Association of Foundation Drilling (ADSC) and the Geo-Institute of the American Society of Civil Engineers.)

Although very rare, soilcrete has been constructed to strengths in excess of 20 MN/m² in clean sands using specially developed grout mixes. An estimate of average strength for ordinary operations can be seen in Figure 6.22.

6.3.6.2 Drilling tolerances

Drilling tolerances are particularly relevant with jet grouting as overlapping of columns is vitally important. Inadequate interlocking not only takes place through drilling deviation which increases the offset from a neighbouring column with depth, but also through penetrating into a neighbouring column that has already set. The latter problem results in jetting within set and rigid material, consequently leading to unsuccessful works as no column is formed, as diagrammed in Figure 6.23. Inadequate interlocking can only be limited by excellent drilling coupled with in-hole survey techniques. Because of this, jet-grouted holes should be surveyed whenever possible to ensure deviation is within acceptable limits.

According to Japan Jet Grouting Association, drilled holes deeper than 30 m must have their inclines measured with an inclinometer or gyroscope. If the incline exceeds the standard of 1:250, holes must be re-drilled. In general, drilling tolerances of up to 1 in 100 can be achieved, but special consideration for the specific risks have to dictate the definite tolerances for the radius in the depth of interest.

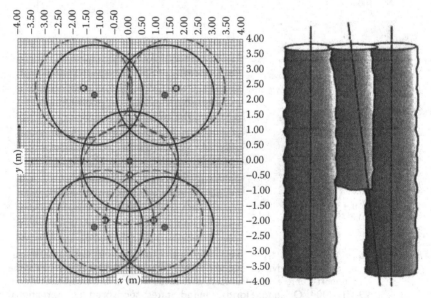

Figure 6.23 Effects and risks related to drilling deviations, and effect of poor drilling tolerance on column construction.

6.3.6.3 Control of jet grout returns

The method of jet grouting uses hydraulic erosion to construct the column geometry. The erosion media is either grout slurry or water, depending on the system deployed. In order to control the *in-situ* erosion environment, the borehole annulus must be an open pathway for return materials.

These return materials can vary greatly depending on the soils being eroded and the erosion media. In some circumstances, it may be desirable to 'pre-cut' or perform an erosion stroke with water only prior to jetting with grout. This could be the case if

- A higher cement content (strength) was needed, feasible by eroding a higher percentage of fine-grained soil from the desired depths.
- Assurance of protecting against surface heave, feasible if the returns pathway is restricted.

Assurance of continuous returns during jet grouting is necessary for the control of jet grouting and is a hallmark of quality jet grout construction. Disregarding this requirement can result in nearby heave or settlement, poor quality soilcrete, lack of geometry control, and impacts to nearby utilities.

Many things can be responsible for a loss of spoil return, such as the following:

- A borehole restriction
  - Too small a drill hole (annulus)
  - Too small a hole through a footing
  - Soft, squeezing clays
  - Gravels that are collapsing
- Loss of air return
  - Open, porous, gravely zones
  - Fibrous peats
  - Very soft clays
- Cohesive soil erosion
  - Erodes in pieces that block the annulus
  - Very thick (viscous) spoil

Spoil return can be enhanced by adjusting aspects of the jet grouting:

- Changing the grout viscosity
- Changing air pressure and flow rate
- Use of casing to reduce up-hole friction
- Pre-cutting measures
- Auxiliary air-lift system
- Changing the borehole size
- Manual reaming of the borehole
- Reducing the jetting energy

6.3.6.4 Sequence of construction

For every jet grouting project, there is a sequence of work that will provide for differences in quality, deformations, returning spoils, and/or geometry. The selection of installation sequence is important to the desired product and, when working beneath or nearby structures, to the effect on them.

This sequence is selected based on experience and the most desirable effect. The sequence for assuring continuity of a wall or base seal against groundwater is different than what might be selected for the highest strength. Similarly, when underpinning a structure or utility, the sequence must be such that loads can be redistributed by arching to adjacent ground until adequate strength can be developed by curing of the soilcrete.

Every case cannot be addressed here, but it is sufficient to say that the sequence is a planned approach that requires attention.

6.3.6.5 Quality control and validation

Section 6.3.6.3 alludes to problems of deviation but column diameter, position, and properties must also be considered. Therefore, it is important wherever possible to record and validate the installation of individual jet grout columns. Most specialists have the instrumentation to record the following parameters during installation:

- Depth
- Withdrawal rate and
  - Step height and step timing (considering rotation speed)
  - Uniform lift rate (considering rotation speed)
- Air pressure and flow rate
- Grout or water pressure and flow rate
- Rotation speed
- Grout density

In addition, some specialists have developed inclinometers built into the jet grout monitor that measure deviation of the drill string. It is also equally important to carry out quality control testing on the grouts used. This normally includes specific gravity, viscosity, and strength by 28-day cube strengths.

The knowledge of all these parameters allows the site engineer to review the column installation and come to a decision as to whether any column is misplaced or incorrectly installed. This is of paramount importance for base slabs or tunnel break-in or break-out where the omission or misplacement of a column can have the most serious effect on performance or safety. A further difficulty is the repair of these jet grout bodies as usually failures are difficult to locate.

Franz (1972), Fritsch and Kirsch (2002), and Kirsch and Sondermann (2002) list standards and publications relating to the control and execution of jet grouting. Burke (2009) reviewed quality control considerations. Eurocode EN 12716 is the European jet grouting standard code for execution.

Validation of jet grouting can be problematic. In order to fully validate a project column diameter, position and strength or permeability must be checked. Techniques typically carried out are as follows:

Column diameter: The most appropriate technique is to construct trial columns and then expose them to measure diameter directly. This is an excellent method but can only be used at shallow depths due to the expense of accessing columns at depth. Coring of columns can be successful but often suffers poor core recovery leading to difficulty in interpretation of diameter or strength. Electronic CPTs have been used to define geometry as they can easily 'feel' the surface of a soilcrete column (Burke et al., 2003). In Japan and Europe, thermocouples are being used to calculate diameter by comparing energy from the binder hydration to the measured in-situ energy (Meinhard et al., 2010). Borehole callipers can be lowered and extended to measure the extent of a column prior to initial set, but only in certain soil conditions by an experienced contractor. Some geophysics companies are developing nondestructive techniques utilising 3D borehole radar, electric resistivity, and sonic response, as yet still remaining at the research stage but offering a promising solution (Burke, 2012).

Column position: Column position relates to measurement of drilling tolerance, and as discussed above this is either accomplished by built-in inclinometers or by survey of the hole prior to jetting.

Column properties: This is the most commonly measured using coring techniques although some companies offer sampling within the column prior to initial set. Some forms of nondestructive techniques can be used as discussed above.

When working beneath or nearby structures and utilities, these items should be monitored during all operations and alarms set such that the operations are temporarily ceased if exceeded.

6.4 APPLICATION OF JET GROUTING

As set out in the introduction, jet grouting is an exceptionally versatile tool when considering ground improvement as part of a project. There are many applications that suit jet grouting but they can be grouped together as follows:

- Groundwater control
- Movement control
- Support
- Environmental

Groundwater control applications include

- Preventing flow either through the sides or into the base of an excavation
- Controlling groundwater during tunnelling
- Preventing or reducing water seepage through a water retention structure such as a dam or flood defence structure
- Preventing or reducing contamination flow through the ground

Movement control applications include

- Preventing ground or structure movement during excavation or tunnelling
- Supporting the face or sides of a tunnel during construction or in the long term
- Increasing the factor of safety of embankments or cuttings
- Providing support to piles or walls to prevent or reduce lateral movement

Support applications include

- Underpinning buildings during excavation or tunnelling
- Improving the ground to prevent failure through inadequate bearing
- Transferring foundation load through weak material to a competent strata

Environmental applications include

- Encapsulating contaminants in the ground to reduce or prevent contamination off site or into sensitive water systems
- Providing lateral or vertical barriers to contaminant flow
- Introducing reactive materials into the ground to treat specific contaminants by creating permeable reactive barriers

These lists show that jet grouting has a multitude of uses, all of which must be understood, designed, and executed accordingly. Some important main applications are now described in more detail.

6.4.1 Groundwater control

The last three decades have seen an increasing number of large excavations constructed in water-bearing soils. The use of conventional groundwater lowering techniques has been reduced as a result of the increasing importance of

- Economic water control
- Environmental aspects of the aquifer

- Observance of existing water rights
- Protection of existing buildings

Conventional chemical-based injection systems have been almost completely replaced by jet grouting techniques where the use of cement-based grouts reduces alkalinity.

Typical waterproofing elements are vertical and horizontal walls with and without an additional structural function in deep excavations, or for dams and dikes, break-in and break-out blocks to assist tunnel-boring machine operations. While with jet grouting columns a permeability of 10^{-5} to 10^{-6} m/sec can normally be expected, the permeability of the system as a whole ranges from 10^{-4} to 10^{-5} m/sec. As a rule, the excavation cannot commence until the allowable flow rate has been achieved and proven by a pumping test. Excess seepage is generally a result of a defect and can have countless additional causes.

The detection and location of leaks is extremely difficult, sometimes even impossible, and full or partial drawdown of the water table and the observation of piezometers or the measurement of the ground temperature during re-establishment of the water table are the most promising methods of leak detection. The necessary remedial works are often time consuming and extremely expensive, so the proper design and execution of jet grouting sealing elements is vital to the success of the project. The design requires the definition of sufficient strength and minimum permeability, homogeneity, and dimensional accuracy. It is essential to remember that water will not forgive any mistakes.

Defects in jet grout bodies can occur as a result of

- Insufficient overlapping of individual jet grout columns
- Jet shadows caused by natural or man-made obstructions
- Inhomogeneities in the ground (hard layers embedded in sand, peat layers)
- Instability and subsequent collapse of jet grout columns before they set
- Process deficiencies and interruptions, errors

To mitigate these risks, a thorough quality assurance plan is essential and, indeed, is state of the art. The plan should include the following elements, also identified in EN 12716:

- Setting out of the jet grout columns by x-y coordinates
- Drilling depth determined by efficient levelling systems
- Definition of drilling and jetting parameters
- Execution of test columns, documentation and evaluation of results
- Definition of the sequence of the works
- Identification of obstacles and countermeasures

- Grout composition and measurement of characteristics by sampling at mixing station and in back flow
- Measurement of drilling accuracy and countermeasures
- Process documentation during execution in real time of
  - Speed of insertion and extraction of monitor
  - Pressure and flow rate of grout, water, and air
  - Drilling and jetting rotation
  - Data secured on memory cards and modem transferred to backup systems

When looking at the evolution of grouting techniques in contractual terms, it is clear how much injection of sediments has departed from rock grouting. It must be remembered that the completed jet grout body is not homogeneous and therefore generally does not exhibit a constant strength or hydraulic characteristic. Design, specifications, and quality control must therefore reflect an uneven distribution of strength and permeability due to the variability of the soil under treatment.

Horizontal jet grout barriers in deep excavations should therefore be designed and executed with the following considerations:

- Minimum slab thickness not to be less than 1.0 m and to be increased by 0.1 m for every metre in excess of 10 m depth for safe uplift slabs
- Large slab areas to be divided into compartments of 2000 m²
- Increase of slab thickness in the immediate vicinity of vertical walls
- Avoid different slab elevations in one compartment
- Avoid location of slab within unsuitable soil conditions
- Time schedule to allow for possible remedial work
- Avoid anchored jet grout slabs
- Prepare emergency plan

Similar recommendations apply for vertical jet grout barriers as structural members:

- Applications with water pressures in excess of 5 m require special attention (redundant design, appropriate checking procedures, emergency plan)
- Identify soils with erosion potential in case of leakages
- Avoid slender construction elements
- Special care required when ground anchors are necessary

6.4.2 Underpinning

Underpinning of structures using jet grout normally involves the construction of a body of improved ground beneath the structure such that the

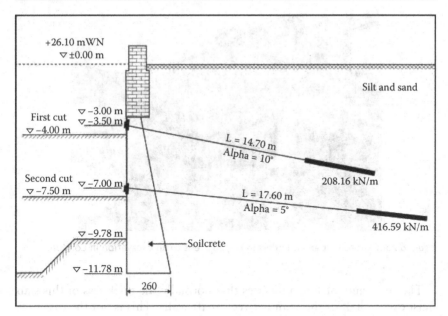

+26.10 mWN
▽ ±0.00 m

Silt and sand

First cut   ▽ –3.00 m
▽ –4.00 m   ▽ –3.50 m

L = 14.70 m
Alpha = 10°

Second cut   ▽ –7.00 m
▽ –7.50 m

208.16 kN/m

L = 17.60 m
Alpha = 5°

▽ –9.78 m

416.59 kN/m

Soilcrete

▽ –11.78 m

260

Figure 6.24 Jet grout underpinning adjacent to an excavation in London.

structural load is transferred to depth. If the underpinning is carried out next to an excavation, then the jet grout body must be designed accordingly and the stability checked for bearing capacity, sliding, and overturning. There is sometimes an economic relationship between the creations of a gravity underpin (i.e., a body that is self-supporting and stable) and a propped or strutted body where overturning or sliding is restrained by props or anchors (as for the case history below).

An example of jet grout underpinning adjacent to an excavation is shown in Figures 6.24 and 6.25. This example is taken from a project in London where a self-supporting underpin was required adjacent to a new basement construction.

The design of a jet-grouted underpin is exactly similar as for any gravity structure except that consideration needs to be taken into account that the strength of the jet grout body is usually significantly lower than brick or concrete.

6.4.3 Tunnels and shafts

6.4.3.1 Bottom slab

Base sealing of the slabs of shafts for tunnelling can be designed for the application of jet grouting to prevent base heave or piping in cohesionless soils saturated with groundwater. As discussed in Section 6.4.1, these constructions are risky if incorrectly executed and require careful design and application.

Figure 6.25 Exposed jet grout underpinning adjacent to an excavation in London.

The Academia of Japan dictates that normally the thickness of this slab must exceed half of the span between shaft walls. This is not the case outside of Japan, where sealing slabs have incorporated tie-down anchors, or were placed well below the excavation depth, to counter the buoyant forces. Thinner slabs are possible by employing circular arc beams on which only compressive stress acts, as illustrated in Figure 6.26. This method of design results in an arch prop 3 m in thickness even at a position of 40 m below ground level.

A tentative calculation gives a maximum value of 1.1 MN/m^2 and a minimal value of 0.95 MN/m^2 as compressive stress on both sides of the arch. As the average unconfined compressive strength of treated soils by jet grouting commonly exceeds 3 MN/m^2, this gives a high assurance of success.

6.4.3.2 Subsurface props

Displacement of walls is always of primary concern in open excavation. Late propping during excavation often causes tilting and/or settlement of not only adjacent buildings but water supply, sewer lines and other underground facilities. Therefore, jet grouting offers the radically different approach of an in-situ soil-mix propping prior to excavation.

A practical case history briefly explains the result. The work required an excavation of 10 m depth in a soft clayey layer for basement construction, but adjacent houses were so close to the site that they were afraid of being largely undermined due to displacement of walls for shoring, as shown in Figure 6.27. Consequently, jet grouting-produced props of just 1-m thickness at the bottom of excavation have proved successful together with a row of conventional strutting at ground level. Adding a row of grouted

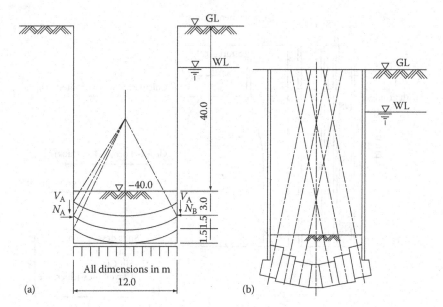

Figure 6.26 Base sealing of a shaft. (a) Conceptual cross-section. (b) Layout of jet-grouted arch.

props enabled the reduction of displacement by approximately 80%, as clearly shown in Figure 6.27.

6.4.3.3 Roof barriers

In starting a tunnel-boring machine (TBM) through a wall of a shaft into an alluvial deposit, the soil surrounding the TBM may be lower in strength due to the loosening effect of the construction of the structure. This could trigger collapse or settlement because of extension of this loosening to the ground surface, especially in the case of shallow tunnelling. Given such difficulties, jet grouting offers theoretical advantages in designing roof barriers. The design geometry is explained by reference to Figure 6.28, which illustrates how to obtain the zone to be treated ($R–a$), the property of which is to be reinforced by jet grouting.

A successful design follows an achieved line of shear strength to exceed a failure envelope of Mohr circle of the original ground. Figure 6.28 also shows that the radial and tangential stresses balance each other on the boundary line of the elastic region from the plastic one, and consequently derives Equation 6.4:

$$\frac{\partial \sigma_r}{\partial r} = \frac{\sigma_\theta - \sigma_r}{r}$$

(6.4)

where σ_r = radial stress, σ_ϕ = tangential stress, and r = variable radius.

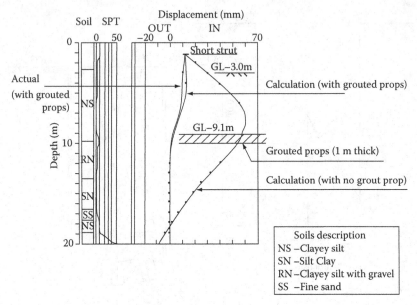

Figure 6.27 Comparison of calculated and actual wall displacements.

Next, since a failure takes place when the failure envelope becomes horizontal and the internal friction angle becomes zero, Equation 6.5 is derived as:

$$\sigma_\theta - \sigma_r = 2c \tag{6.5}$$

where c = cohesion.

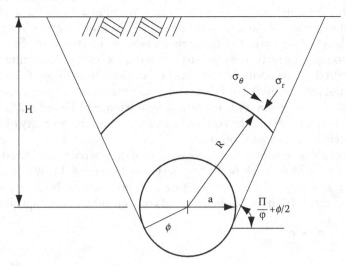

Figure 6.28 Roof barrier of a tunnel.

Then, substituting boundary conditions into the above simultaneous equations to obtain the plastic region leads to the following equation:

$$ln\left(\frac{R}{a}\right) = \frac{\gamma_t}{2c}(H - R) \tag{6.6}$$

where R = plastic region, γ_t = average unit weight of the soil, H = depth to the centre of the tunnel, and a = radius of the tunnel.

6.4.4 Environmental applications

One of the more interesting uses of jet grouting is in the environmental field. There are many applications based on the ability of jet grouting to form bodies at considerable depth while only requiring small penetrating drill holes. The main uses can be classified as follows.

6.4.4.1 Encapsulation

Achieving encapsulation of contaminants at depths where conventional excavation would be difficult, as for the example shown in Figure 6.29. Additionally, the grouted body is usually more impermeable than with conventionally grouted ground, leading to more security in contaminant control.

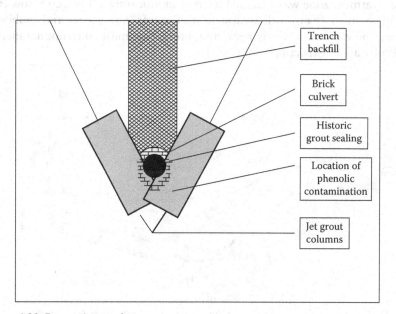

Figure 6.29 Encapsulation of contaminants at depth.

6.4.4.2 Containment barriers

In difficult ground, jet grouting can provide an effective method of creating a barrier as was achieved in a sand seam at a landfill site in Dundee, Michigan, United States in 2005 (Burke, 2007). Its main advantage is the ability to be selective in which zone is to be cut off. This is especially advantageous for deep applications. The grout or water jet (depending on system utilised) scours weak and loose material, penetrating into fissures and fractures and replacing the permeable infill material with relatively impermeable grout. The effective distance penetrated will depend on the system and rock type but has been shown on a number of projects to be effective up to 1 m from the hole position. Lateral barriers are typically specified in terms of permeability, and for rock it is usually possible to achieve 5–10 times lower permeability than using conventional rock grouting. For soils, the reduction when compared to permeation grouting can be as high as 10–50 times. As with all jet grout projects consisting of barriers or bodies constructed from interlocking columns, care must be taken during construction to minimise deviation from design locations, and this should always be taken into account when designing the scheme.

At the Dundee, Michigan project, an industrial manufacturer needed a groundwater containment barrier on the down gradient side of a disposal area (Figure 6.30). A 360-m-long wall was needed, with a requirement for thickness = 0.9 m, strength = 1,034 kPa minimum, and permeability < 1 × 10^{-6} cm/sec. Although the focus was on a sand seam at a depth of 6–9 m, the treatment zone was specified to treat significantly above and below the sand seam to ensure full treatment and cutoff. Site access and mobility along the wall was severely restricted, preventing most conventional methods of wall construction.

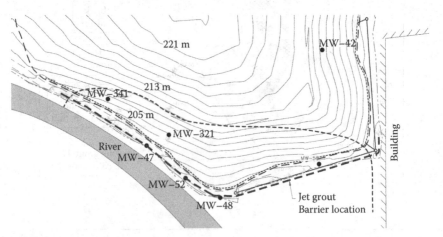

Figure 6.30 Plan view of jet-grouted soilcrete barrier wall.

Figure 6.31 Dual axis jet grouting operation for the jet-grouted soilcrete barrier wall.

The double fluid system of jet grouting was used, utilising a dual axis rig (Figure 6.31). This enabled two columns to be constructed concurrently from the horsepower of a single grout pump. Pairs of columns were constructed 'fresh-in-fresh', meaning that set was intentionally not required so to ensure jetting energy connected with adjacent work to preclude leaving windows in the wall. This raised the certainty of closure of the barrier and left only drilling verticality as a potential problem for the continuous wall. Slowed drilling penetration and high speed rotation ensured vertical drilling after setup.

6.4.4.3 Active barriers

In recent years, jet grouting has been used in the construction of permeable reactive barriers (PRB). These barriers contain materials that react with specific contaminants such that they are rendered harmless or less dangerous. Typical materials are zero valent iron (ZVI), granulated active carbon (GAC), or biologically active (BA). Design of these barriers is beyond the scope of this chapter. A guide was published by the United Kingdom Environmental Agency in 2002. To construct these barriers, the reactive material is either introduced in place of the grout, or a cavity is created by jetting and is tremie filled with a prepared material, as was the case in Memphis, Tennessee, United States in 2006. The process is illustrated in Figure 6.32 and described below.

ZVI was installed in a PRB to reduce levels of chlorinated volatile organic compounds (CVOCs) in a deep fluvial aquifer at the former Defense Distribution Depot known as the Memphis Depot (Endo, 2009). The bottom fluvial aquifer is approximately 24.3 m below ground surface and is 1.8–2.4 m thick at the PRB location. The fluvial aquifer is composed of sand, sandy gravel, and gravelly sand. Groundwater in the fluvial aquifer

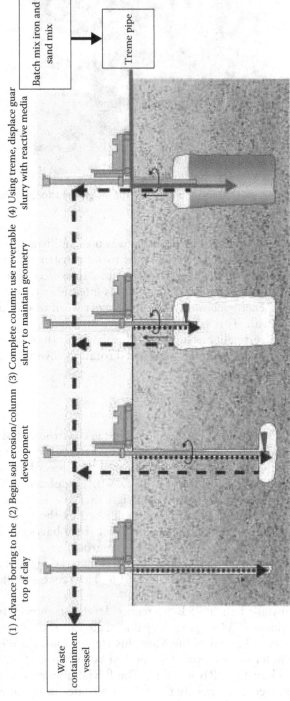

Schematic of ZVI PRB installation

(1) Advance boring to the   (2) Begin soil erosion/column   (3) Complete column; use revertable   (4) Using treme, displace guar
top of clay   development   slurry to maintain geometry   slurry with reactive media

Batch mix iron and sand mix

Treme pipe

Waste containment vessel

Figure 6.32 ZVI-PRB installation processes.

beneath the Memphis Depot has historically contained 1, 1, 2, 2-tetra-chloroethane (1, 1, 2, 2-PCA) up to 41,000 micrograms per liter (μg/l) and trichloroethane (TCE) up to 7,100 g/l. The PRB at Memphis Depot was expected to be capable of treating the CVOCs to drinking water standards.

The jet grouting at the Memphis Depot was performed in two phases. Phase I involved the construction of the geometry of the PRB using conventional jet grouting methodologies with a water and a revertible (biodegradable guar drilling fluid), which was used to erode and remove the soils. Phase II involved mixing the iron with sand and placing it down the hole via a tremie pipe; the enzyme required for breaking the guar slurry was also added during this phase. Because of its higher specific gravity, the sand and iron mix displaced the guar/water/soil mix within the jetted geometry.

6.4.5 Waterfront structures

With aging waterfront structures and the need for deep water berthing, jet grouting has emerged as an economic solution. Jet grouting is a system that can work around the many buried features (sheetpiling, anchors, deadmen, piling) to reduce the loads on existing walls and improve the stability of the *in-situ* system (Figure 6.33). It can target specific locations and depths to overlap and offer vertical and horizontal support.

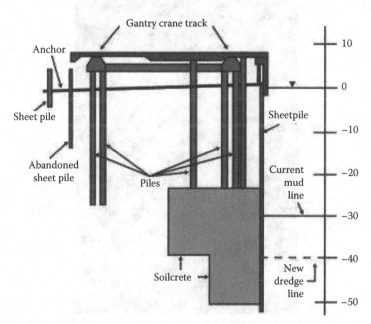

Figure 6.33 Typical port cross-section and solution for berth deepening.

6.5 CASE HISTORIES

6.5.1 Seismic remediation: Wickiup Dam, La Pine, Oregon, United States

SuperJet grouting provided seismic remediation of liquefiable soils layers within a dam embankment in western Oregon. Wickiup Dam is a zoned, rolled, earthfill embankment with a main river embankment section height of 30.48 m and a crest elevation of 1,324.97 m. The main embankment transitions into a 4.83-km-long, 12.19-m-high dike section on the left abutment. The dam is founded on bedrock while the left abutment dike is founded on deep and bedded fluviolacustrine deposits.

Analysis of the foundation materials in the left abutment dike indicated that two separate layers of diatomaceous silt and one layer of volcanic ash are likely to liquefy if the dam is subjected to the design earthquake. Superjet grouting allowed the liquefiable foundation soils to be targeted for stabilisation in situ, allowing normal reservoir operations and reducing the inherent risks associated with an excavate-and-replace alternative (Figures 6.34 and 6.35). Additionally, the jet grouting programme significantly reduced the environmental impacts of construction operations on the pristine location.

Figure 6.34 SuperJet grouting in progress at Wickiup Dam.

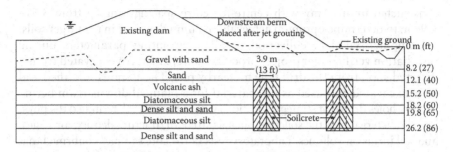

Figure 6.35 Section view of left abutment dike at Wickiup Dam.

A total of 854 soilcrete columns were constructed along a 685.80 m length of the dam toe. The columns were spaced on 3.96 m centres and each had a diameter of 4.27 m. The treatment was performed within the depth interval of 13.41–26.52 m at the northwest end of the alignment, and sloped upward to the depth interval of 2.44–6.10 m at the southeast end. Over 68,152,238 l of grout slurry was pumped, and the volume of the soilcrete exceeded 153,675 m³.

A preproduction test programme verified soilcrete geometry and quality prior to final design and construction. The test SuperJet columns were

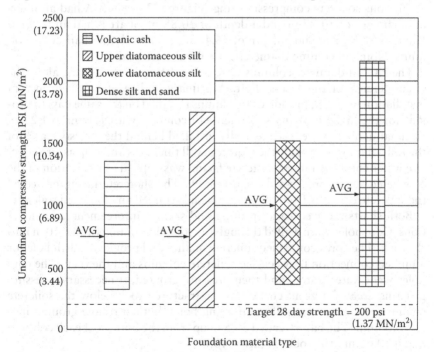

Figure 6.36 Unconfined compressive strength test results from Wickiup Dam. The owner required a minimum specific gravity, which kept the strength higher than what the specification called for.

constructed in an array with centre-to-centre spacing ranging from 3.51–3.96 m to determine optimum achievable column diameter in the target soils.

Computerised data collection of all jet grouting parameters during installation verified that project procedures were met. An excavated trough directed spoil from the drill hole to nearby pits. These spoils solidified and were excavated and used as fill material for a planned downstream berm.

Core holes drilled at strategic locations within the array determined the physical characteristics of the soilcrete, such as strength, fracture density, air vesicle and soil inclusion volumes. Laboratory tests of the core samples confirmed that soilcrete strengths exceeded the target 28-day q_u of 1.37 MN/m^2 (Figure 6.36).

6.5.2 Compression ring access shaft: Grand & Bates Sewer Relief, St. Louis, Missouri, United States

Triple fluid jet grouting was used to construct two access shafts in clayey soils for the Metropolitan St. Louis Sewer District's Grand & Bates Sewer Relief project (Figure 6.37) (Camper, 2002). The close proximity to roadways and homes was a primary concern.

The shafts were composed of interlocking soilcrete columns forming a continuous soilcrete compression ring (Figure 6.38). Shaft A had an internal diameter of 10.97 m and a depth of 20.88 m. Shaft B had an internal diameter of 9.75 m and a depth of 16.46 m, using full- and half-circle columns. Nominal column diameters of 1.07 m were achieved.

The 1.07-m-diameter columns were on a 0.91-m-centre spacing along the circumference of the access shaft. A primary and secondary sequence of installation was used (no adjacent columns installed on the same day) to provide for the highest quality soilcrete. Half columns with 1.07-m to 1.22-m nominal diameters were strategically installed behind the interstice area of the full columns to provide the required wall thickness and help assure effective water cutoff. An inner shotcrete facing was applied as excavation of the completed shafts progressed (Figure 6.39). The shotcrete facing protected the soilcrete columns from the elements and retained any loosened soilcrete.

Both shafts received rock grouting and shear reinforcement at 31 locations. Core holes were drilled through the soilcrete columns and 6.10 m into the rock. The cores confirmed soilcrete quality and provided a drill hole for a pinned connection between the soilcrete columns and the rock. The core holes were water tested, and then pressure grouted, if necessary. Pressure grouting created a grout curtain in the fractured rock below the soilcrete columns. A 9.14-m-long #8 reinforcing bar was then tremie grouted into each column. The bar extended 3.05 m up from the bottom of each column, and 6.10 m into the rock (Figure 6.40).

Two full test columns and one half-test column were installed at production column locations prior to the start of production work at each shaft location. These tests helped to confirm the jet grouting parameters necessary to create the

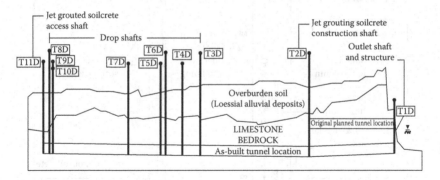

Figure 6.37 Section view of tunnel alignment with soilcrete access shaft locations.

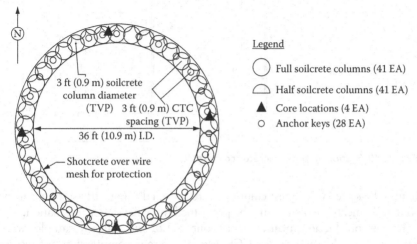

Figure 6.38 Plan view of soilcrete access shaft.

Figure 6.39 Shaft excavation in progress.

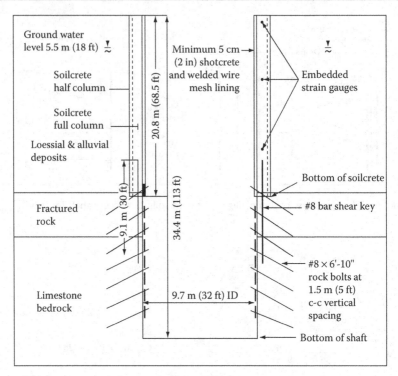

Figure 6.40 Section view of soilcrete access shaft.

required geometry. The test columns were installed using early setting cement so that verification coring could be performed three days after installation.

During production, neat cement grout and *in-situ* soilcrete samples were taken once daily and cast for UCS testing. Average compression test results for both shafts were 6.13 MN/m² for wet soilcrete samples at 28 days. Once the columns achieved adequate strength, four continuous 85-mm-diameter soilcrete core samples were taken from columns equidistant around the shaft circumference. The first core hole was taken at the interstice area of the test columns. The three remaining core holes were taken from production columns at their interstice areas. A total of 32 core samples from each shaft were tested. Eight samples were tested at 28 days for each core-hole location. The average UCS test result of the cores from each shaft was 5.86 MN/m² at 28 days, offering good correlation with the wet samples. All core-hole locations were filled with neat cement grout after this testing was performed.

Vertical ground movements of the surrounding area were monitored before and after each production shift. No ground heave was detected. Twelve vibrating wire strain gauges were installed into each full column at both shafts immediately after grouting. The strain gauges were installed at three elevations for a comprehensive reading (Figure 6.41).

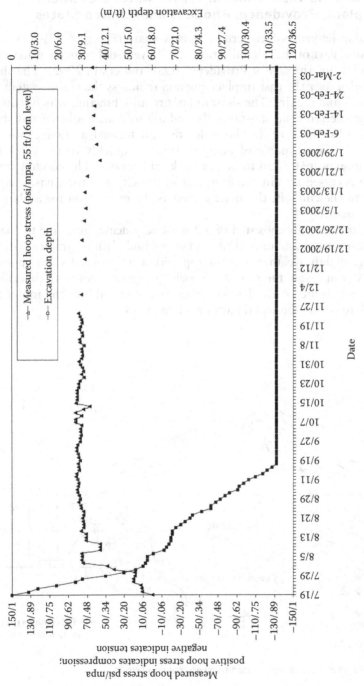

Figure 6.41 Measured hoop stress provided from strain gauges.

6.5.3 Groundwater control: Waste-water treatment plant, Providence, Rhode Island, United States

A circular jet-grouted cofferdam access shaft (Figure 6.42) was constructed as part of the rehabilitation of a 75-year-old primary water supply conduit for the city of Providence (Oakland et al., 2002). The shaft was needed over the cast-in-place portion to access the entire system for inspection and repairs. The shaft did not require bracing, which reduced the construction time substantially and allowed an unobstructed shaft for access to the tunnel. The soilcrete wall formed a stable, erosion-resistant surface capable of being disinfected quickly in the event the tunnel unexpectedly had to be put back into service. The conduit could only be accessed within the footprint of the city's settling basin. After repairs to the conduit, the shaft served as the excavation for a planned valve structure.

The targeted soils consisted of 5.2 m of very dense, gray fine to coarse sand with various amounts of silt and gravel underlain by granite bedrock that was slightly weathered at the top with a 0.3 m sand seam approximately 0.3 m below the top of the rock. Jet grout holes were predrilled, and the jet-grouted cofferdam columns were doweled into the top of the bedrock to ensure the seal (Figures 6.43 and 6.44).

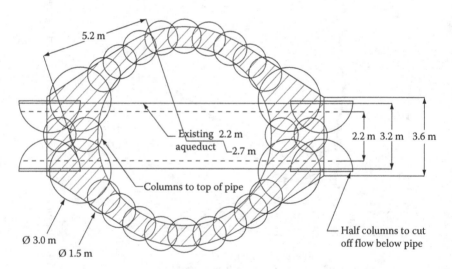

Figure 6.42 Plan view of jet-grouted soilcrete column layout.

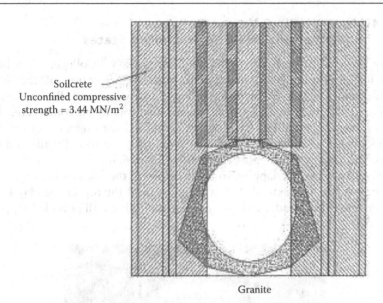

Soilcrete
Unconfined compressive
strength = 3.44 MN/m^2

Granite

Figure 6.43 Cross-section of jet-grouted soilcrete cofferdam around and below the existing conduit.

Figure 6.44 Soilcrete seal around and below the conduit with the bulkhead visible within the conduit.

6.5.4 Underpinning: Vassar College, Poughkeepsie, New York, United States

Avery Hall, a 140-year-old historic building at Vassar College, was scheduled for demolition and replacement, with the exception of the West Façade, which was to be incorporated into a new building (Burke, 2007). A new mat foundation, requiring excavation, was to be constructed for the replacement building. Jet grouting combined with temporary tieback anchors was chosen for both the underpinning of the West Façade and for temporary excavation support (Figures 6.45 and 6.46).

By constructing overlapping soilcrete columns, the footprint of the existing foundation was extended to within 0.3 m of the top of the clay layer located 6 m below grade. The soilcrete underpinning wall provided support

Figure 6.45 Excavated Avery Hall façade underpinned with jet-grouted soilcrete columns.

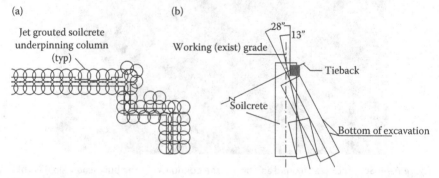

Figure 6.46 (a) Partial plan view of jet-grouted soilcrete underpinning columns beneath Avery Hall façade. (b) Section view.

by bearing below the excavation of the new mat foundation, and provided the necessary temporary excavation support along the façade.

The grouting extended to a minimum of 0.6 m above the bottom of the foundation in order to bind the rubble foundation together. Work was sequenced such that any movements would be minimised. Before production work, a test programme consisting of the installation of two soilcrete columns was performed to enable the grouting contractor to confirm or adjust the jet grouting parameters to ensure column geometry.

6.5.5 Environmental: Philadelphia Airport, Runway 8-26, Pennsylvania, United States

A new commuter runway (Runway 8-26), was planned for construction at the Philadelphia International Airport. A 300-m portion of the runway was to be constructed over the Enterprise Avenue Landfill (Furth et al., 1997). The landfill mass consists of 6 m of incinerator ash and concrete, asphalt, rock, metal, and hazardous materials. The site was clay capped in the 1980s, but additional remediation was needed to meet US Environmental Protection Agency final closure requirements before the runway could be constructed. One component of the closure plan included installation of a low-permeability horizontal barrier above a very thin (approximately 0.61–0.91 m) natural clay stratum which underlies an approximately 1,020 m² area of the landfill footprint so as to ensure that a minimum 1.52-m-thick low-permeability barrier exists beneath the entire 150,000 m² landfill (Figure 6.47). The new barrier was constructed using double fluid jet grouting to achieve remote excavation and replacement of the bottom 0.91 m of the waste mass with a low-permeability soilcrete. The grout slurry was formulated to meet

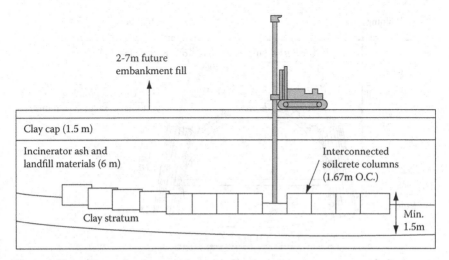

Figure 6.47 Jet-grouted soilcrete column profile view.

the low-permeability (1×10^{-9} m/sec), low-elastic modulus (124,100 kPa) and compressive strength requirements (900 kPa) for the project design.

6.5.6 Settlement control: Japan

A tunnel was required to be constructed beneath a street under which were buried numerous services. In addition, the adjacent buildings were sensitive to movement. The solution adopted was to construct a heading from spiles (horizontal piles) supported on jet grout columns toed into competent ground. In this way the jet grouting supported the tunnel drive and reduced settlements to acceptable levels. The small diameter holes required to install the columns were also of benefit in penetrating between the services. The crossjet system was chosen as the ground conditions were variable and with this system the column diameter could be guaranteed (Figure 6.48).

In this project, an arrival shaft of a shield tunnelling machine was supported with many anchors, which had the potential to hinder the machine's arrival at the shaft, and therefore needed to be removed. At part of the wall where the shield reaches the shaft, it was decided to remove the anchors prior to the arrival of the machine and improve and stabilise the soil behind the shaft wall. Figure 6.49 shows the exposed soilcrete at the opened wall.

6.5.7 Cofferdam sealing: New Orleans, Louisiana, United States

At two locations where levee walls were breached during Hurricane Katrina, the repair required that new sheetpile walls be constructed to

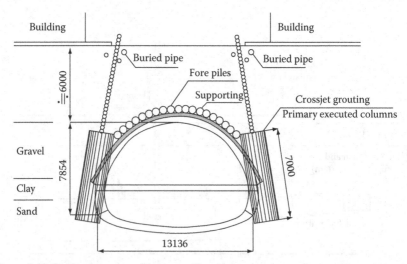

Figure 6.48 Settlement control for tunnelling, Japan.

Figure 6.49 The exposed soilcrete at the opened wall.

enable removal and replacement of the emergency fill (Burke, 2007). In each location several groups of jet grout columns were constructed to connect and seal the space between the old and the new sheetpiling. In each location, the double fluid system of jet grouting was used to create the soilcrete columns, which hardened to provide an excellent groundwater seal (Figure 6.50).

Jet grouting was employed for its ability to easily access these locations on hastily constructed soil berms, and the ability to cleanly erode the soft organic soils and encapsulate the sheetpile sections with a high-strength low-permeability soilcrete.

6.5.8 Waterfront structure: Battery Park City Authority, New York City, United States

Battery Park City Authority, on the Hudson River, is a combined residential/commercial development built on land 'created' from material excavated during the construction of the World Trade Center (Boehm, 2004). Further development in the 1970s included the construction of a 21.3-m-wide riverfront esplanade consisting of a reinforced concrete relieving platform supporting sand fill. Parallel to the river, the esplanade supports vertical timber sheeting to retain up to 1.8 m of soil. Recent improvements in Hudson River water quality resulted in an increase in the Teredo Navalis mollusk population. These worm-like, marine borers were attacking and destroying the timber sheeting.

Because borer activity would eventually result in loss of soil and surface subsidence, replacing or supplementing the timber sheeting was imperative. However, extensive development of the area, limited workspace, and difficult subsurface conditions precluded conventional construction methods.

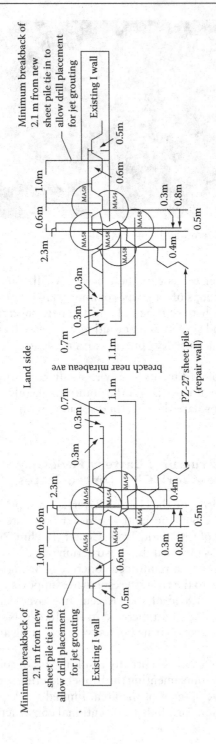

Figure 6.50 Plan view of jet-grouted soilcrete locations for sheetpile seal.

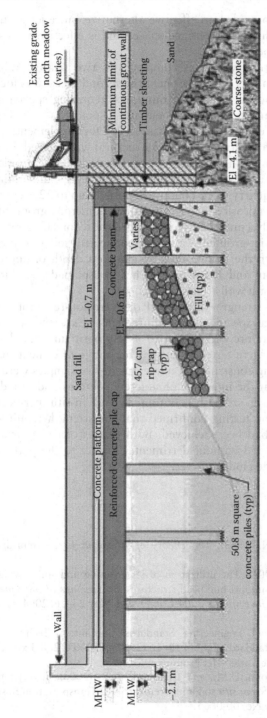

Figure 6.51 Profile of jet-grouted soilcrete wall at Battery Park City Authority.

Jet grouting provided an effective alternative, since jet grouting can be readily accomplished in confined spaces and is effective across the widest range of soil types.

The jet grouting work was completed in two phases. While the first phase work area was relatively open, the second phase was located within extremely restrictive, urban surroundings, requiring special attention to site conditions and spoil containment and disposal.

Project requirements on each phase called for supplementing the timber sheeting with an in-situ, jet-grouted structural wall, placed directly behind and in contact with the timber sheeting. The subsurface profile consists of sand backfill placed over filter stone. This in turn is underlain by a layer of crusher-run quarry stone containing cobbles up to 22.8 cm in diameter. This very high–porosity material required numerous grout additives and a specific, tightly controlled work procedure to preclude excessive grout loss. For each jet-grouted wall, interconnected soilcrete columns were constructed with the double fluid system to a depth of approximately 6 m along 243.8-m and 152.4-m stretches of esplanade, creating effective, 0.9-m-thick in-situ walls (Figure 6.51).

A very high–strength (20.68 MPa), corrosion-resistant soilcrete was needed to meet specification requirements. Extensive preconstruction testing was therefore carried out to assess optimum mix design. Eleven different mixes were tested, using a wide range of cement materials and additives. During construction, numerous in-situ samples were retrieved at close intervals at the interstice of soilcrete columns and tested for unconfined compressive strength, continuity, and in-situ permeability. This post-construction testing confirmed that the strength requirement in the soilcrete walls had been achieved. Both phases of jet grouting were successfully completed without detrimental impact to the park, the existing structures, or the Hudson River.

REFERENCES

Abramovich, G.N. (1963). *The Theory of Turbulent Jets*, Cambridge, MA: MIT Press.

Boehm, D.W. (2004). The utilization of jet grouting and soil mixing methods to repair and support bulkhead structures, *Proceedings of the American Society of Civil Engineers Ports 2004 Conference*, May 23–26, 2004, Houston, Texas, United States.

Building and Civil Engineering Standards Committee, Technical Committee. European Standard DIN EN 12716 (2001). CEN/TC 288, Execution of special geotechnical works – Jet grouting, 05.18.07.

Burke, G.K, Cacoilo, D.M., and Chadwick, K.R. (2003). SuperJet grouting new technology for *in situ* soil improvement, TRB Transportation Research Record 1721, Paper No. 00-0665.

Burke, G.K. (2004). Jet grouting system: advantages and disadvantages, *Proceedings of Sessions of the GeoSupport Conference: Innovation and Cooperation in the Geo-Industry*, Jan. 29–31, 2004, Orlando, Florida, United States; sponsored by International Association of Foundation Drilling (ADSC) and the Geo-Institute of the American Society of Civil Engineers.

Burke, G.K. (2007). Vertical and horizontal groundwater barriers using jet grout panels and columns (GSP 168), *Proceedings of Geo-Institute's Geo-Denver 2007: New Peaks in Geotechnics*, February 18–21, 2007, Denver, Colorado, United States.

Burke, G.K. (2007). New methods for underpinning and earth retention, grouting for ground improvement: Innovative concepts and applications (GSP 168), *Proceedings of Geo-Institute's Geo-Denver 2007: New Peaks in Geotechnics*, February 18–21, 2007, Denver, Colorado, United States.

Burke, G.K. (2009). Quality control considerations for jet grouting, *Geotechnical News*, December 2009, BiTech Publishers Ltd.

Burke, G.K. (2012). The state of the practice of jet grouting, *Proceedings 4th International Conference on Grouting and Deep Mixing*, February 15–18, 2012, New Orleans, Louisiana, United States; sponsored by the Geo-Institute of the American Society of Civil Engineers.

Carey, M.A., Fretwell, B.A., Mosley N.G., and Smith, J.W.N. (2002). Guidance on the use of permeable reactive barriers for remediating contaminated groundwater, National Groundwater and Contaminated Land Centre Report NC/01/51, Bristol, England: Environment Agency.

Camper, K.E. (2002). Safe passage: jet grouted columns on Rand & Bates tunnel, *Tunnel Business Magazine*, December 2002.

Endo, W.K. (2009). The use of jet grouting techniques in permeable reactive barriers, presented to Chemical Grouting Co., Ltd., Japan, July 29, 2009.

European Standard Amendment Draft 4093 to DIN EN 12716.

Franz, N.C. (1972). Fluid additives for improving high velocity jet cutting, *Proceedings of the 1st International Symposium on Jet Cutting Technology*, British Hydromechanics Research Association, April 5–7, 1972, Coventry, England, A7–93.

Fritsch, M. and Kirsch, F. (2002). Deterministic and probabilistic analysis of the soil stability above jet grouting columns, *5th European Conference on Numerical Methods in Geotechnical Engineering*, NUMGE, Paris, Presses de LENPC.

Furth, A.J., Burke, G.K., and Deutsch, W.L. (1997). Use of jet grouting to create a low permeability horizontal barrier below an incinerator ash landfill, *Proceedings of the 1997 International Containment Technology Conference and Exhibition*, February 9–12, 1997, St. Petersburg, Florida, United States.

Hermans, J.J. (1953). *Flow Properties of Disperse Systems*, Amsterdam, The Netherlands: North Holland Publishing Company.

Kirsch, F. and Sondermann, W. (2002). Zur Gewölbestabilität über Soilcrete-Korpern. 9. *Darmstädter Geotechnik Kolloquium*, Technische Universität Darmstädt, March 14, 2002, Darmstadt, Germany.

Kirsch, K. (1997). Contractor's view on the risks involved with deep excavations in water bearing soils. *Proceedings of the 14th International Conference on Soil Mechanics and Foundation Engineering*, September 6–12, 1997, Hamburg, Germany.

Lichtarowicz, A. (1995). Future of water jet technology basic research, *Proceedings of the 4th Pacific Rim International Conference on Water Jet Technology*, April 20–22, 1995, Shimizu, Japan, pp. 13–26.

Meinhard, K, Adam, D. and Lackner R. (2010). Temperature measurements to determine the diameter of jet-grouted columns, *11th DFI and EFFC International Conference on Geotechnical Challenges in Urban Regeneration*, May 26–28, 2010, London, England.

Noda, H., et al. (1996). Case of jet grouting for 10.8 m diameter shield. In: Yonekura R., Shibazaki, M., and Terashi, M. (Eds.), *Grouting and Deep Mixing*, Rotterdam, The Netherlands: Balkema.

Oakland, M.W., Ashe, J.M., Wheeler, J.R., and Blake, R.C. (2002). Design-build of a jet grout access shaft for tunnel rehabilitation, *Proceedings of North American Tunnelling 2002*, May 18–22, 2002, Seattle, Washington, United States.

Pollath, K. (2000). Baugrube Schleuse Uelzen II, *Deutsche Gesellschaft für Geotechnik*. Baugrundtagung, September 18–21, 2000, Hanover, Germany.

Reichert, D., et al. (2002). Baugrube Domquarree. *Deutsche Gesellschaft für Geotechnik*. Baugrundtagung, September 25–28, 2002, Mainz, Germany.

Shavlovsky, D.S. (1972). Hydrodynamics of high pressure fine continuous jets, *Proceedings of 1st International Symposium on Jet Cutting Technology*, British Hydromechanics Research Association, April 5–7, 1972, Coventry, England.

Shibazaki, M. (1996). State of the art grouting in Japan, *Proceedings of the 2nd International Conference on Ground Improvement Geosystems*, May 14–17, 1996, Tokyo, Japan, pp. 857–862.

Shibazaki, M. and Ohta, S. (1982). A unique underpinning of soil solidification utilizing super high pressure liquid jet, *Proceedings of the Conference on Grouting in Geotechnical Engineering*, American Society of Civil Engineers, February 10–12, 1982, New Orleans, Louisiana, United States, pp. 685–689.

Shibazaki, M., Ohta, S. and Kubo, H. (1983). *Jet Grouting*, Tokyo, Japan: Kajima Publishing Company, pp. 15–20.

Wachholz, T., et al. (2000). Planung, Konzeption der Schleuse Uelzen II und Berechnung der Baugrube. *Deutsche Gesellschaft für Geotechnik* Baugrundtagung, September 18–21, 2000, Hannover, Germany.

Wichter, L. and Kugler, M. (2002). Fehlstellen in Dusenstrahlkörpern durch Inhomogenita ten im Baugrund. *Deutsche Gesellschaft für Geotechnik*. Baugrundtagung, September 25–28, 2002, Mainz, Germany.

Wittke, W. et al. (2000). Neues Konzept für die Schildeinfahrt in eine Baugrube im Grundwasser. *Taschenbuch für den Tunnelbau 2000*, Essen, Germany: Verlag Glückauf.

Yahiro, T. (1996). *Water Jet Technology*, Tokyo, Japan: Kajima Publishing Company, pp. 15–29.

Yahiro, T., Yoshida, H. and Nishi, K. (1983). *Soil Modification Utilizing Water Jets*, Tokyo, Japan: Kajima Publishing Company, pp. 7–20.

Yanaida, K. and Ohashi, A. (1980). Flow characteristics of water jets in air, *Proceedings of the 5th International Symposium on Jet Cutting Technology*, British Hydromechanics Research Association, June 2–4, 1980, Hannover, Germany, A3–A33.

Yoshida, H. and Saito, K. (2009). Mechanism of cross jet and application, *Proceedings of the International Symposium on Ground Improvement 2009*, December 9–11, 2009, Singapore, pp.159–164.

Yoshida, H. (2010). The progress of jet grouting in the last 10 years in the Japanese market, *Proceedings of the Deep Foundation Institute 35th Annual Conference*, October 12–15, 2010, Hollywood, California, United States, pp. 503–514.

Chapter 7

Soilfracture grouting

Eduard Falk and Clemens Kummerer

CONTENTS

259

7.1 INTRODUCTION

The phenomenon of hydraulic soil fracturing was initially observed as an undesirable side effect of traditional grouting measures. The uncontrolled propagation of soil fracs that were rapidly filled with grout suspension did not achieve the objective of homogenously filled voids in granular soils. In the past, high pumping rates that no longer allowed the grout to penetrate the pore system of the soil continuously were regarded as a grouting defect. It was crude oil technology using hydraulic soil fracturing for increasing the permeability and thus the yield of oil fields that provided the impetus for systematically applying geotechnical methods for using deliberately produced voids in the soil. In the meantime, fracture grouting has been used for systematically improving soil properties. The load-bearing capacity and permeability of both granular and cohesive soils can be modified by incorporating a cement or solid matter skeleton. The repeated application of this method also allows for the controlled raising of buildings with very different foundation systems. The most spectacular use of the method is found in connection with the complex tasks of compensating for settlements which as a result of tunnel excavation threaten the structural integrity of buildings above the tunnel.

7.1.1 Characteristics of fracture grouting

High-viscosity grout is introduced through valves installed in the ground
in such a way that the sum of the reachable voids in the surrounding soil
accommodates only a small percentage of the amount of fluid grout introduced. As far as equipment is concerned, this method requires mixers for
producing suspensions rich in solid content and pumps that achieve a sufficiently high pressure increase of the suspension which is accumulating in
the soil. After the fracturing pressure in the soil has been exceeded, cracks
open up in the soil which are widened immediately by the subsequent grout.
By injecting small amounts of solid substance per grouting operation and
by repeatedly pressurising individual grouting valve positions, it is possible
to achieve a grout framework of hardened solid veins and lamellae (see
Figure 7.1).

The lamellae as described have an irregular shape and a median thickness that can range from just a few millimetres to several centimetres. Soil
with large voids may need to be pre-treated in order to achieve a pressure
increase and, subsequently, to be able to carry out the process of hydraulic
fracturing. In order to control the development of 'fracs' in the ground
regarding their length, grout volumes are strictly limited and the flow characteristics of the grout are controlled by use of additives.

7.1.2 Construction and technical aspects

The use of hydraulic fracturing in construction technology has to meet
some important preconditions:

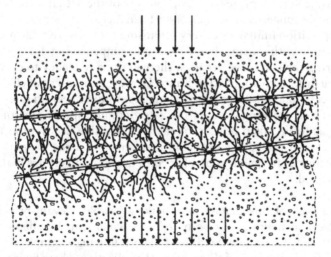

Figure 7.1 Supporting framework consisting of solid veins and lamellae for transferring
concentrated stresses.

(1) Basic project assumptions: The achievable improvement of the soil parameters has to be assessed just as realistically as the geometric relations of foundations subjected to load, which allows for systematic lifting of a structure. Two- to five-fold increases in stiffness can be achieved.

(2) Performance description: The performance description on which the works contract is based has to take into account realistic anticipated progress, with observation periods and work interruptions necessarily playing a part for organisational reasons.

(3) Time factor involved in the course of the construction work: A permanent soil improvement by hydraulic fracturing is achieved when the number of grouting phases is large. Consequently, the desired effect only occurs after the passage of a considerable amount of working time. For compensation grouting it is particularly important to include in the time schedule a suitable period for installing the grouting system and pre-treating the soil up to the point in time when the structure is ready to be lifted.

(4) Measuring technology: The method is controlled by means of well-developed measuring technology, which makes it possible to observe both surface movements as well as any deformation in the subsoil. The reliability of the measured values and their early evaluation are essential preconditions for a successful application of the method.

(5) With hydraulic fracturing the soil is improved in small stages and with the objective of achieving a permanent increase in the lateral soil resistance. Consequently, only those tests which, both in terms of time and space, reflect the geometry of the actual application can provide conclusive evidence of performance.

(6) Application limits: After very promising results were achieved in connection with raising buildings and compensating for settlements during the last three decades, there exists a tendency to exceed previously known application limits. However, an essential element of a successful application is observing maximum injection rates per soil unit and working day. As far as economic aspects are concerned, it has to be taken into account that a sensible decision on the usefulness of the application can be taken only if the prevented damage and its frequently complex consequential effects are realistically assessed.

The purpose of this chapter is to make available relevant principles on which decision making can be based and which can be adapted to the respective specific project conditions. The following examples should be regarded as an overview of those areas of application where hydraulic fracturing has been used in recent years.

7.1.3 Basic applications of soilfracture grouting

(1) Modification of soil stiffness, such as soil homogenisation underneath a machine foundation (see Figure 7.2).
(2) Stabilisation of long-term or current differential settlements. For example, differential settlement of a church which had continued over centuries and eventually became a safety problem (see Figures 7.3 and 7.4).
(3) Compensation of existing settlements, such as levelling of structures affected by settlements in the vicinity of deep excavation pits or tunnel excavations (see Figure 7.5).
(4) Active compensation of absolute and/or differential settlements as they occur. For example, compensating for settlements above cavity structures, which are often built in several phases by means of traditional excavation (see Figures 7.6, 7.7, and 7.8).
(5) Conditioning and compensation for settlements that occurred as a result of excavation with tunnel-boring machines (see Figures 7.9 and 7.10).

7.1.4 Existing bibliography

It should be noted that because of the spectacular results, many cases of the application of compensation grouting have found access to international

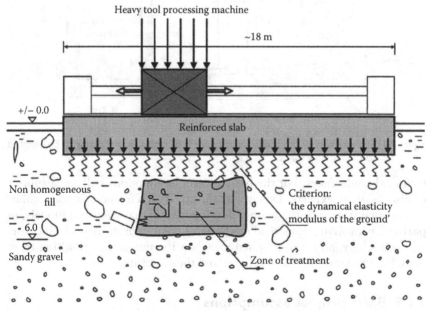

Figure 7.2 Consolidation of a demolished, inadequately refilled basement underneath the foundation of a precision machine, with the objective of homogenising the dynamic deformation modulus of the soil.

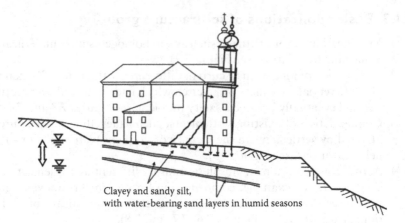

Clayey and sandy silt,
with water-bearing sand layers in humid seasons

Figure 7.3 Shrinkage processes caused by cyclic drying of the soil lead to differential settlement occurring in stages.

Longitudinal section   Plan view

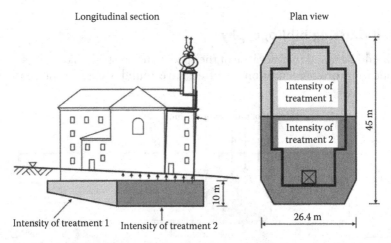

Figure 7.4 The intensity of the stabilisation process by hydraulic fracturing to be adapted to the existing loads.

technical conferences, but the basic project assumptions still are a subject of discussion by experts. This is why the attached references contain publications that deal with the further development of mathematical models, in particular involving numerical methods, current improvements in measuring technology, and well-founded views on the interaction resulting from forced deformation between the soil and the structure.

7.1.5 Basic project assumptions

In principle, it has to be stressed that hydraulic fracturing consists of imposing soil deformation by injecting grout rich in solid components.

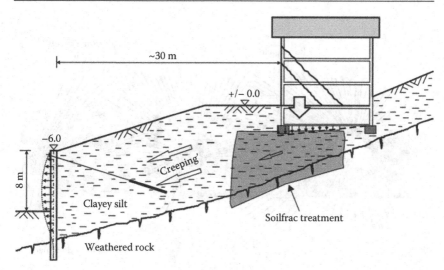

Figure 7.5 Stopping creep movements after deformation in the retaining structures had led to cracks in existing structures.

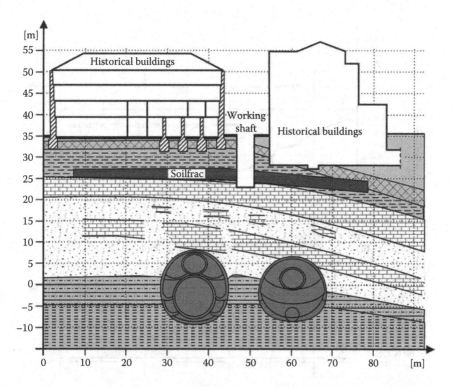

Figure 7.6 Compensating for settlements above two station buildings for an underground railway which was built in stages over a period of approximately two years.

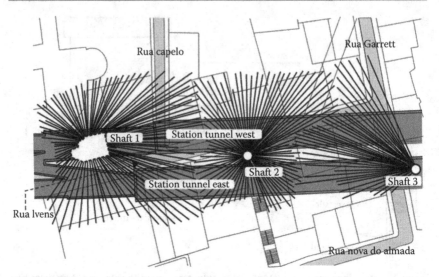

Figure 7.7 Arrangement of shafts and drilling for the installation of grouting pipes for settlement correction in densely built-up area of approximately 15,000 m².

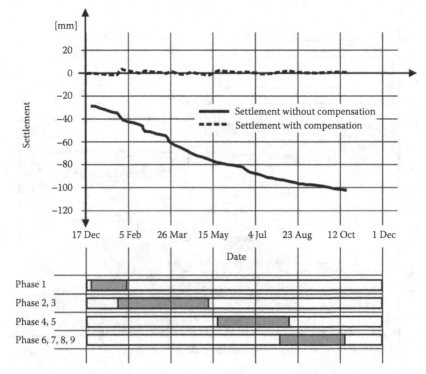

Figure 7.8 Illustration of the average compensation success, indicating the compensation procedure in small stages during the individual excavation phases.

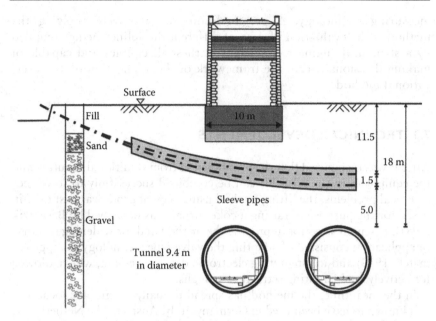

Figure 7.9 Compensation grouting for the protection of a railway bridge with 9.4-m twin tunnel excavation and differential settlement limits of 1/3,000.

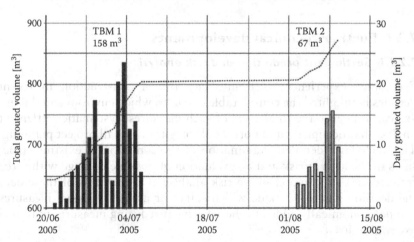

Figure 7.10 Grouted quantities for first and second TBM drive. The mean advancement rate was 18 m/day.

The criterion for effectively applying the method consists of being able to control the movements in the soil as well as the interaction between such movements and the structure concerned. The more or less complex requirements demand close interdisciplinary cooperation between geotechnology, structural analysis, structural process technology, and measuring technology. A pre-condition for effectively applying the method is the evaluation of data of different disciplines being compiled by a structural engineer competent in these disciplines and capable of making decisions within the framework of the application of an observational method.

7.2 TECHNICAL DEVELOPMENTS

After it was recognised that the phenomenon from traditional grouting and the equipment in the oil fields can be combined successfully to solve geotechnical problems, the lifting method using cement grout was first used in Essen for the purpose of raising a coke furnace, as described by Bernatzik (1951). Essential steps for improving the method and for widening its range of application consisted of adapting the valve pipe technology (in approximately 1970) and in integrating electronic data processing, which allowed for actively compensating settlement troughs.

In the meantime, the method has spread to many geographic locations. It is known to have been used in Germany, Italy, Austria, The Netherlands, Portugal, Spain, Belgium, United Kingdom, United States, Canada, Australia, and Puerto Rico.

7.2.1 Further technical developments

7.2.1.1 Settlement prediction and risk analysis

Ever since experience with inner-city tunnel construction in recent decades highlighted the considerable extent to which environmental damage can influence the total costs of such measures, systematic settlement analysis has occupied an important place within overall project planning. This includes a detailed examination of the condition of existing structures in the area at risk and an evaluation of possible damage within the framework of a comprehensive risk analysis. On the basis of these data the decision has to be made which stages or areas additional measures can be economically added to the settlement-reducing measures in tunnel excavation itself.

7.2.1.2 Sleeve pipe technology

The improvement in the grouting system consisting of long-life valve pipes and double packers made it possible to grout individual sleeves many times; to use long pipe lengths and, if necessary, to carry out controlled drilling operations. High-quality pump control devices do not only allow a large number of decisions to be made on parameters, but also enable these data

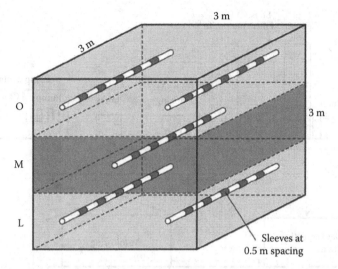

Figure 7.11 Sleeve pipes can be arranged on several levels and can serve different purposes. This figure shows a detail of a lifting mat used in compensation grouting with two horizons: 'O' and 'L' for the purpose of stress distribution, plus a central lifting layer 'M'.

to be recorded and automatically presented. High-performance drilling methods and the use of flexible drilling assemblies gives the option of the use of shafts and also the adaptation of existing working areas such as excavations or tunnels, which permit the method to be used in confined inner-city spaces as well (Figure 7.11).

7.2.1.3 Soilfracturing combined with other methods

To meet difficult requirements and to find economical solutions, the method has already been developed and used in combination with several geotechnical methods: compaction grouting, jet grouting, pipe roofs, hydraulic jacking, floating pile foundations.

7.2.1.4 Data processing

Recording and storing of data obtained in grouting itself and from the monitoring-based observations are not sufficient for ensuring a professional application of the method. Only the use of professional visualisation programmes and the combination of data by means of individual software modules makes it possible for the site manager to take real-time decisions on grouting programmes that have to be modified continuously (Figure 7.12).

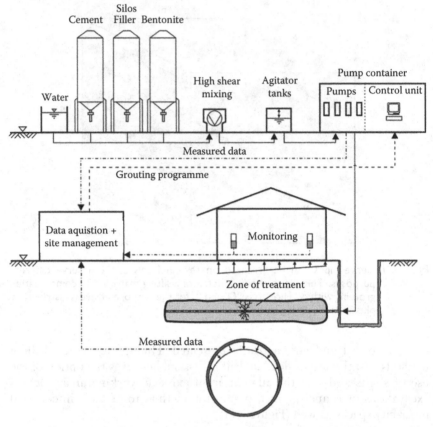

Figure 7.12 Data on deformation measurements and parameter recording are used by the site manager for determining actual grouting programmes.

7.3 EQUIPMENT

7.3.1 Installation of grouting system

7.3.1.1 Drilling technology

The sleeve pipes (also: Tube à Manchettes, TAM) can be installed in boreholes stabilized with mud-flush or by a casing. Both the rotary drilling method using a down-the-hole hammer and methods driving the casing with an external hammer are used. In soft stable soils it is also possible to apply auger drilling methods. The resulting spoil material is conveyed by air pressure or water flushing. In special cases it is necessary to use directional drilling where long bores with limited deviation tolerances are required. Moreover, directional drilling enables curved drilling. The curvature that can be achieved must be larger than 120 m.

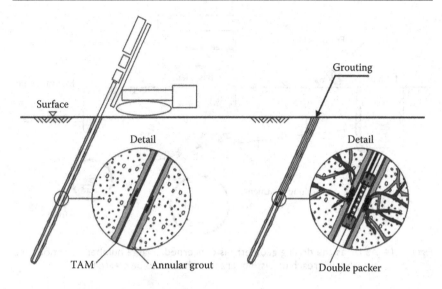

Figure 7.13 Sleeve pipes (TAMs) are installed in a stable (cased) borehole and sealed by an annular grout. Individual valves can be pressurised by double packers.

In the case of double-packer technology, it is advisable to limit the pipe length to approximately 50 m, although borehole lengths exceeding 75 m have been successfully achieved. It should be noted that in the case of extended drilling length the general application problems increase disproportionally. Directional drilling can be used for these cases to limit drilling deviation. In the case of controlled 100 m boreholes, the double-packer technology can be replaced by 'no-return' valve technology (Figure 7.13).

7.3.1.2 Borehole setups

Efficient drilling masts for producing boreholes in excess of 20 m comprise lengths of at least 3 m, but preferably approximately 5 m. Any drilling shaft, trench/excavation, and working tunnel should be designed to be at least 1 m wider than the length of the drilling masts. Arrangements for orderly discharge of flushing water containing cement have to be included in the works schedule, just like the safe introduction and lifting of the drilling and grouting equipment. All safety requirements have to be observed during the different operating phases and are normally combined in a safety schedule which is pointed out to those participating in the project prior to the commencement of work. If shafts have to be located in the direct vicinity of the area where the lifting operations take place, it may be necessary to provide a dilatation joint, if necessary even underneath the water table (Figure 7.14). Typical drilling equipment for a shaft is shown in Figure 7.15.

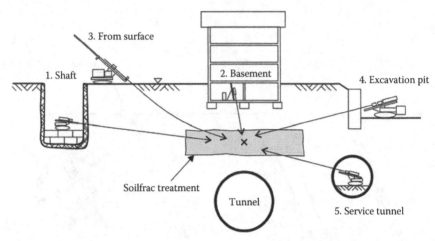

Figure 7.14 As far as the drilling geometry is concerned, a large number of options are available for reaching distinct areas in the soil to be treated.

Figure 7.15 Drilling rig for a 3.5-m diameter shaft.

7.3.2 Preparation of grout suspensions

7.3.2.1 Storage system

The storage system must facilitate separate storage of the major components required for mixing suspensions. It is necessary to provide steel or plastic tanks for storing water in drinking-water quality, silos with a capacity of at least 20 tonnes for bonding agents (cement, lime, fly ash) and filler material (limestone, slag, bentonite) as well as small containers for additives.

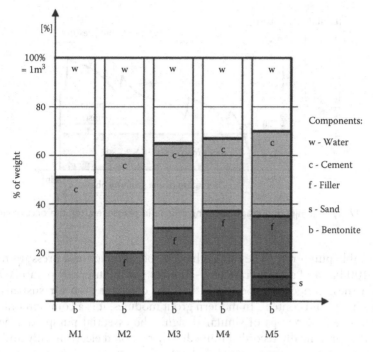

Figure 7.16 Examples of suspension compositions that can be used in different soils and operations.

7.3.2.2 Mixing technology

It involves the use of so-called colloidal mixers or high-frequency mixers which allow the homogeneous mixing of suspensions with a water/solid ratio of 0.45. Even with grouts with a high solid content and bentonite added, the mixing capacity has to ensure an adequate supply of material to the proposed number of pumps (Figure 7.16).

7.3.2.3 Stored quantity of suspension

Limited quantities of 500 – 1000 litres are stirred in agitator tanks with electronic control devices ('multirangers') to ensure that grouting units are supplied continuously and that larger suspension quantities are prevented from being 'stirred dead' due to any interruption in the grouting operation.

7.3.3 Grouting technology

7.3.3.1 Grouting pumps

Modern grouting pumps provide different control options, thus permitting an automatic reaction on the development of pressure at a constant

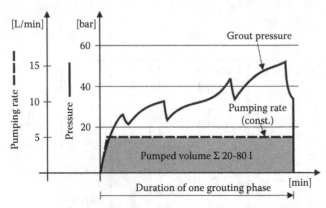

Figure 7.17 Typical injection graph showing a drop in pressure after the occurrence of 'fracs'.

or variable pump rate. The pumps have to be suitable for a pressure range of 0–100 bar and a pump rate of 1–20 litres per minute; any parameter set within these ranges is expected to be kept constant even for suspensions with a high solid content. In modern grout modules, it is a common practice to combine 2–8 pumps of similar design. The essential pump parameters are either graphically recorded immediately or stored electronically and, via software programmes processed in databanks and printed out. In any case, the type of data saved has to ensure that in each individual grouting operation the pressure and quantity ratios must be clearly associated with the respective location of the grouting operation. This is the reason why modern grouting data recording and visualisation programmes are coupled; they permit an early interpretation of data (Figures 7.17, 7.18, and 7.19).

7.3.3.2 Sleeve pipes

Pipes typically available consist of PVC, fibre glass, or steel. Their diameter ranges from 1–4 inches and the distance between valves amounts is 0.33–1 m. In special cases, the steel pipes are reinforced and the rubber valves are protected by steel rings or special covers in such a way that either the pipes can be driven directly or that the function of the reinforcement of the soil, for example the shape of the pipe umbrella, is supported. In principle, by surrounding the grouting pipes with a so-called skin-forming grout (stable, low-strength, but stiffened sleeve grout) a direct connection between the individual grouting valves is prevented.

7.3.3.3 Packer system

Double packers are used to limit the grouting operation to one specific valve of the grouting pipe (Figure 7.20). The packer elements consist of

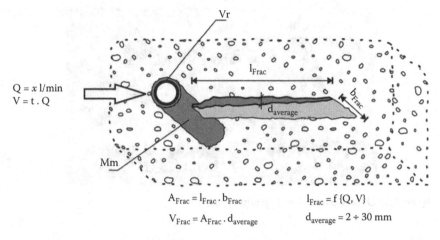

$$A_{Frac} = l_{Frac} \cdot b_{Frac} \qquad\qquad l_{Frac} = f\{Q, V\}$$
$$V_{Frac} = A_{Frac} \cdot d_{average} \qquad d_{average} = 2 \div 30 \text{ mm}$$

Figure 7.18 A frac usually has a flat, oblong shape and starts from an initial crack in the sleeve grout. However, frequently there occur secondary fracs whose geometric description can be given in statistical form only.

a wire mesh and have a rubber surface, and are inflated either with compressed air or water.

The pressure applied has to be clearly higher than the maximum grouting pressure expected. After completion of the grouting operation, the grouting pipes are cleaned by high-pressure hoses integrated into the packer system or by separate cleaning systems. The quality of the individual components and their careful use are the key to an efficient process application and their successful use over longer periods of time.

Figure 7.19 Coloured frac sample deriving from grouting in cohesive soils.

Figure 7.20 Double packer for 2" grouting pipes.

7.3.3.4 Data recording

A compensation grouting operation generates a huge amount of data, which has to be suitably administered and made available to the site management in order to make appropriate decisions. For instance, there have been applications in the past with more than 100.000 individual grouting operations and the continuously measured values of more than 100 measuring elements had to be recorded over a period of more than one year.

Associating the data in terms of time and place is just as important as combining the effects, which should start from the excavation stages—from the grouting measures and the influence of individual regions of the structures positioned above. The objective is therefore to generate a simplified presentation of the structures and the soil in a form that shows the interaction between excavation, grouting operation, and structure. When designing the measuring system, care has to be taken to ensure that the actual measuring accuracy and the frequency of data recording indicate movement tendencies. Important external influences such as temperature variation have to be filtered out.

Valuable tools in the application of compensation grouting are graphical display systems which, for example, at any time show the development of pressure ratios during grouting or the distribution of quantities inside geometrically limited units.

7.4 THEORY AND DESIGN

Injecting solid material into the soil leads to deformation on all sides. The directions in which individual injections spread largely depend on the homogeneity conditions in the soil. From a statistical point of view, it has to be assumed that the greatest part of the volume introduced into the soil

leads to deformations whose amounts are distributed proportionally relative to the respective stress conditions. Only a small part of the movement
rates is caused by the compression of the existing soil, as the highest effective injection pressure cannot greatly exceed the magnitude of the highest
existing standard stress. However, locally and over a short period of time,
higher forces can become effective if tensile forces within the grout skeleton
which have already been set and cohesion forces in the soil are activated
(Figure 7.21).

Experience has shown that if a limit rate with respect to the injected
quantity per unit time within a limited soil volume is exceeded, said forces
are overcome and, in consequence, the time-dependent deformation resistance is clearly reduced. This conclusion is based on the observation that
the efficiency of injected quantities is reduced in the course of a lifting
operation, if the quantity injected per working day is increased excessively
in order to try to achieve a greater lifting speed. The above-mentioned limit
balance between effective injection pressures applied (not the injection
pressure measured at the pump) and the annularly acting pressure forces
in the surrounding soil depends on the respective soil characteristics to
such an extent that global recommended values for reliable injection quantities cannot reasonably be given. It is advisable to monitor limited injection
areas by making use of the available measuring technology and to determine the achievable 'grout efficiency factor' (defined as the ratio between
'heave volume' monitored on surface to grouted volume) and lifting speed
by varying the injection parameters (Figure 7.22).

Clear information on the deformation rates and their directions as a
result of the injection of solids cannot reliably be obtained by only observing surfaces. Information from the soil is essential, with extensometers and
inclinometers able to provide useful service. For the purpose of checking

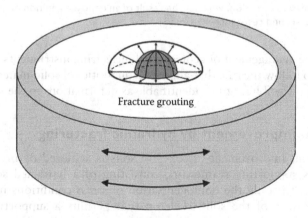

Fracture grouting

Figure 7.21 The model of the 'confining' ring comprises the sum of all forces that allow
for a central lifting injection.

Phase	Horizontal (H)	Vertical (V)
Pre-treatment	100%	0%
Multi-stage injection before actual heave	95 ÷ 100%	0 ÷ 5%
Heaving phase	75 ÷ 95%	5 ÷ 25%

$$V_{Heave} = (0.05 \div 0.25) \times V_{inj.}$$

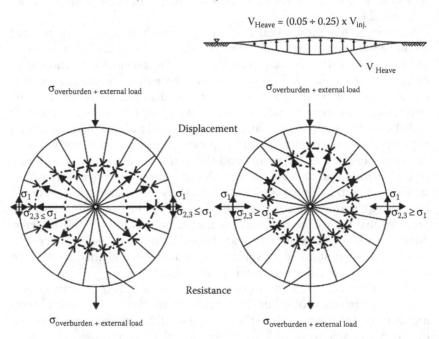

Figure 7.22 Heterogeneities in the soil and the stress distribution influence the deformation direction, which is the result of an imposed addition of material and all-round resistance.

the accurate arrangement of geotechnical measuring instruments, it is conservative to follow the rule that almost all quantities of solid material introduced into the soil have to be identifiable as deformations in the soil.

7.4.1 Soil improvement by hydraulic fracturing

A permanent improvement in cohesive soils is achieved by producing a continuous supporting framework consisting of a hardened solid skeleton. Since it is only the homogenisation of stress conditions in the soil and the closure of the solid skeleton that permits a supporting effect independent of the original soil, the improvement curve of a multiphase injection application is not to be regarded as being linear by any means.

The stresses acting on foundations of up to approximately 1000 kN/m²
typically occurring in structures can often only be transferred into the
ground if soils are improved (e.g., by soil fracturing). It is possible to
transfer stresses in the range of 2.0–3.0 MN/m² with negligible deformation rates implementing fracture grouting measures. Borderline cases are
soft soils. Whereas effective consolidation has already been achieved in
peat layers, there are no examples yet for very soft and structureless types
of soil, although in such cases, the selection of a suitable injection agent
promises success, too.

Soil improvements by hydraulic fracturing are always carried out with
the intention of generating a controlled stress flow in the ground. Because
of the low material-strength values of the soil and the intention to include
said soil in a supporting system, it is frequently necessary to increase the
initial soil stiffness by 2–5 times the existing value.

As the deformation method described involves the displacement of large
soil masses by small individual amounts, the respective plans must take
account of the principle that slim deformation elements should not be considered in planning such methods. Figure 7.23 shows basic geometric relations that result from experience with different lifting operations. In such
considerations the different load intensities of adjacent foundation inter-
faces and the different depths at which they are located also play important
parts.

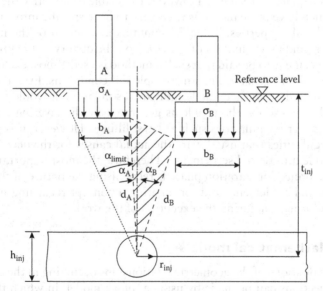

Figure 7.23 Recommendations for geometrical relations and load intensities to permit
controlled levelling: α_{limit} = 10–25°; $d_A/d_B \leq 1.5$ with $\alpha_A + \alpha_B \leq 20°$; $b_A/d_A \geq 1$
resp. $b_B/d_B \leq 2$; $\sigma_A/\sigma_B \leq 3$ with $\alpha_A + \alpha_B \geq 20°$. Situations exceeding these
indications need additional verification or practical trial.

Table 7.1 List of parameters with significance for project considerations

Designation	Short symbol	Unit	Importance
Grain-size distribution	–	mm	1
Coefficient of non-uniformity	U	–	1
Moisture content	W	–	1
Porosity	n	–	1
Void ratio	e	–	1
Relative density	D	–	1
Consistency index	I_c	–	1
Liquidity index	I_L	–	1
Coefficient of permeability	k	m/s	2
Young's modulus	E	kN/m²	2
Shear strength parameters	φ', c', φ_r	°, kN/m², °	2
Undrained shear strength	C_u	kN/m²	3
Model of rheology			3

1 – Absolutely necessary
2 – Important for calculations
3 – Significant for numerical models

7.4.2 Soil description

In principle, hydraulic fracturing is suitable for improving all types of soil with an adequate consistency. However, to be able to quantify the requirements which have to be met, it is necessary to have specific information on the initial soil properties. Table 7.1 contains a selection of the necessary parameters and an evaluation of the effect on the quality of the prediction. In principle, it has to be said that such a method is used almost exclusively in difficult nonhomogeneous and anisotropic soil conditions. Even if it would be possible to describe accurately individual soil layers, it would be almost impossible to describe the problems generated by the layering effect and by even greater irregularities such as karst fillings. However, it is precisely these irregularities that usually are the actual causes for the occurrence of considerable differential settlements. Therefore, the most important objective of the project preparation phase is recognising the nature of the causes of differential settlements and formulating a concept regarding necessary further information during the execution of the work.

7.4.3 Mathematical models

Simple evaluations of the geometric relationships occurring in the course of lifting injections can be made by using a 'block model' in which the zones of different treatment intensities are modelled in the form of an idealised 'confining' stress ring, and the lifted zone in the form of a centrally positioned lifted piston.

The presently available numerical calculation methods allow the modelling of many deformation phenomena in the soil and their interactions with structures positioned above. In the past, a considerable impediment regarding the simulation of the lifting injections has been found by the continuous change of the input parameters as the soil properties change during the treatment of the soil.

Recent publications describe two- and three-dimensional analytical and numerical solutions for modelling the effect of compensation grouting. With these models, case histories have been investigated matching the real situation in a considerable manner.

A back analysis of fracture grouting applications was presented by Schweiger et al. (2004). The numerical model adopted simplified analytical and numerical models according to the finite element method for the settlement correction simulation for the central station of Antwerp. The Finite Element Mesh for the calculation is represented in Figure 7.24.

With the proposed model, it was shown in a reasonable manner that FEM calculation can be valid for the prediction of the compensation effect (Figure 7.25). However, the prediction of exact grouting quantites is still difficult as numerical models do not take into account various effects of the practical grouting behaviour (filling of voids, variable grout efficiency, effects of the grouting parameters, and so on).

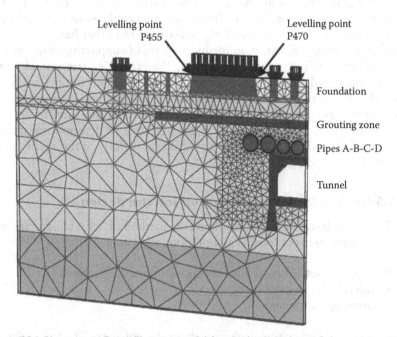

Figure 7.24 Plane strain Finite Element model for the back analysis of the construction stages for the tunnelling at Antwerp Central station.

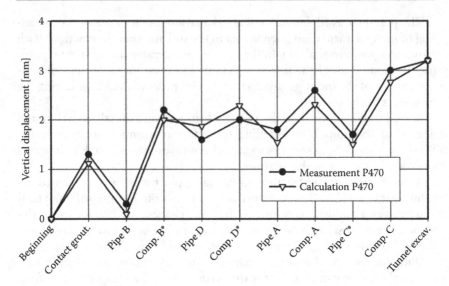

Figure 7.25 Comparison between field measurements and FEM back analysis for the compensation grouting work of the Antwerp Central station.

7.4.4 Monitoring

The concept for comprehensive deformation measurements always includes the two completely different sets of problems of the structure on the one hand and of the soil underneath on the other hand. As far as structures are concerned, monitoring the level of supporting components is of primary importance. In addition, floors and existing cracks can be provided with instruments whose values are read automatically or visually. Measuring instruments such as inclinometers, extensometers, incremental extensometers, settlement piezometers, and similar devices can be arranged in the ground with the objective of determining spatial deformations and their directions in the direct vicinity of the area treated (Figures 7.26 and 7.27).

7.4.5 Basic information required for the design

To be able to design a project involving the method of hydraulic fracturing, it is necessary to make reliable assumptions regarding

- Expected settlements, when excavations are encountered
- Tolerable settlements
- Grouting efficiency

Moreover, the pretreatment phase, where the voids are filled in order to condition the soil, and preheaving have to be taken into account.

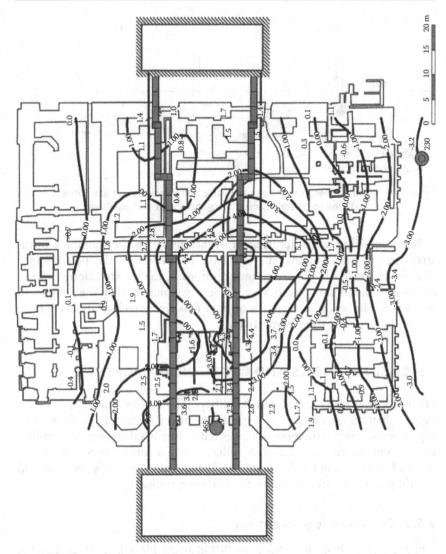

Figure 7.26 Example illustrating the deformation of a historic railway station which was continuously monitored by means of a complex measuring system.

7.4.6 Performance elements

The plan for the application of hydraulic fracturing is normally followed by a performance description which is used as a basis for the works contract to be concluded with the contractor. Table 7.2 contains a list of normally occurring working phases and a selection of necessary specifications. In addition, there are proposed units according to which the very different services have to be evaluated, depending on the type of project.

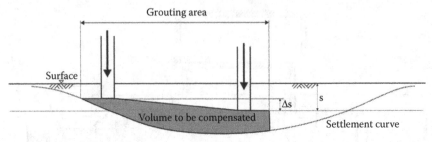

s...Allowable absolute settlement
Δs...Allowable differential settlement

Figure 7.27 Actual settlements and the allowable deformation state basically determine the volume to be considered during the compensation phase.

When applying soil fracturing in connection with very complex infrastructure measures, it may be advisable to include in the performance description regularly occurring interruptions in the construction work; in addition, a distinction has to be made between phases when the equipment is ready for injection and phases of pure measuring technical observation and equipment idle time, respectively.

7.5 CASE HISTORIES AND LIMITS OF APPLICATION

Hydraulic fracturing can be used for different objectives and types of settlement reduction operations. The spectrum ranges from pure prevention or repair works after settlement damages have already taken place to simultaneous compensation of excavation induced settlements for inner-city areas. A number of examples illustrate the opportunities for this method and should provide creative ideas for the planning engineer.

7.5.1 Settlement prevention

Soil improvement by hydraulic fracturing takes place to homogenise the stress conditions in large soil volumes or to increase the load-bearing capacity of soils to enable them to support additional loads. It is important to point out the economic nature of this method compared to deep foundations systems because a sufficient degree of soil improvement has been reached for the load acting on the soil-structure interface when initial heave is achieved.

7.5.1.1 Increasing the load-bearing capacity of soils

In the case of already over-loaded strip foundations of a three-story building it was decided to improve the soil down to 8 m depth to enable the

Table 7.2 Performance elements of compensation grouting

Performance element	Important specifications
Licences for drilling areas and construction site equipment, access to structures for exploration purposes	Duration of use/number of measuring points to be provided
Exploring installation situation	Type of exploration Safety measures
Recording details of structures	Type of structure
Preparing and maintaining drilling surfaces	Shaft depth and diameter Types of excavation/expenditure for adapting existing spaces Access situation
Setting up drilling equipment	Mean and maximum drilling length Drilling method, borehole diameter Diameter of valve pipe to be built in, spaces available
Execution of drilling operations installation of sleeve pipes	Borehole diameter, diameter and wall thickness of pipes, material properties, distance between valves, valve characteristics
Filling the annular space	Suspension mixture data
First injection	Number and sequence of individual injections, quantity injected per injection operation
Mobilisation of grout station	Number of injection units to be operated separately
Multiple injection; preconsolidation and 'conditioning'	Number of injection operations and quantity to be introduced per injection operation, minimum and maximum pump rates
Prelifting and injection up to point when the structure is ready to lift the structure, 'contact'	Number of injection operations and quantity to be introduced per injection operation, minimum and maximum pump rates
Lifting 'compensation injection'	Listing lifting stages and lowest and highest settlement rates to be compensated for per shift
Installing an automatic measuring system	Type and number of measuring elements to be installed
Providing and maintaining the measuring system for the duration of the project and/or for the period of continued observation	Assessing the required functional periods, required accuracies, measuring frequencies
Data recording and visualisation	Providing, in a standard form, the injection parameters and data from monitoring, frequency and type of data sets to be handed over
Technical site management	Interpreting the measured data jointly with designer, setting up individual injection programmes/evidence of experience with similar construction situations

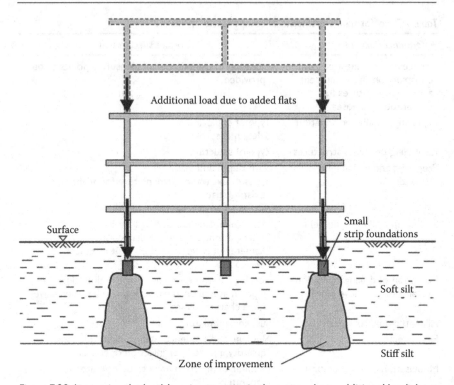

Figure 7.28 Increasing the load-bearing capacity in the areas where additional loads have to be carried due to addition of two storeys.

building to support, free of undue settlements, additional loads resulting from the construction of additional storeys (Figure 7.28).

7.5.2 Passive settlement reduction

If differential deformations are expected, a pretreatment with the objective of improving the soil can considerably reduce the absolute extent of such a deformation and, if necessary, allows an active correction of actually occurring movements. Normally, the costs of such a measure are significantly lower if the installation of the sleeve pipes is already included in the construction design.

7.5.2.1 Deformation reduction for deep excavation pits

The most economical variant of securing a construction pit by means of an anchored diaphragm wall was likely to result in an extent of horizontal deformation that was unacceptable for the adjacent buildings. As an alternative to secondary underpinning of foundations by jet grouting, it

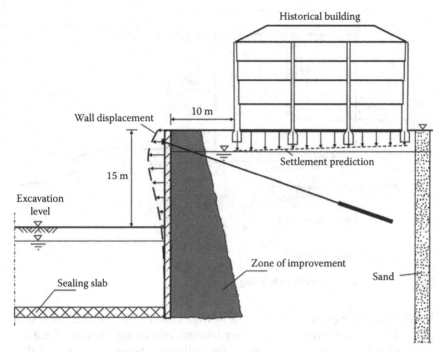

Figure 7.29 Passive settlement reduction by carrying out specific soil improvement in the area subject to deformation.

was considered to improve the block of soil behind the diaphragm wall by means of soil fracturing which, if needed, would also allow for volume losses to be compensated during the excavation phase (Figure 7.29).

7.5.3 Back levelling of structures

There are several traditional methods for performing repairs of settlement-damaged buildings, including the use of piling systems, jet grouting, and hand digging. Most methods have complications connected with the use of working areas inside the building. In addition, lifting of buildings by hydraulic jacking is a very complicated operation that even if applied accurately requires a very sophisticated control technology and a sufficiently intact and stable structure. By adapting the soil fracturing technology to small-scale applications, it is possible to obtain a technically and economically interesting alternative and the access to the building is limited for observation purposes only.

7.5.3.1 Stabilisation of a peat layer

A wedge-shaped layer of peat existing underneath the foundation slab has not been properly identified during the construction of a multi-storey

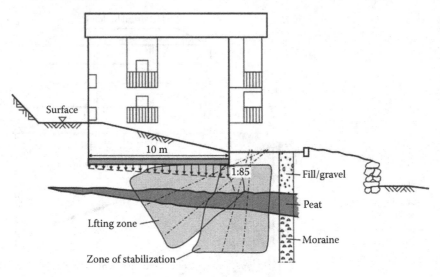

Figure 7.30 Lifting an apartment building.

building and as a consequence a shallow foundation with a concrete slab was executed. The resulting differential settlements led to a maximum inclination of 1:85, and thus the serviceability of the building was questionable. On this occasion it was successfully shown that for the first time organic soil can also be improved by hydraulic fracturing, although at higher costs (Figure 7.30).

7.5.4 Concurrent and observational compensation grouting

Ever since electronic measuring systems have allowed a real-time deformation measurement, it is possible to actively compensate for concurrently developing settlement troughs by means of soil fracturing. In this context, it is essential to assess realistically the total settlement to be expected as well as the highest settlements occurring during a limited period of time. Obviously, minimising settlements by excavating a tunnel in a suitable manner constitutes the essential element of an environmentally acceptable construction method in inner-city areas.

7.5.4.1 Compensation for settlements above conventional underground excavations

Over a period of 18 months, historic structures had to be protected from differential settlements caused by station tunnels with internal diameters of up to 20 m. Total settlements in excess of 100 mm were compensated in lifting stages of 2 mm per lifting operation, so that the position of the

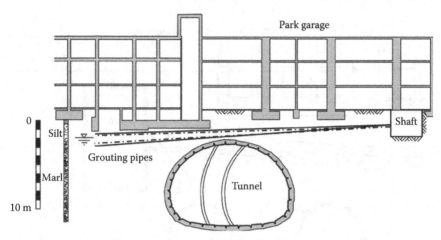

Figure 7.31 Settlement compensation for a car park with very small cover.

building foundations prior to the tunnel driving operation remained almost unchanged (Figures 7.6, 7.7, and 7.8).

Another case study was the excavation of a large tunnel underneath a car park where the cover between the foundation and the shotcrete lining had a minimum distance of less than 1 m. The 45-m-long grouting pipes were installed from a shaft inside the building without interference by means of the grouting operation the tunnel excavation was performed with limited settlements of about 7 mm (Figure 7.31).

7.5.4.2 Compensation for settlements above tunnels excavated with tunnel-boring machines

Excavating tunnels with tunnel-boring machines differs fundamentally from using mining methods, mainly with respect to time characteristics. Because of higher driving speeds up to 25 m/day and occasionally abrupt interferences in the zone of the earth pressure support, greater attention has to be paid to passive effect of compensation grouting. As a result of the number and types of sleeve pipes, it is possible to additionally use their reinforcing effect or the soil. Furthermore, an extensive monitoring system extended over a sufficient area contributes considerably towards controlling shield machines when installing a suitable communication system.

One of the major compensation grouting jobs in Europe was the protection of approx. 22,000 m² of urban area in the centre of Leipzig, Germany. The twin tunnel with a diameter of 9 m had a cover ranging from 7.5 to 15.6 m. The geological profile was varying from soft soils with boulders to rock. In total, 13 shafts were constructed with a diameter from 3.5–6.5 m, with the level of grouting pipes at 10 m depth from ground surface and water pressures of 8 m. For the 60 buildings, an allowable settlement of 10 mm

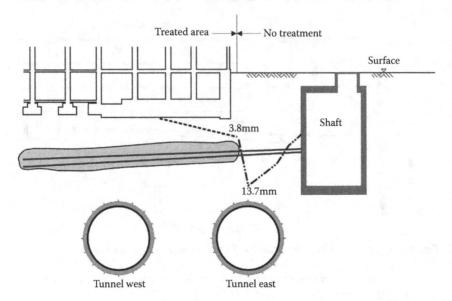

Figure 7.32 Cross section of the City Tunnel Leipzig project with settlement curve showing the difference between nontreated and grouted zone.

was specified. More than 30,000 m of drilling and 1,100 liquid levelling points were installed. The settlement reduction can clearly been seen in Figure 7.32. In the areas outside the soilfrac treatment, a settlement of 13.7 mm occurred wheareas in the treated areas only 3.8 mm were registered.

A recent case of compensation grouting for a TBM excavation was made for a Metro Line in Rome. Fracture grouting substituted the initially considered jet grouting operation, as less impact in terms of occupied public space and discharge of spoil was given and in addition an active control of the building protection was possible. Two EPB-tunnel-boring machines passed underneath masonry and concrete buildings with a mimimum cover of less than 3 m. More than 6,000 m of directional drillings were made, settlements introduced due to drilling were about 1.5 mm although the distance to the foundation was only 1 m in certain cases. The measured settlements were less than 5 mm, significantly below the limit of 10 mm, see Figures 7.33 and 7.34.

7.5.5 Application limits

The large number of practical applications shows that soil fracturing can be applied under very different conditions with technical and economic advantages. As far as improving soft soils is concerned, the application limit is reached in very soft and organic soils. If the method is to be applied in waste disposal, it is necessary to carry out a detailed preliminary chemical investigation. Regarding the geometrical conditions, it should be pointed out once again that the treatment mainly modifies the stress flow in the soil. The concept of the design should therefore be formulated in the sense of an improved and controlled stress transmission and should not be reduced to the limited dimension of the foundation elements. The assumption that the

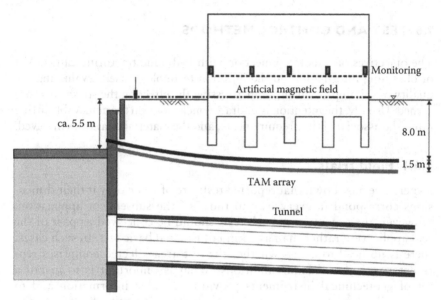

Figure 7.33 Cross section of the Metro Rome settlement protection by means of soilfracturing.

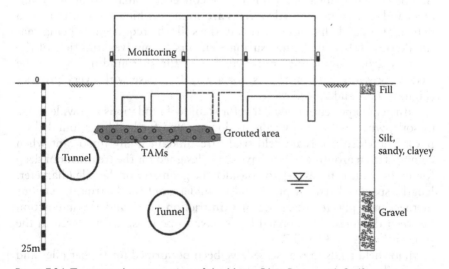

Figure 7.34 Transversal cross section of the Metro B1 in Rome with Soilfrac area.

dimension of the treated soil body must permit the deflection of the stress occurring can be the simple guide principle.

If there is doubt, large-scale tests have always been useful. These can be integrated into subsequent main works and thus do not incur substantial additional costs.

7.6 TEST AND CONTROL METHODS

The objectives of tests in connection with hydraulic fracturing can consist of obtaining information on the injection technology itself, evaluating the quality of soil improvement, or in assessing the effect on the structures concerned. Usually, the intention is to draw conclusions on the achievable lifting rates, the associated time requirements, and the material quantities involved.

7.6.1 Field trials

Experience has shown that reported tests are of value only if their dimensions correspond in every way to those of the subsequent application. However, the execution of such tests should be conceived as part of the real application rather than as a kind of isolated basic test. In such cases, there is no need to assess whether the test area selected comprises representative soil and load conditions. What is important is to install a set of geotechnical instruments providing reliable information and to record all relevant parameters. The documentation of such a test has to allow the evaluation of all parameters in terms of their time sequences. As far as geotechnical instruments are concerned, apart from inclinometers and extensometers, all systems are suitable which are able to record deformation and changes in stress in the soil, the requirement being that the deformability of the measuring elements themselves and that of the annular grout should be similar to that of the surrounding soil; in the case of pressure gauges, this is not always the case with every product (Figures 7.35 and 7.36).

Sufficient experience is available for shallow foundations in a wide range of soils, whereas less knowledge is documented for buildings founded on piles. Therefore real-scale field trials are fundamentally important when compensation grouting works have to be designed for the treatment of deep foundations. It is necessary to consider the geometry of the pile (diameter, length, spacing between piles) and to understand the bearing behaviour of the single pile (end-bearing of skin-friction type) and the interaction between the piles. The applied loads have to be those encountered in the project (Figures 7.37 and 7.38).

Major field trials in real scale have been performed for timber piles and bored piles (Figures 7.37 and 7.38).

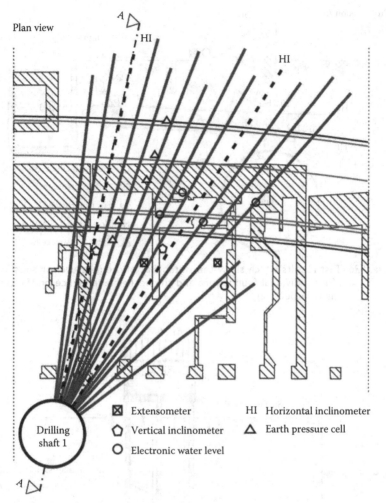

Plan view

HI

HI

Drilling
shaft 1

⊠ Extensometer

⬡ Vertical inclinometer

◯ Electronic water level

HI Horizontal inclinometer

△ Earth pressure cell

Figure 7.35 Layout of a large-scale field test which was planned as part of the subsequent settlement compensation measure.

7.6.2 Laboratory tests

The suspensions provided as a skin-forming mixture, and for the injection itself can vary considerably with respect to the number and type of components used and their composition. In view of the different reactions of similar bonding agents with different production origins, it is absolutely essential to carry out basic tests at the start of each project. The effect of additives can especially be influenced by the local properties of cement and filler materials. The tests are carried out with the objective of determining the flow properties, the bleeding of water and the setting behaviour as well as the stability of suspensions.

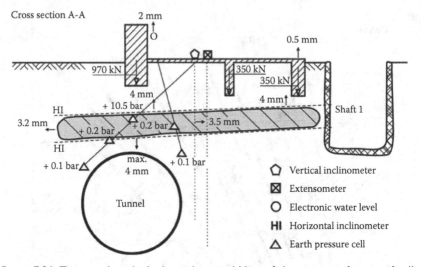

Figure 7.36 Test results which show the suitablility of the measure for specifically lifting individual foundations and the negligible influence on the tunnel shell, respectively.

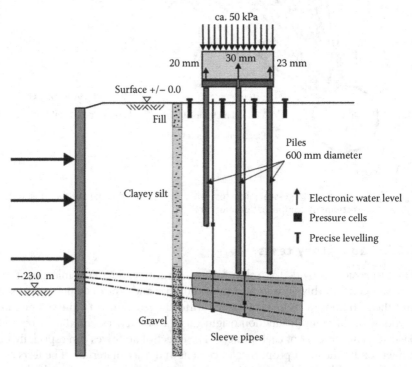

Figure 7.37 Cross section of the field trial performed to prove that piled foundations can be lifted with soil fracturing in a controlled manner.

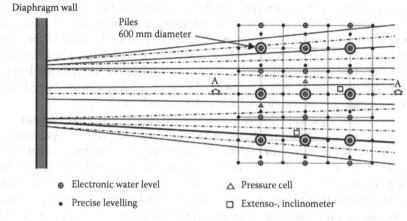

Diaphragm wall

Piles
600 mm diameter

⊕ Electronic water level   △ Pressure cell

• Precise levelling   □ Extenso-, inclinometer

Figure 7.38 Plan view of the field trial performed with the monitoring system.

Although it is important to determinate special technical characteristics, the most important premise remains the processibility of the suspensions using the equipment available on site. While it is a common practice to vary the suspension composition in the course of the staged execution of the soil fracturing work, it is advisable to set up a site laboratory. The objectives of suitability tests can be the production of 'soft mixtures' to achieve low-end strength values of granular soils as well as the production of 'harder mixtures' with the release of small amounts of water for the purpose of improving soft soils.

7.6.3 Monitoring technology

The measuring technology used must allow the measures applied in the soil and their effects on the soil to be very clearly associated with the structures concerned. It is important to ensure that any blurred measured value does not feign any movement tendencies which could lead to premature changes in the works programme. Therefore, in case of doubt, reference arrangements have to be used under controlled conditions. Such calibrations under realistic conditions can concern all types of electronic and visual settlement-measuring systems such as water level systems, automatic levelling instruments and theodolites, rotary lasers, precision levelling instruments, and special types such as floor-level measuring instruments, inclination instruments, and crackmeter devices.

7.7 ENVIRONMENTAL AND CONTRACTUAL ASPECTS

Although hydraulic fracturing mainly manifests itself in the form of movements in the soil and on structures, it is its ability to control the modification of the soil and of the respective condition of the structure on which attention is focused. When agreements are drafted in connection with complex structural methods, it is important to clarify at an early stage ownership conditions and the interests of those who may be affected. The chemical environmental compatibility of the individual injection components has to be proved. In cases of doubt, additional tests of the actually used combination of materials and products have to be carried out.

Agreements regarding the realisation of projects involving hydraulic fracturing are concluded on the basis of mutually agreed projects. As some of the effects of the measures applied can often not be specified at that particular point in time, the agreement must permit the consistent use of the observation method. The type of reaction to possible scenarios in the individual construction phases has to be planned and contractually permitted, of course with the intention of safeguarding the rights of third persons and the economic execution of the project in the interest of solving an existing problem. While taking into account the measures associated with strategies against undesirable effects, it should be noted that a performance schedule can list additional measures for limiting or warding-off injection effects. Equally, the acceptance of crack formations and arching taking place in the foot path region while a building is being lifted can be included in the working agreement as part of the overriding project objective.

REFERENCES

Bernatzik, W. (1951). 'Anheben des Kraftwerkes Hessigheim am Neckar mit Hilfe von Zementunterpressungen', *Der Bauingenieur* 26(4).

Brandl, H. (1981). 'Stabilization of excessively settling bridge piers', Florence: *Proceedings of the Xth ICSMFE*, pp. 329–336.

Chambosse, G. and Otterbein, R. (2001). 'Central Station Antwerp Compensation grouting under high loaded foundations', Istanbul: *Proceedings of the XVth ICSMFE*.

Droof, E. R., Tavares, P. D., and Forbes, J. (1995). 'Soil fracture grouting to remediate settlement due to soft ground tunnelling', San Francisco: Rapid Excavation and Tunnelling Conference.

Falk, E. (1997). 'Underground works in urban environment', Hamburg: *Proceedings of the XIVth ICSMFE*.

Falk, E. and Schweiger, H. F. (1998). 'Shallow Tunneling in Urban Environment – Different Ways of Controlling Settlements', *Felsbau* 16(4).

Haimoni, A. M. and Wright, R. H. (1999). 'Protection of Big Ben using compensation grouting', *Ground Engineering* 32(8):33–37.

Harris, D. I. (2001) 'Protective measures', London: Response of buildings to excavation–induced ground movements, CIRIA Conference.

Jakobs, M., Otterbein, R., and Dekker, H. (2001). 'Erfahrungen beim Einsatz der Druckschlauchwaage zur Höhenüberwachung setzungsempfindlicher Bauwerke', *Zeitschrift Bauingenieur* (76).

Kaalberg, F.J., Essler, R.D., and Kleinlugtenblet, R. (2011). 'Compensation grouting of piled foundations to mitigate tunnelling settlements', *Proceedings of 7th International Symposium on Geotechnical Aspects of Underground Construction in Soft Ground.*

Kudella, P. (1994). Mechanismen der Bodenverdrängung beim Einpressen von Fluiden zur Baugrubenverfestigung, Karlsruhe. *Veröffentlichungen des Institutes für Bodenmechanik und Felsmechanik der Universität Fridericiana.*

Kummerer, C. and Passlick, T. (2008). 'Kompensationsinjektionen als besondere Schutzmaßnahmen von Bauwerken bei Tunnelunterfahrungen'. *Tagungsband zum 23. Ch. Veder Kolloquium,* Institute for Soil Mechanics and Geotechnical Engineerging, Graz University of Technology, 131–144.

Mair, R. J., Harris, D. I., Love, J. P., Blakey, D., and Kettle, C. (1994). 'Compensation grouting to limit settlements during tunnelling at Waterloo Station, London.' *Tunnelling* 94:279–300.

Raabe, E. W. and Esters, K. (1993). 'Soilfracturing techniques for terminating settlements and restoring levels of building and structures.' In: *Ground Improvement,* Chapman & Hall, Glasgow.

Sagaseta, C., Sánchez-Alciturri, J. M., González, C., López, A., Gómez, J., and Pina, R. (1999). 'Soil deformations due to the excavation of two parallel caverns.' Rotterdam: *Proceedings of the Twelfth European Conference on Soil Mechanics and Geotechnical Engineering.*

Samol, H. and Priebe, H. (1985). 'Soilfrac – ein Injektionsverfahren zur Bodenverbesserung', San Franciso: *Proceedings of the XIth ICSMFE.*

Schweiger, H. F., Falk, E. (1998). 'Reduction of settlements by compensation grouting – Numerical studies and experience from Lisbon underground', Brazil: *Proceedings of the World Tunnel Congress 1998 on Tunnels and Metropolises, São Paulo.*

Schweiger, H. F., Kummerer, C. Otterbein, R. (2004). 'Numerical modelling of settlement compensation by means of fracture grouting', Soil and Foundation, *J. of Japanese Geotechnical Society,* 11(1): 71–86.

Sciotti, A., Desideri, A., Saggio, A. Kummerer, C. (2011). 'Mitigation of the effects induced by shallow tunneling in urban environment. The use of "compensation grouting" in the Underground Line B1 work'. *Proceedings of the Seventh International Conference on Geotechnical Aspects of Underground Construction in Soft Ground, Rome* (in print).

Van der Stoel, A. E. C., Haasnoot, J. K., Essler, R. D. (2001). 'Feasibility of Compensation Grouting of Timber Pile Foundations to Mitigate TBM Settlements, Full Scale Trial North/South Line Amsterdam', London: CIRIA Conference.

Chapter 8

Compaction grouting

James Hussin

CONTENTS

8.1 INTRODUCTION

Compaction grouting involves the subsurface injection of a stiff mortar grout. Since its inception in the mid-twentieth century, its use has evolved to address many subsurface problems. This has produced discussion within the industry as to the proper definition of compaction grouting, components and characteristics of the grout, and the proper procedures that should be followed.

Injection of thick mortar grout was initially performed in the 1950s to fill relatively small voids beneath structures (Warner, 1982). It was soon discovered that after a void was filled, additional pumping resulted in 'jacking' of the overlying structure, which made the process a valuable relevelling tool. It wasn't until the late 1960s that the side effect of soil compaction was identified (Graf, 1969). This is when the term 'compaction grouting' was coined.

The low mobility grout became known as 'compaction grout' and the process of injecting it, 'compaction grouting'. However, as when it was originally developed, the grout's low mobility characteristic made it attractive for other applications such as filling of subsurface voids, either man-made (such as mines) or naturally occurring (such as karst conditions), or for constructing high modulus columns to reinforce soft wet soils that were not compacted by the process. These applications often did not involve soil compaction, and some in the industry felt the term 'compaction grouting' was being improperly applied. To help resolve this controversy, the term 'low mobility grout' (LMG) was proposed in the late 1990s (Byle, 1997) to describe the grout and the broad process of using LMG. Compaction grouting would be a subcategory of LMG.

In 2010, the Grouting Committee of the Geo-Institute of the American Society of Civil Engineers (G-I ASCE) published the 'Compaction Grouting Consensus Guide' (ASCE/G-I 53-10). In this guide, compaction grouting is defined as follows:

> Compaction Grouting is a ground improvement technique that improves the strength and/or stiffness of the ground by slow and controlled injection of a low mobility grout. The soil is displaced and compacted as the grout mass expands. Provided that the injection process progresses in a controlled fashion, the grout material remains in a growing mass within the ground and does not permeate or fracture the soil. This behavior enables consistent densification around the expanding grout mass, resulting in stiff inclusions of grout surrounded by soil of increased density. The process can be applied equally well above and below the water table. It is usually applied to loose fills and loose native soils that have sufficient drainage to prevent buildup of excess pore pressure.

Although change is slow, it appears that the industry as a whole is in the process of adopting the term 'low mobility grout (LMG)' to describe the grout itself and its broad use, and the term 'compaction grouting' only when the goal is to compact soils (Figure 8.1). Since this transition is still in progress, this chapter shall discuss the general use of LMG, both when

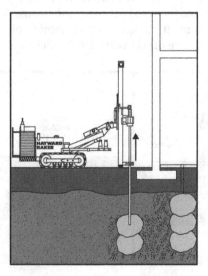

Figure 8.1 Compaction grouting to densify soil beneath existing foundation. (Courtesy of Hayward Baker Inc.)

used to compact soils and for other applications. The term 'low mobility grout (LMG)' shall be used for the grout.

8.2 HISTORY

Compaction grouting is the only major grouting method originating in the United States. The following timeline presents a general history of the development of the LMG industry:

Early 1950s A small contractor in California first pumped a stiff mortar-type grout to fill voids beneath distressed structures and to relevel settled foundations and slabs.

Late 1960s Compaction of the soils surrounding injected mortar grout was identified and the term 'compaction grouting' was first used.

Late 1970s Compaction grouting was first used to compact soils loosened during soft ground tunnelling for the Baltimore subway to prevent settlement of overlying structures (Figure 8.2). The process was so successful that it was also used for other tunnelling projects, including the subways for Boston, Seattle, and Los Angeles.

Early 1980s Compaction grouting was used for large-scale preconstruction soil improvement to densify loose sands beneath a planned power plant in Jacksonville, Florida (together with dynamic compaction), and beneath many structures for the planned nuclear submarine base in Kings Bay, Georgia (together with dynamic compaction and vibroreplacement) (Hussin, 1987) (Figure 8.3).

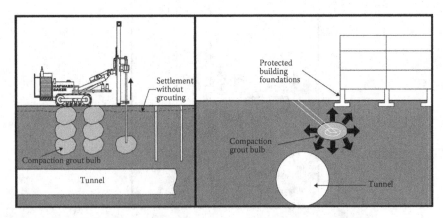

Figure 8.2 Compaction grouting to prevent settlement of overlying foundations during soft ground tunnelling. (Courtesy of Hayward Baker Inc.)

Mid-1980s   Prior to this time, slurry grout (Portland cement and water) was used to stabilise sinkhole conditions in karst regions. However, since the grout was very fluid and the limestone contained many cracks and crevasses, the required grout quantity was very difficult to predict, resulting in frequent large cost overruns. A specialty contractor in Florida (Henry, 1986) realised that LMG offered the benefit of filling the larger voids and compacting loosened soil while at the same time, its stiff nature would limit grout takes in the thin cracks of the limestone formation (Figure 8.4). The process was very successful and essentially replaced slurry grouting to stabilise sinkholes, compact resulting loosened sands, and relevel overlying settled structures. Soon after, the technique saw its first use to treat sites in sinkhole prone areas prior to construction to reduce the potential for future sinkholes. This application has since been adopted in karst geology regions throughout the United States. By the late 1980s the technique began to be applied in Europe, notably the UK, mainly for ground improvement and foundation remedial works (see Crockford and Bell, 1996).

Circa 1990   Compaction grouting was exported to Japan where it began to gain acceptance to raise structures and was later used to relevel structures that settled as a result of the Kobe earthquake (1995).

Figure 8.3 Compaction grouting combined with vibro-replacement and dynamic compaction beneath planned structures at Georgia submarine base. (Courtesy of Hayward Baker Inc.)

Figure 8.4 LMG to stabilise large sinkhole at a central Florida mining facility. (Courtesy of Hayward Baker Inc.)

The news of the applications and successes spread and LMG is now used in many parts of the world. The term 'compaction grouting' is still used outside the United States to mean the whole process used in the full range of applications including ground densification, relevelling, or compensation using LMG.

8.3 APPLICATIONS

The low mobility characteristic of LMG makes it suitable for modifying subsurface conditions both beneath existing structures and prior to construction. Common applications include

- Preconstruction soil improvement to permit shallow foundations
- Compaction of loose fills
- Creation of grout columns and filling voids within loose or deteriorating natural soil conditions (i.e., organic degradation, etc.) or voided fills
- Compaction of loose soils resulting from adjacent excavation activity, tunnelling (Figure 8.5d), sinkhole activity (Figure 8.5b), improper dewatering, broken utility lines, and so forth
- Compaction to increase bearing capacity or reduce potential settlement beneath existing foundations when modifications to the existing structure increase the foundation loading (Figure 8.5c)
- Compaction of deep loose zone (static or dynamic settlement potential)

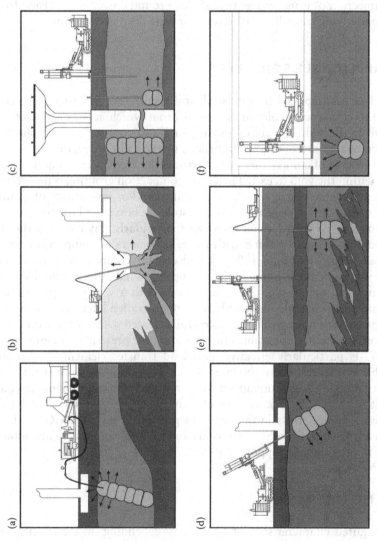

Figure 8.5 Low mobility grout applications. (Courtesy of Hayward Baker Inc.)

- Improvement of ground between pinnacled limestone to avoid deep piling within slots (Figure 8.5e)
- Injection of grout beneath settled structures to heave and relevel the structures (Figure 8.5a and f)
- Increasing lateral support for existing deep foundations (Figure 8.5c)
- Injection of grout to compensate for ground movements arising from underlying tunnelling operations or near deep excavations

8.4 APPLICABLE SOIL TYPES

LMG has been applied in almost all soil types. The process can be controlled to construct a column of mortar grout which acts as a reinforcing element in the soil. The ability of the process to compact the surrounding soils (compaction grouting) is influenced by the soil's properties.

The process of compacting soil constricts the void spaces, requiring any water within the void to exit. Therefore, compaction grouting is most effective in soils which have a high permeability and/or low degree of saturation. Loose, highly permeable granular soils are best suited for compaction grouting. As the fines content increases (particularly clay content), the soil permeability decreases along with the effectiveness of compaction grouting. Sands with less than 10% silt and no clay compact well above and below the water table. Some limited compaction has been accomplished in nonsaturated, nonplastic, fine-grained soils with a very slow grout injection rate and a carefully designed grout mix (detailed later in this chapter). Collapsible soils have also been successfully treated with compaction grouting. Soils best suited for compaction grouting are presented in Figure 8.6.

LMG is particularly effective as a load transfer element in noncompactable soil layers where the layer thickness is only several feet thick. In this situation, a compaction grout column with a minimum diameter equal to half of the layer thickness can effectively support the load that would otherwise induce stress in the layer. As previously stated, LMG has been effectively used to fill subsurface voids and to build columns in abandoned mines to provide roof support (Figure 8.7).

8.5 LMG MIX DESIGN

The required characteristics of LMG vary depending on the application. LMG used for compaction grouting is generally most restrictive and will first be discussed. Afterwards, other applications for which the requirements can be eased will be discussed.

LMG has what may appear to be the conflicting requirements of being pumpable yet immobile. The grout is most effective if it has enough fines to

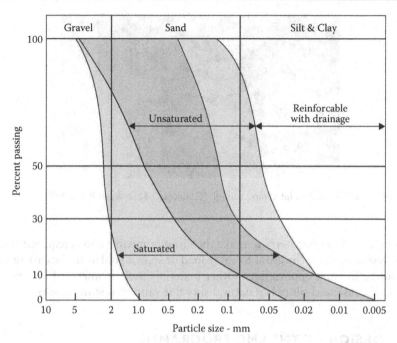

Figure 8.6 Range of soil gradation curves best suited for compaction grouting. (Courtesy of Hayward Baker Inc.)

maintain its pumpability but be sufficiently permeable to give up its water once it exits the end of the pipe. The gradation of the grout's aggregate is most influential in achieving this property. A commonly used gradation is a uniformly graded aggregate, including gravel to silt size material with the 100% finer than the 5–20 mm range and the 0% finer than the range of 0.001–0.03 mm. This gradation will generally maintain the grout's internal friction and permeability, resulting in an immobile grout once it leaves the end of the pipe in the ground.

Care should be used when adding bentonite clay to LMG. Although bentonite makes the grout more pumpable, the percentage should be kept low (<1% by total dry weight of solids) to avoid excessively reducing the grout's permeability and possible fracturing within the soil. Bentonite has been successfully used when compacting low fines content sands and injecting the grout at a very low rate (~2 cubic feet or 0.06 cubic metres per minute.) The use of bentonite in LMG is less of an issue in void-filling and karst applications.

The amount of water included in the mix should be the minimal amount required for the grout to be pumpable. This generally results in a very stiff mix.

For compaction grouting applications, the primary goal is generally compaction of the surrounding soils, and the grout only needs to be as strong as the soil. Therefore, cement is not a requirement. However, cement is often used to

Figure 8.7 LMG used to fill a mine tunnel. (Courtesy of Hayward Baker Inc.)

provide the fines necessary to make the grout pumpable and is required when the design requires the grout to carry load or span a void as in the case of cap grouting in karst applications. Other materials used as fines in LMG mixes include flyash and silt (when natural silty fine sand is used in the mix).

8.6 DESIGN OF THE LMG PROGRAMME

Several important steps are involved in the design of a LMG programme.

8.6.1 Subsurface conditions

It is critical to understand the subsurface conditions in order to design the LMG programme to achieve the desired results. The stress conditions in the ground prior to grouting will influence the design quantities and injection pressure. Identifying variations in soil density with depth allows the programme to target loose strata and avoid wasted effort in dense strata. Soft strata may only require a certain diameter grout column for reinforcement. The rate of injection may be able to be varied based on the permeability of different strata. The quantity of grout to be injected will depend on any subsurface voids or hard inclusions.

8.6.2 Constraints

The above- and below-ground constraints must be understood prior to designing the grouting programme. Some areas of the site may not be accessible or may require limited access equipment or hand-held equipment. Both above- and below-ground utility locations must be carefully identified. Subsurface structures or adjacent retaining walls must be identified to avoid damage during the high-pressure injection of the grout.

8.6.3 Requirements

The requirements of the programme are important to understand before the LMG programme can be designed. The requirements could involve densification of the soils, reinforcement with grout columns, relevelling of foundations or compensation for loss ground due to tunnelling.

If densification is the target, the minimum required post-treatment density must be determined. Relevelling of an existing structure could be necessary. In the scenario of compaction grouting above soft ground tunnelling to avoid excessive settlement of overlying structures, the maximum allowable settlement must be defined.

8.6.4 Criteria

The criteria that define success or failure must be defined and understood by all involved parties. Common acceptance criteria include injected grout volume, injected grout pressure, final elevation of lifted structures and test results (SPT, CPT, etc.) of soil between injection locations. Often, the same test is performed both before and after grouting to allow an accurate determination of the improvement (see Figure 8.8).

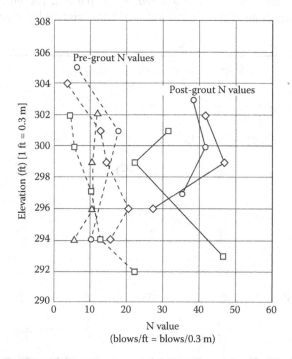

Figure 8.8 Sample SPT results before and after compaction grouting. (Courtesy of Hayward Baker Inc.)

8.6.5 Layout

The design of the grout injection pipe layout is based on the subsurface conditions, site constraints, programme requirements, and the criteria to be achieved.

8.6.5.1 Spacing

The spacing of the pipes in plan depends primarily on the soil type and the criteria to be achieved. Spacing generally varies between 3–15 feet (0.9–4.6 m) on centre. These are extreme examples. The spacing is on the lower end of the range for shallow treatment (low overburden or overlying structural load) and in fine-grained soils. Spacing is on the higher end of the range for deeper treatment (greater overburden or heavier overlying structural load), in granular soils, or when limited improvement is required. Most compaction grouting is performed at a spacing of 5–7 feet (1.5–2.1 m) (see Figure 8.9).

8.6.5.2 Injection sequence

The sequence by which the injection is performed is important to maximise the improvement achieved. For a given layout pattern, the best results are achieved by injecting at every other location first (primary locations) followed

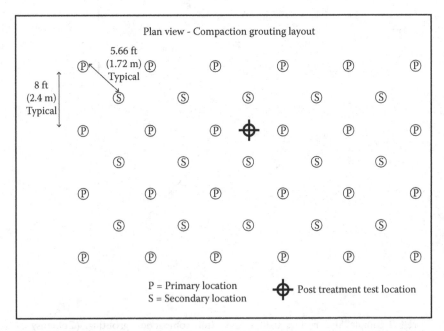

Figure 8.9 Sample compaction grouting layout plan. (Courtesy of Hayward Baker Inc.)

by the skipped locations (secondary locations), see Figure 8.9. The primary injections provide confinement of the soils during the secondary grouting.

The vertical sequence at a specific injection location can also be varied. LMG may begin at the bottom of the treatment zone and proceed upward (upstage) or at the top of the treatment zone and proceed downward (downstage). The upstage process involves installation of the grout pipe to the bottom of the treatment zone and then slowly extracting the pipe as the LMG is injected. The upstage process is generally easier, quicker, and less costly. The downstage process involves installation of the grout pipe to the top of the treatment zone, injecting the LMG at that depth, waiting for the grout to set, drilling through the grout, and then pumping grout beneath the previously injected grout. The process is repeated until grout has been injected into the full treatment depth. This process is more effective when lifting structures from a shallow depth and shallow weak soils.

8.6.6 Procedures

8.6.6.1 Injection pressure

Injection of LMG requires relatively high pressures, possibly in excess of 1,000 psi (6.99 MPa). It may require 100–200 psi (0.7–1.4 MPa) just to pump the stiff LMG through the grout hose and injection pipe before the pipe is inserted into the ground. Once the pipe is inserted into the ground, the additional pressure required to displace and compact the surrounding soils increases with depth and with soil density. The pump injection pressure will typically increase as the injection continues at a particular location due to the increase in density of the surrounding soils.

8.6.6.2 Injection rate

The maximum effective injection rate for LMG depends on the permeability of the soils being treated. As the fines content increases, the injection rate should be lowered. A clay content of 1%–2% can greatly reduce the permeability of granular soil and lower the maximum effective injection rate. In low permeability soils or low confinement situations (shallow depth and no overlying structures) an injection rate of 0.5–1 ft³ (14.2–28.3 litres) per minute is common. In well-draining soils, dry soils, or soils at depths, 4–12 ft³ (113–340 litres) per minute is appropriate. The injection rate is not limited in void-filling applications.

8.6.6.3 Injection volume

The injection volume depends on the density of the soil and the amount of improvement required. The volume reduction from a granular soil's minimum density to its maximum density may be 15%–20%. However, typical injection volumes are in the range of 8%–12% of the soil to be compacted. The volume may require adjustment as the programme progresses.

8.6.6.4 Heave

Heave typically occurs when the resistance to grout displacement is less above the grout than it is laterally. It can also be an indication that fracturing has occurred. Heave is a limiting factor in that when it occurs, little additional compaction is occurring in the surrounding soils. Heave can also damage overlying structures. However, if the objective is to restore the levels of an overlying structure which has previously experienced settlement, some heave may be desirable to relevel the structure.

Heave can occur when grouting at shallow depths. It is difficult to densify soils at depths shallower than 10 feet (3 m) because the limited overburden pressure is less than the lateral pressure of the soils. Heave can also occur during deeper grouting when the surrounding soils have been compacted such that they resist further displacement. Once heave is detected, grouting is generally discontinued at that depth. Compaction of overlying loose zones can still be achieved.

The heave criteria should be established before the grouting programme begins. It is important to realise that heave is often cumulative. Therefore, the maximum heave criteria at any particular depth should be less than the maximum allowable for the overlying structure to allow continued grouting in soils above the depth where the heave criterion is first observed. For example, if 0.5 inch (12.7 mm) is the maximum total allowable heave, a reasonable refusal criterion for any particular depth might be 0.1 inch (2.5 mm).

8.7 LMG PROGRAMME PROCEDURES

The LMG programme procedures are established prior to the programme beginning and assist in making sure that the parameters established during the design are followed. A close working relationship between the specialist contractor and the client's field inspector is beneficial in achieving the programme's goals. Monitoring of the work and surrounding structures is an integral part of the procedures.

8.7.1 Layout and sequence

The plan layout of the injection locations and sequence in which they are performed should be carefully followed. Sequence includes the order that the locations are performed, including if a primary/secondary sequence is to be followed. Also, the sequence includes either an upstage or downstage procedure at each location.

8.7.2 Pipe installation

The main concern when installing the grout pipe is that the surrounding soil is in tight contact with the exterior of the pipe. This is important to avoid grout travelling up the annulus between the soil and pipe exterior as well as to provide resistance against the pipe jacking up out of the ground during pumping. Installation methods include driving, flushing, and drilling. Problems associated with driving the pipe include encountering refusal prior to reaching the target depth, damaging the pipe during driving, and negatively affecting existing structures due to the driving vibrations. Flushing can be external or internal. External flushing should only be used with care since it can cause an open annulus. It has been used successfully in granular soils in limited amounts since the granular soils collapse around the pipe at the end of flushing. Internal flushing is generally acceptable. Drilling is the most common installation procedure since it avoids these problems. When flushing is used in conjunction with drilling, these comments concerning flushing apply. Upstage and downstage sequencing is discussed above in Section 8.6.5.2. When performing downstage grouting, it is common to have to wait 8 to 10 hours for a stage to achieve initial set before advancing the pipe through it to grout the underlying stage.

During pipe installation the location and angle of insertion should be carefully monitored and documented so that the grouting is performed in the planned location and no locations are missed or repeated. The difficulty of installation should also be monitored and documented as an indication of the soil density with depth prior to grouting.

8.7.3 Grout injection

The drilling and grouting operations can be performed by the same crew and equipment, completing both at each location before proceeding to the subsequent location, or they can be performed by separate crews; one installing the pipes and another following behind performing the grouting and pipe removal.

Pumping water through the grout hoses prior to pumping grout prevents the loss of water from the first stroke of grout mix as it is pumped through the hose, avoiding possible plugging. It is also important that all hose and pipe connections are water-tight to avoid water loss and possible grout plugging at these locations. Before connecting the grout hose to the top of the injection pipe, grout should be pumped to fill the hose.

Often the volume of grout is determined by counting the number of pump strokes. Prior to beginning the programme, the grout pump should be calibrated. One simple method of calibration consists of counting the number of strokes required to fill a container of fixed volume, such as a 55-gallon (208-litre) drum.

During grout injection the grout pressure and volume should be monitored and recorded versus depth. The pressures in the hose should be monitored and recorded both near the pump and near the top of the injection pipe (requiring pressure gauges at these locations). The pressure at the top of the pipe is a close indication of the injection pressure into the soil. A comparison of the two gauges reveals if high pressures or grout refusal is a result of grout plugs within the line.

The three parameters that often are specified as controlling when a grouting stage is complete are grout injection pressure, grout volume, and ground heave. The reasoning behind specifying a maximum pressure is that when achieved, the soils surrounding that injection stage have been sufficiently compacted. A maximum volume is specified when the design requires a grout column of a defined diameter to carry a load or possibly at primary locations to provide confinement of the soils during the subsequent secondary location grouting. A maximum heave is specified to avoid excessive grouting without additional compaction and to avoid damage to overlying structures. Grouting is often continued at a stage until one or the other of these is first observed.

8.8 EQUIPMENT

Proper equipment is integral to the success of a LMG programme and particularly to a compaction grouting programme.

8.8.1 Batchers and mixers

The batcher and mixer requirements vary depending on the daily grout volume required and the grout mix requirements, both of which depend on the application. When LMG is used in void-filling or karst applications, the internal friction of the grout is not as critical and often is more easily mixed (i.e., may simply contain a fine sand aggregate and, possibly, bentonite). These mixes may be provided by a ready-mix plant and mixed during transport in the revolving drum of a common concrete truck (Figure 8.10). However, when performing compaction grouting, the internal friction is often required and this mix does not mix well in a truck. Also, the lower volumes typical of compaction grouting do not suit ready-mix plants and concrete trucks because of the significant time each load might be delayed in the truck while waiting to be pumped.

For compaction grouting applications, on-site batching is typical. Pugmill mixers usually are slow and labor intensive (adding materials by hand) and are only used when small quantities are expected (less than 5 cubic yards [3.8 cubic metres] per shift). For projects requiring larger quantities, a continuous mix batch plant is often used. These batch plants can vary in size and capacity and are often mounted on a truck. Typically, the components are metered onto

Figure 8.10 Concrete truck delivers LMG to grout site. (Courtesy of Hayward Baker Inc.)

Figure 8.11 Examples of on-site LMG batch plants. (Courtesy of Hayward Baker Inc.)

a belt which travels beneath the material storage bins. The materials are then mixed with water in a continuous screw auger which then feeds the grout into the pump. These batch plants can produce as much as 50 cubic yards of grout (38 cubic metres) per hour. Sample batch plants are depicted in Figure 8.11.

8.8.2 Pumps

LMG is typically pumped with modified concrete pumps. These are piston pumps which pump a defined volume per stroke (the volume of the piston). The typical concrete pump generally requires modification of the piston diameter and control mechanisms to be capable of pumping at the high pressures (1,000 psi or 6.9 MPa or greater) and slow rates of less than 2 cubic feet (57 litres) per minute required to perform compaction grouting.

8.8.3 Injection pipe installation equipment

The method of installing the grout injection pipe can vary depending on site constraints and the application.

Figure 8.12 Hand equipment to install LMG injection pipe. (Courtesy of Hayward Baker Inc.)

8.8.3.1 Hand equipment

Many LMG projects are in tight environments with injection location barely accessible to workers alone (Figure 8.12). In these situations, the pipes can be installed with hand-held equipment. These can consist of hand-held rotary percussion drills, or small hydraulic or pneumatic driving equipment.

8.8.3.2 Tracked and wheeled drill rigs

When more room is available, tracked or wheeled equipment may be suitable. The size of this equipment varies from very small that can fit through a standard doorway, to very large and crawler-crane mounted, see Figure 8.13. Generally, equipment with the longest mast (single stroke) is most productive in that it can install longer sections of pipe at a time. Augering a hole and then inserting the grout pipe is generally not desirable since it does not result in a tight fit between the pipe and the soil.

8.8.4 Miscellaneous

Several miscellaneous items are also important to consider.

Figure 8.13 Small-wheeled and large-tracked drills to install LMG injection pipe. (Courtesy of Hayward Baker Inc.)

8.8.4.1 Extraction of pipe

Extraction of the pipe during grouting can be performed by the same equipment that inserted it (drill or crane) or could be performed as a separate operation. When handled separately, purpose-built hydraulic rams are often used (Figure 8.14).

8.8.4.2 Hoses

High-pressure hoses or steel pipes with flush joints and a minimum diameter of 2 inches (51 mm) are required to avoid excessive resistance to pumping the grout through the lines. The fittings (couplings) between sections of hose or pipe should be tight to avoid water leaks which could lead to blockages in the lines.

8.8.4.3 Pressure gauges with gauge savers

Pressure gauges are necessary to monitor the grout injection pressure. As noted in Section 8.7.3, gauges should be located both near the pump and near the top of the injection pipe (Figure 8.15).

8.9 QC & QA

A properly specified quality control (QC) and quality assurance (QA) programme is essential to the successful performance of a LMG programme.

Figure 8.14 Hydraulic ram to extract LMG injection pipe. (Courtesy of Hayward Baker Inc.)

Figure 8.15 Pressure gauge and gauge saver near top of grout injection pipe. (Courtesy of Hayward Baker Inc.)

The programme should be designed to specify, monitor, and measure the parameters necessary for the specific application.

8.9.1 LMG characteristics

Depending on the application, specific aspects of the LMG should be monitored and documented. A laboratory mix design is typically performed prior to the grouting work to ensure that the grout will achieve the required parameters (strength, slump, etc.). The results of a previously performed mix design programme could apply if the same components were used. If the grout is batched on site, the ratio of the components should be accurately measured and documented to assure a consistent product.

If the design requires minimum grout strength, then test specimens, typically cylinders, should be cast on a regular interval. This interval may be based on time (e.g., twice a day) or grout volume (e.g., every 100 cubic yards [76 cubic metres]). If grout slump is important, it is common for the slump measurements to be performed when the UCS samples are cast (Figure 8.16).

8.9.2 Test programme

For many larger projects, a test programme is performed to demonstrate that the planned materials and procedures will achieve the required results. A test programme may be performed within or separate from the production area. If densification of soils is required (compaction grouting), penetration testing of the target soils before and after treatment is often performed. If the LMG is planned to reduce settlement potential in soils which may not show significant improvement in penetration test values (fine-grained soils), a load test may be performed of an individual grout column or of a treated area.

Figure 8.16 Casting test cylinders and measuring LMG slump. (Courtesy of Hayward Baker Inc.)

8.9.3 Soil improvement testing

When soil improvement is required, a performance specification is often selected. Typically, the acceptance criterion requires that post-treatment penetration testing of the soil midway between the grout injection locations achieves a minimum and/or average value. Common post-treatment penetration testing could include Standard Penetration Tests (SPT), Cone Penetrometer Tests (CPT), and Dilatometer Tests (DMT). The specified value should be based on an analysis of the required performance (bearing capacity, settlement, liquefaction, etc.). An example of testing location and results is depicted in Figures 8.8 and 8.9.

8.9.4 Quality control during injection

Several parameters should be monitored and recorded during grout injection.

8.9.5 Grout injection pressure

The grout pressure during injection should be measured in the grout hose both near the pump and near the top of the grout injection pipe (as explained in Section 8.7.3). The LMG will require a significant pumping pressure to overcome the line friction and passive resistance of the soil at the bottom of the injection pipe. An initial injection pressure in the range of 100–200 psi (0.7–1.4 MPa) is common. The pressure required to inject additional grout increases as the soil is displaced and compacted. Final injection pressures may exceed 1,000 psi (6.9 MPa).

8.9.6 Grout injection rate

The grout injection rate should be monitored and recorded. The common method is by calibrating the pump piston (as mentioned in Section 8.7.3) and then counting the number of piston strokes during injection. Controlling the rate of grout injection is important when performing compaction grouting. If the rate is too great, the soil pore water may not have time to drain and result in an increase in pore water pressure. The pore pressure increase may result in an increase in grout pumping pressure, giving a false indicator of soil improvement. The typical grout injection rate is between 1–5 cubic feet (28–142 litres) per minute. A gradual increase in injection pressure indicates controlled densification. A sudden drop in pressure indicates hydraulic fracturing.

8.9.7 Volume of grout injected

The grout volume injected should be monitored and recorded. Depending on the initial density of the soil being treated, the grout injection volume typically required to compact the soil is between 5%–15% of the volume of the soil being treated. The volume of the soil being treated extends halfway to the adjacent injection locations. If significantly more grout is injected without a significant increase in injection pressure, the location should be explored for subsurface voids or utilities. When a primary/secondary injecting sequencing is planned, the volume injected at the primary locations is typically restricted to a maximum of 15% with the goal of even grout distribution. This is necessary to determine payment in unit price contracts.

8.9.8 Heave

During grout injection, the grout will displace the ground in the direction of least resistance. In loose soils and when injecting at depth, the overburden weight provides more resistance than the lateral passive resistance, resulting in the grout displacing the soil horizontally. As the soils become denser and when injecting at depths shallower than about 10 feet (3 m), the path of least resistance may be towards the ground surface. This may result in heave of the ground surface and overlying structures. Little additional soil compaction is occurring during heave. Many different instruments are used to monitor when heave occurs and the magnitude of the heave. Examples include crack monitors, tilt meters, plumb bobs, and spirit levels.

8.9.9 Instrumentation, electronic monitoring, and computer data acquisition

Computers have become more capable of withstanding the vibrations and dust associated with construction, allowing their incorporation into many aspects of construction. Computers can be connected to instrumentation that measures grout injection pressure, volume and depth. The computer along with the instrumentation is referred to as a data acquisition (DAQ) system. This is not only useful for documentation of the work, but can also allow the operators to monitor their work and make adjustments as necessary, since measurements are displayed on a screen in real time (Figure 8.17).

8.10 CASE HISTORIES

The following case histories present several applications for LMG.

8.10.1 Harlem Hospital, New York, New York

A 48,000-square-foot (4459-square-metre) site, containing liquefiable soils within the depth range of 10–50 feet (3–15.2 metres) below grade, was selected for the construction of a six-story structure at Harlem

Figure 8.17 Compaction grout DAQ system with grout line under table where grout pressure and volume is measured. (Courtesy of Hayward Baker Inc.)

Hospital in New York City. A combination of spread footings and a large structural mat foundation on compaction grout–improved ground was selected during a peer review as a value-engineered alternative to deep foundations.

The grout injection pipe was installed and extracted in a continuous operation utilising a large-tracked drill-mounted vibratory hammer (Figure 8.18). The geotechnical contractor designed and installed real-time data acquisition systems on two rigs to record depth, grout pressures, grout volumes, and grout injection rates. The plots of the data were produced in real time to assist in QC. The operators had similar screen displays to allow them to determine if criteria were being met and if they could advance to the next stage. In addition, a three-dimensional visualization package was created to assist in evaluation of the grouting process (Figure 8.19). The three-dimensional rendering of grout volume and pressure was completed automatically, based on data acquired from the field.

Over 130 CPTs were performed to verify post-treatment soil improvement. As each was performed, the electronic data was imported into a programme which calculated the factor of safety against liquefaction, the static settlement, and the seismically induced settlement (Figure 8.20).

8.10.2 Interstate 5 at Hasley Canyon, California

Compaction grouting was selected to densify liquefiable sands and gravels around existing and newly installed piles supporting Interstate 5 where it

Figure 8.18 Large track-mounted drill rigs with telescopic masts installed the injection pipes in a single stroke to the maximum required treatment depth of 50 feet (15.2 meters). (Courtesy of Hayward Baker Inc.)

crosses the Castaic Creek approximately 30 miles (48 kilometres) north of Los Angeles, since it is capable of being performed from low headroom working conditions. The creek is an ephemeral creek that is normally dry. The soils in the creek bed are typical fluvial sediments composed of mostly cobbles, gravels and sand. SPT blow counts ranged from 8–15. These soils would be prone to liquefaction during an earthquake. The liquefiable soils were as deep as 30 feet (9.1 m) below ground surface.

The geotechnical contractor performed multiple stages of compaction grouting to accommodate traffic requirements, sensitive habitat, limited/restricted access, and construction sequencing along a major interstate highway. There were approximately 100 injection locations to depths up to 59 feet (18 metres) grouted for abutment 1, 225 locations to depths ranging from 52.5–59 feet (16–18 metres) grouted for pier 2, and more than 300 locations to depths of 16.4 feet (5 metres) for abutment 4.

A small crawler drill was used in the creek bed to drill grout pipes to the depths required (Figure 8.21). Holes were drilled to depths ranging from 15–59 feet (4.6–18 metres) below ground surface. Primary holes were drilled on approximately 20-foot (6.1-metre) spacing, and secondary holes were drilled between these holes leaving 10-foot (3-metre) spacing

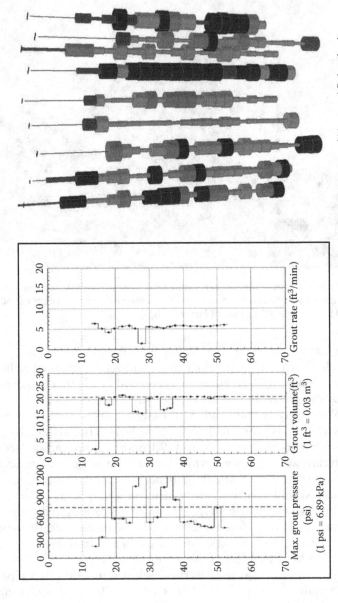

Figure 8.19 Real-time plots of grouting data and three-dimensional visualization graphic. (Courtesy of Hayward Baker Inc.)

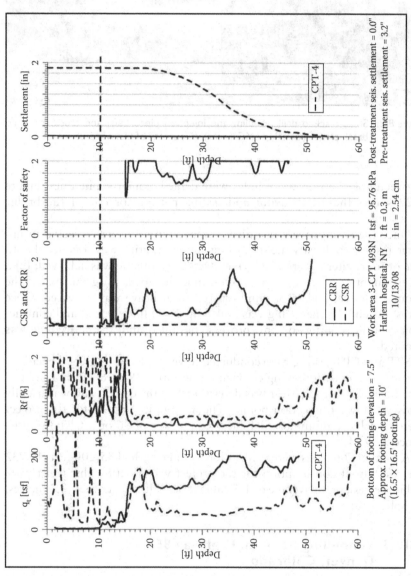

Figure 8.20 Sample of pre- and post-treatment CPT and analysis results. (Courtesy of Hayward Baker Inc.)

Figure 8.21 Small-tracked drill installing the injection pipes (left) and truck mounted batch plant producing LMG (right). (Courtesy of Hayward Baker Inc.)

between holes. Five rows of holes were drilled along the pile lines. Figure 8.22 shows the grout layout and testing pattern for one of the bridge abutments.

Once the casings were installed, compaction grout was injected in up-stage fashion to densify the soil. Compaction grouting was conducted until one of four criteria were met: (1) a given overpressure was achieved; (2) a given quantity of grout was injected into the zone; (3) grout exited the ground surface; and (4) ground or structural heave was observed. Once refusal occurred, the casing was pulled up 2 feet (0.6 metres) and compaction grouting continued. This process continued until the entire hole was grouted.

SPT and CPT testing were conducted to verify that the soil had been sufficiently densified following compaction grouting (Figure 8.23). The location of the SPT or CPT test was determined by the owner, and was usually midway between injection points. The acceptance criteria for the project were to achieve minimum SPT blow count N_{160} of 36 or a CPT tip resistance > 19 MPa.

A total of 766 holes were drilled, for a total length of 35,200 feet (10,732 metres). Grout consumption for the project was 53,208 cubic feet (1,507 cubic metres), or an average of 1.5 cubic feet (42 litres) per foot (0.3 metres) of hole.

8.10.3 Tunnelling beneath Highway 85, Denver, Colorado

A sinkhole opened up during rush hour on Highway 85 north of Denver due to sloughing during a tunnelling operation. Within the hour, compaction grouting crews were on site. Within hours the situation had been resolved using grouting techniques.

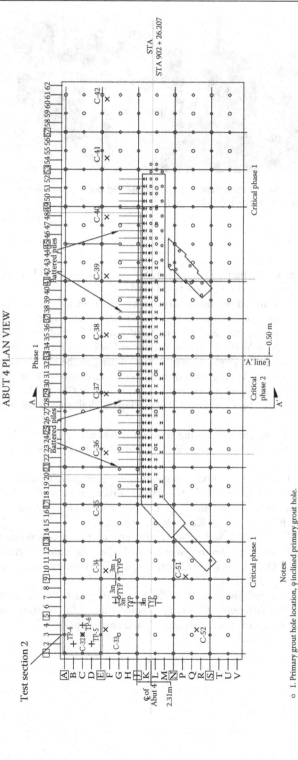

Figure 8.22 Grout injection and testing layout pattern. (Courtesy of Hayward Baker Inc.)

Figure 8.23 Compaction grouting densifies soils loosened during tunneling operations. (Courtesy of Hayward Baker Inc.)

REFERENCES

American Society of Civil Engineers Standards 53-10-Consensus Guide Committee. (2010). Compaction grouting consensus guide.

Byle, M.J. (1997). Limited mobility displacement grouting: When 'compaction grout' is not compaction grout. In: Vipulanandan, C. (ed.), *Grouting: Compaction, Remediation and Testing*, Geotechnical Special Publication No. 66, New York, New York: American Society of Civil Engineers, pp. 32–42.

Crockford R. and Bell A.L. (1996). Compaction grouting in the UK – a review. In: Yonekura R., Terashi M., and Shibazaki M. (eds.), *Grouting and Deep Mixing*, *Proceedings of the Second International Conference on Ground Improvement Geosystems*, May 14-17, 1996, Tokyo, Japan: Balkema Publishers, pp. 279–284.

Graf, E.D. (1969). Compaction grouting technique, *Journal of the Soil Mechanics and Foundations Division*, ASCE, Vol. 95, No. SM5, Proceedings Paper 6766, pp. 1151–1158.

Henry, J.P. (1986). Low slump compaction grouting for correction of central Florida sinkholes, National Water Well Association, *Proceedings of the Conference on Environmental Problems in Karst Terranes and Their Solutions*, October 28–30, 1986, Bowling Green, Kentucky, United States.

Hussin, J.D. and Ali, S. (1987). Soil improvement at the Trident submarine facility. In: Welsh, J.P. (ed.), *Soil Improvement – A 10-Year Update, Proceedings of a Symposium, ASCE Convention*, April 28, 1987, Atlantic City, New Jersey, United States, pp. 215–231.

Warner, J. (1982). Compaction grouting – the first thirty years, *Proceedings of the ASCE Specialty Conference, Grouting in Geotechnical Engineering*, February 25–28, 1982, New Orleans, Louisiana, United States, pp. 694–707.

Chapter 9

In-situ soil mixing

Michał Topolnicki

CONTENTS

9.1 INTRODUCTION

The use of soil mixing (SM) to improve the engineering and environmental properties of soft or contaminated ground has increased widely since its genesis. Growing interest for SM mainly results from the high flexibility of this method, which can be purposely adapted to specific project requirements and site conditions, as well as from cost-to-performance efficiency of respective geotechnical solutions. In this method of ground improvement, soils are mixed in situ with different stabilising binders, which chemically react with the soil and/or the groundwater. The stabilised soil material that is produced generally has a higher strength, lower permeability, and lower compressibility than the native soil. The improvement becomes possible by cation exchange at the surface of clay minerals, bonding of soil particles, and/or filling of voids by chemical reaction products. The most important binders are cements and limes. However, blast-furnace slag, gypsum, and ashes as well as other secondary products and compound materials are also used. For environmental treatment, binders are replaced with chemical oxidation agents or other reactive materials to render pollutants harmless.

Soil mixing can be subdivided into two general methods: the deep mixing method (DMM) and the shallow mixing method (SMM). Both DMM and SMM include a variety of proprietary systems.

The more frequently used DMM is applied for in-situ stabilisation of the soil to a minimum depth of 3 m (a limit depth introduced by EN 14679:2005). The binders are injected into the soil in dry or slurry form through hollow mixing shafts tipped with various cutting tools and equipped with auger flights, mixing blades, or paddles to increase the efficiency of the mixing process. The shafts, mounted in single or multiple arrangements, rotate about the vertical axis and produce individual or overlapping soil-mix columns. In the case of special cutting/mixing arms equipped with cutter wheels mounted on horizontal axes to create panels or with a revolving cutting chain to create continuous walls, the slurry is pumped through injection pipes and outlets mounted along the arm. In some methods, the mechanical mixing is enhanced by simultaneously injecting fluid grout at high velocity through nozzles in the mixing or cutting tools.

The complementary SMM has been specially developed to reduce the costs of improving loose or soft superficial soils overlying substantial areas, including land disposed dredged sediments and wet organic soils a few metres thick. It is also a suitable method for in-situ remediation of contaminated soils and sludges. In such applications, the soils have to be thoroughly mixed in-situ with an appropriate amount of wet or dry binders to ensure stabilisation of the entire volume of treated soil. Therefore, this type of soil mixing is often referred to as 'mass stabilisation'. Mass stabilisation can be achieved by installing vertical overlapping columns with up and down movements of rotating mixing tools, as in the case of standard DMM, and is most cost-effective when using large-diameter mixing augers or multiple shaft arrangements. For shallow depth applications, however, generally limited to about 5 m, another very efficient method of mass stabilisation is usually implemented, and the mixing process is carried out repeatedly in vertical and horizontal directions through the soil mass using various cutting and mixing tools that are different from the tools developed for DMM (e.g., mixing drums). Consequently, in the classification scheme used in this chapter the SMM includes both systems of mass stabilisation.

In-situ soil mixing is a versatile ground improvement method. It can be used to stabilise a wide range of soils, including soft clays, silts and fine-grained sands. Stabilisation of organic soils such as gyttja (sedimentary organic soil), peat, and sludges is also possible, but is more difficult and requires carefully tailored binders and execution procedures. However, the engineering properties of the stabilised soil will not only depend on the characteristics of the binder. They will also depend, to a large extent, on the inherent characteristics of each soil and the way it has been deposited, as well as on mixing and curing conditions at a particular worksite. Therefore, a thorough understanding of chemical reactions with these factors is necessary in order to ensure successful application of this ground improvement technology.

In this chapter, the current status of in-situ soil mixing is outlined, taking into account recent execution and design practice, international literature, and experience. General application areas are identified and discussed, and a few case histories selected from international projects are included for illustration. The focus is on civil engineering applications of DMM, and, to a lesser extent, of SMM. Some specialised soil mixing issues in relation to environmental projects, such as mass treatment of subsurface hazardous wastes by various processes including solidification, stabilisation and chemical treatments, reactive barriers, etc., are only touched upon, therefore the cited literature should be referred to for more information. Furthermore, overly extensive descriptions of the complicated chemical processes occurring in the stabilised soil when mixed with various binders have been excluded from the contents. This choice, however, should not undermine the importance of this aspect of soil mixing. It may rather reflect the fact that in spite of considerable knowledge about basic reaction mechanisms, identified and described for instance by Babasaki et al. (1996) for soils stabilised with lime or cement, it is still not possible to predict the strength of in-situ mixed soil with a reasonable level of accuracy. As a consequence of this fundamental deficiency, which we are challenged to overcome, it is believed that the development of SM will be continued along a somewhat erratic experimental path, and will be to a large extent dependent on accumulated experiences. Therefore, the scope of this chapter instead concentrates on the characteristics of equipment in current use, execution procedures with reference to selected operational methods, applications, merits, and the limitations of the technology. Design aspects as well as quality control and quality assurance issues of DMM are also considered. The design approach outlined herein follows the practice established in Japan, the US, and Europe, assuming that the treated soil is practically an impermeable material. The approach used with respect to DM columns stabilised with unslaked lime or lime and cement, which may act as vertical drains, is covered in Chapter 10.

9.2 HISTORICAL DEVELOPMENT AND CLASSIFICATION

The historical development of SM was extensively covered in the second edition of this book, taking into account the status of this technology about 10 years ago. Anticipated and observed growth of applications worldwide has significantly changed this situation, however, and SM can now be regarded as well-established ground improvement technology. Consequently, relevant information on historical development has been shortened to include only the most important and stimulating achievements.

The roots of deep soil mixing go back to the mid-1950s, when the mixed-in-place (MIP) piling technique was developed by Intrusion-Prepakt Inc. (FHWA, 2000). In this method a mechanical mixer was used to mix cementitious grout into the soil for the purpose of creating foundation elements and retaining walls. The grout was injected from the tip of a mixing tool consisting of a drilling head and separated horizontal blades. Modern deep mixing techniques reflect, however, mainly Japanese and Scandinavian efforts over the last five decades as well as recent European achievements in cutter wheel and trench-type mixing systems.

The level of research and development activity in Japan in relation to DMM remains the highest in the world. The first commercial application of the deep lime mixing (DLM) method, utilising a mechanical binder feeding system, was conducted in 1974 by Fudo Construction Co. Ltd. using the Mark IV machine to improve reclaimed soft alluvial clay in Chiba Prefecture in Japan. The first marine use of DLM was in 1975 at Tokyo Port (Terashi, 2002a). In an effort to improve the uniformity of the stabilised soil, a new concept using cement mortar and cement-water slurry as binders was implemented in the mid-1970s, with CMC and DCM (deep cement mixing) methods developed by Kawasaki Steel & Fudo and Takenaka Group, respectively, with close supervision from Port Harbour Research Institute. The first on-land and marine applications of CMC and DCM were conducted in 1976. Also that year, the Seiko Kogyo Co. developed and introduced the soil-mixed wall (SMW) method using discontinuous augers and paddles positioned at discrete intervals, usually along three shafts arranged in a row. This method was applied primarily for excavation support and groundwater cut-off walls, with the possibility for installation of reinforcing steel sections within fresh columns to increase bending stiffness of the supporting DM elements.

Major marine ground improvement works at Daikoku Pier, beginning in 1977 and continuing for about 10 years, contributed to important developments of the wet method of deep mixing (e.g., DCM, DECOM, POCOM, and others). These developments included the elaboration of design standards and construction control procedures, slowly hardening binders and new positioning systems for offshore applications (Terashi, 2002a).

A general method using a variety of stabilising binders in slurry form (wet method) has been named cement deep mixing (CDM) method. In 1977, the CDM Association was established in Japan to promote and improve the CDM method via a collaboration of general contractors, marine works, and foundation works contractors, as well as industrial and research institutes. As a result, new efficient machines were developed, such as CDM-Mega, CDM-LODIC, CDM-Land4, and CDM-Column 21. For marine applications the CDM method has mainly been used to improve the foundations of revetments, as well as quay wall and breakwater foundations. The diameter of the mixing blades ranges from 1.0–1.6 m and the maximum depth of improvement is about 70 m below water. For land applications, the CDM method has been mainly applied for slide and liquefaction prevention, settlement reduction, and to improve the bearing capacity of foundations. The standard CDM machines have two shafts, mixing blades with a diameter of 1 m, and a penetration depth limited to about 50 m. Typical machines for marine and on-land use are shown in Figures 9.1 and 9.2, respectively.

Another remarkable development conceived in Japan in 1993 is the TRD (trench re-mixing and cutting deep wall) method. In this system, a continuous soil-mix wall is created in situ by lateral motion of vertical 'chain saw' in a one-phase process that involves simultaneous full-depth cutting and mixing of soils with binders in slurry form. This method has been applied in Japan for more than 400 projects, and in the US since 2006 (Figure 9.3). Approximately two thirds have been structural retaining walls and one third were cut-off walls (Garbin et al., 2010).

The development of the wet method in Japan includes successful attempts to combine mechanical mixing with high-velocity injection. In 1984, the spreadable wing (SWING) method was introduced. In this unique system a retractable mixing blade mounted on a single drilling shaft allows treatment of specific depths with large diameters (0.6 m with blade retracted and up to 2 m after expansion). Following that, jet grouting was incorporated into SWING and its first application was in 1986. With additional jetting during withdrawal, mechanically mixed and jet mixed concentric zones are produced with a total diameter up to 3.6 m (Kawasaki et al., 1996).

(a)   (b)

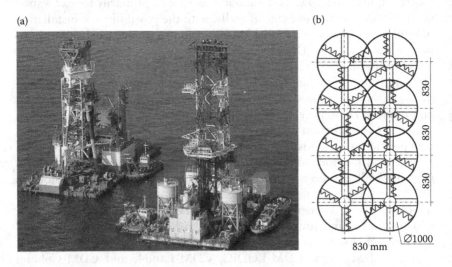

Figure 9.1 (a) CDM barges for marine deep mixing works using the wet method, Japan. (Courtesy of CDM Association, Japan.) (b) Possible arrangement of eight mixing shafts.

(a)   (b)   (c)

Figure 9.2 CDM machine for on-land deep mixing works using the wet method, Japan (note auger screws along the shafts enabling partial extrusion of disaggregated soil).

(a)   (b)

Figure 9.3 (a) Execution of a cut-off wall with the TRD method (Herbert Hoover Dike rehabilitation project in Florida), (b) close-up of the cutting/mixing chain. (Courtesy of Hayward Baker, Inc.)

Moreover, in 1992, the jet churning system management (JACSMAN) method using combined mechanical mixing and modern cross-jet grouting systems was developed by Fudo Construction Inc. and Chemical Grouting Co. (Kawanabe and Nozu, 2002). Another innovative development of Fudo is the CI-CMC method. With this method it is possible to disperse solidifying materials to the entire improvement area by injecting cement slurry into the air stream with an 'ejector discharge' device incorporated into the stirring blade. Consequently, penetration loads are reduced and the soil becomes easier to mix due to the air lifting effect, resulting in better homogeneity, reduced lateral displacements and lower costs of mixing (Fudo Inc., 2011).

In 1978 the development study of Japanese dry method designated for on-land applications utilising a pneumatic binder feeding system known as dry jet mixing (DJM) was initiated. The constructed machine was first applied in 1981. It was subsequently improved, with a landmark project in 1985 on San-yo Motorway, where a 10-m-high embankment was constructed on 10 m of sensitive soft clay (Terashi, 2002a). In the early 1980s the DJM Association was established in Japan, with a role similar to that of the CDM Association for the wet method. The current standard DJM equipment has two mixing shafts, with blades of 1.0 m diameter and a maximum penetration depth of 33 m, as shown in Figure 9.4.

(a)   (b)

Figure 9.4 (a) DJM machine for on-land deep mixing works using the dry method, Japan. (b) Two mixing tools, diameter 1.0 m.

The development of DMM in Scandinavia was initiated in Sweden in 1967 when laboratory and field research began for a new method of stabilisation of soft clays with unslaked lime. The first light wheel-mounted mixing equipment was manufactured by Linden-Alimak AB in cooperation with Swedish Geotechnical Institute, BPA Byggproduktion (presently LCM AB, Keller Group), and Euroc AB. In Finland, research was initiated during that time as well. Commercial use of the lime column method started in Sweden in 1975 for support of excavation, embankment stabilisation, and shallow foundations near Stockholm. Other types of dry binders, like cement and two component binders composed of unslaked lime and cement, have been subsequently investigated and put into practice. The first commercial project with the lime cement column method in Finland was conducted in 1988 and then 1990 both in Sweden and Norway under Swedish guidance and using Swedish contractors. This type of DMM, generally with two component binders tailored for various soft soils, has been consequently developed over the years. It is now referred to as the 'Nordic dry deep mixing method' (Holm, 2002). With an increasing number of proven applications, especially since 1989, by the mid-1990s this method had become the predominant technology of ground improvement in Scandinavia. Relatively light and mobile equipment with one shaft is typically used to produce columns of 0.5–1.0 m in diameter to a maximum depth of about 25 m depending on the soil conditions. The application focus remains on ground improvement to reduce settlement and enhance stability of road and railroad embankments, and soil/column interaction solutions for very soft, highly compressible clayey and/or organic soils. The first application for mitigation of vibrations induced by high speed trains took place in Sweden in 2000 (Holm, 2002a).

In European countries outside Scandinavia, the Nordic method has been mostly used in Poland since 1995, with a recent major application during construction of road S7 near Elbląg, involving about 743,000 lin. m of dry DM columns (Figure 9.5). Increasing number of projects were also conducted in the UK (first in 2001), and field trials in The Netherlands and Germany.

In Finland another dry shallow mixing method for stabilisation of superficial layers of peat, mud, sludge, or soft clay to a depth of about 5 m has been developed since 1992 and applied for road and land reclamation projects. The mixing tools of this method of mass stabilisation have different shapes and are typically attached to the arm of a conventional excavator. They may be constructed as mixing/cutting heads, equipped with blades rotating about a vertical or a horizontal axis, or as mixing drums. The mixing process is conducted repeatedly in vertical and horizontal directions through the soil mass in order to obtain a homogeneous soil-binder mixture. The first commercial project was conducted in Sweden in 1995 in connection with renovation works along Highway 601 Sundsvägen, where about 10,000 m³ of peat were treated.

Figure 9.5 Dry method deep mixing near Elbląg (Poland) using the Nordic method. (a) Keller/LCM equipment. (b) Dry mixing in winter.

The application and development of the contemporary DMM in the United States started in the mid-1980s and was comprised initially of the wet method. In 1986, SMW Seiko Inc. began operations under license from Japanese parent Seiko Kogyo Co. The SMW method was subsequently used in 1987–1989 in a landmark liquefaction mitigation and seepage cut-off project at Jackson Lake Dam, Wyoming, US, where 130,000 lin. m of column were installed to a maximum depth of 33 m (FHWA, 2000).

Following their cooperation with SMW Seiko on the Jackson Lake Dam project, in the late 1980s Geo-Con Inc. developed the first US soil mixing technologies, the DSM and SSM (shallow soil mixing) methods. The DSM method uses 1–6 shafts with discontinuous augers of 0.8–1 m in diameter. The SSM method uses a large-diameter single mixer to economically treat weak superficial soils and contaminated sites to a depth of about 12 m. The SSM has since been extended to accommodate binders in a dry form. This variant using dry form binders was applied in 1991 to stabilise large lagoons containing contaminated sludge residues from a water treatment plant at a refinery near Chicago (Jasperse and Ryan, 1992).

The Japanese SCC method was introduced in the US by SCC Technology Inc. The single-axis system of Hayward Baker (Keller), with diameters of 0.5–2.5 m, typically 2.1 and 2.4 m, began development in the 1990s and has been applied since 1997 (Burke, 2002). Mass stabilisation with the wet method (shallow soil-cement mixing) was also applied, using an excavator equipped with a shallow mixing bucket (Druss, 2002). The bucket contained mixing blades that rotate about a horizontal axis.

The largest DM works in North America to date were conducted between 2009 and 2011 for the New Orleans East Back Levee (LPV-111). The project consisted of ca. 1.3 million cubic metres of wet DM to a depth of up to 20.5 m. Over 17,000 single and double axis elements of diameter 1.6 m were combined in different patterns to create transversal panels, approximately 4.7 m apart, along the 8.53 km extension of the levee. Eight batching plants were used to prepare the grout using over 460,000 tonnes of slag/cement binder (Schmutzler et al., 2011).

A combined mechanical and hydraulic mixing method called GEOJET (Condon Johnson and Associates, Halliburton) has been developed and modified in the US since the early 1990s. GEOJET equipment includes the soil processor equipped with specially designed cutting blades and multiple jetting nozzles which jet mix at pressures up to 35 MPa. The first commercial application was in 1994, followed by some major retaining wall works and installations of pipe piles in soil.

The first commercial project that used the Nordic method in the US was conducted in 1996 in Queens, NY, by the Stabilator Company (Skanska). Subsequent application for settlement reduction at I-15 in Salt Lake City, Utah, took place in 1997. Since 1998, other dry methods, like DJM (Raito Inc.) and TREVIMIX (Treviicos Corp., with Hercules), have been available.

In Europe, the earliest wet DM activities that took place in the 1980s were oriented towards development of a potentially cheaper alternative to jet grouting. In France, Bachy Soletanche developed the COLMIX method in the mid-1980s, in conjunction with the French Railway Authority (SNCF) and the French National Laboratory for Roads and Bridges (LCPC). The method features twin, triple, or quadruple contra-rotating and interlocking augers, generally 3–4 m long and driven via hollow stem rods coupled to a single rotary drive. Several road and rail embankment stabilisation projects have been completed with this method in France, UK, and Italy, as summarised by Lebon (2002).

In Germany, the first application of the mixed-in-place (MIP) system developed by Bauer Spezialtiefbau GmbH, which was based on the rotary-auger-soil-mixing (RASM) method utilising single shafted crane and wet binder, took place in Nürnberg in 1987. MIP piles were executed to create panels of mixed soil filling up a 'Berlin'-type temporary retaining wall constructed in sands (Herrmann et al., 1992). Subsequently, a more advanced triple auger wet mixing system has been developed since the early 1990s. This system has been in use since 1994, primarily for construction of temporary and permanent panels supporting excavations, cut-off walls, ground improvement, and environmental purposes (Außenlechner et al., 2003, Schwarz and Seidel, 2003). For walls with shallow depth, typically 6–15 m, the Bauer soil mixing wall method, which uses three adjacent slightly overlapping augers and mixing paddles, was later developed. Keller Grundbau GmbH developed their first system based on a single paddle shaft equipped with a short auger and mixing blades above the drill bit. Their commercial ground improvement applications for this system have been ongoing since 1995. More advanced mixing tools with twin and triple shafts as well as combined systems involving mechanical and jet assisted mixing have been also introduced.

Another high-capacity specialised wet mixing system developed in Germany in 1994 is the FMI method (Fräs-Misch-Injektionsverfahren = cut-mix-injection). It was applied for the first time in 1996 in Giessen (Pampel and Polloczek, 1999). The FMI machine is comprised of a special cutting tree, along which cutting blades are rotated by two chain systems. The cutting tree can be inclined up to 80 degrees, and is dragged through the soil behind the power unit. With this method it is possible to treat the soil in deep strips, with a mean capacity of 70–100 m³/h. The width of treatment is 1 m down to a depth of 6 m, or 0.5 m down to a depth of 9 m. Initial applications mainly covered ground improvement works along railways. A similar system, called TRENCHMIX, was jointly developed by Bachy Soletanche and Mastenbroek and first applied in France in 2005.

In the United Kingdom wet DM for ground improvement was employed in early 1990s by Cementation Piling and Foundations for construction of a few temporary shafts of approximately 4 m internal diameter and up to 15 m deep (Blackwell, 1994). The columns were installed with a simple auger-type mixing tool, using five passes of the tool over a 1 m withdrawal length. Around 1995, soil mixing was introduced for geoenvironmental applications, with growing importance since 1997. Currently the UK is leading Europe in the research and application of wet mixing to the containment and encapsulation of contaminated soils, including cut-off walls and reactive barriers (Lebon, 2002, Al-Tabbaa et al., 2009). In 2001, the UK saw its first use of the dry Nordic method (Keller), and in 2010 the TRENCHMIX machine was first used for flood defence works in Nottingham (Bachy Soletanche).

In Italy, the Trevi SpA developed in the late 1980s a dry mixing method named TREVIMIX. The equipment has more similarities with the Japanese DJM method than with the Nordic method. In this system one or two (more common) shafts with mixing paddles of 1.0 m (or 0.8 m) in diameter are arranged at variable spacings of 1.5–3.5 m and are used to disintegrate soil structure during penetration with air. The distinction of this system lies in its ability to operate in dry or semi-dry conditions by adding a controlled amount of water to the soil in order to ensure a hydrating reaction. First applications in Italy have been reported by Pavianni and Pagotto (1991). Another development is the TURBOJET wet mixing system that uses a tubular Kelly with drilling bit and mixing blades, and combines mechanical mixing and single fluid jet grouting technology.

In Poland, the wet method of DM was first introduced in 1999 by Keller Polska, initially using single-axis equipment and later twin-shaft tools. The first project involved execution of intersecting columns forming a cut-off wall along an old dam of the Vistula River in Kraków. Since then, the use of wet DMM in Poland is probably the highest in Europe (except for environmental applications), including the first worldwide applications of DM for the foundations of highway bridges (first in 2002) and modern wind turbines (first in 2007).

In Belgium Smet-Boring NV has developed a modified DM system, called TSM (Tubular Soil Mixing). TSM uses a mixing auger inside an outer casing, diameter 43–63 cm, equipped with a set of nozzles for high-pressure jetting. The casing contains holes at its bottom section in order to avoid blockage in the ground if pressure in the soil-slurry mixture builds up. The casing also reduces the lateral displacement of soil/slurry and improves homogeneity of soil-mix. The major applications are for excavation support, with a rather small column overcut (about 5 cm) because of a high accuracy of vertical alignment.

Another important European achievement is the cutter soil mixing (CSM) system, derived from the cutter diaphragm walling technique, jointly developed by Bauer Maschinen GmbH and Bachy Soletanche since 2003. The soil is broken down and mixed in situ with slurry by two sets of cutting/mixing wheels rotating about horizontal axes (cf. Fiorotto et al., 2005). The CSM system offers significant advantages over other walling techniques, and has been successfully transferred to countries outside Europe, including Japan and the US.

Deep SM is also very popular in China and Southeast Asia. In China DMM and DJM were introduced in 1977 and 1983, respectively (Zheng and Liu, 2009). Both methods are now widely used for a variety of applications and proved to be competitive in terms of costs and time of execution. To improve the performance of DM columns various types of composite columns have been also developed (CDMC). They may include installation of a precast concrete pile, reinforced concrete pile, or a steel pipe inside a fresh DM column, or even execution of DM column inside a sand column of comparatively large diameter. In Korea DCM has been developed since mid-1980s and the use is increasing, especially for marine and harbour works using special barges equipped with multiple mixing shafts (Kim et al., 2009).

The hitherto development of different technologies and equipment used in SM is difficult to follow without a certain generic classification system. Several similar systems have already been developed for this purpose (FHWA 2000, CDIT 2002, and EN 14679). The classification format adopted herein is based on three fundamental operational characteristics. The distinction between wet and dry technologies with respect to the form of binder introduced into the soil is the most straightforward, and hence the most widely used format. In the dry mixing methods the medium for binder transportation is typically compressed air, while in the wet mixing methods the medium of transportation is typically water. The second key characteristic is related to the method used to mix the binder (i.e., by mechanical action of the mixing tool with the binder injected at relatively low velocity), hydraulic action of the fluid grout injected at high velocity (jet grouting), or by a combination of both aforementioned techniques (so-called hybrid mixing). The third basic characteristic reflects the location of mixing action at the end or along a specific tool. The elaborated classification chart with the allocation of several operational methods, split with account for the difference between systems involving rotation of the cutting/mixing tool about vertical or horizontal axes or around the whole cutting arm, is shown in Figure 9.6.

When comparing technical features of recently used DMM and SMM machines and operational systems, it should be kept in mind that the aforementioned methods have been developed while taking into account various demands and constraints of regional markets, as well as soil conditions prevailing in areas of potential application. Moreover, various operational systems also reflect different objectives of ground improvement and design approaches. Consequently, not all SM methods can be regarded as equivalents, although all are based on the same overall concept of in-situ soil stabilisation. Despite these variations, the main technical goal of any SM method is to ensure a uniform distribution of binder throughout the treated soil volume, with uniform moisture content, and without significant pockets of native soil or binder.

9.3 EQUIPMENT AND EXECUTION

9.3.1 Dry method deep mixing

Typical dry method DM construction equipment consists of a stationary or movable binder storage/premixing and supply unit, and a mixing machine for the injection of binder material and installation of the columns. The binder is delivered to the mixing machine by compressed air. The equipment components generally include: silos with stabilising agents, pressurised tank with binder feeder system, high-capacity air compressor, air dryer, filter unit, generator, control unit, and connecting hoses. The two major techniques for dry mixing are the Japanese DJM and the Nordic method.

The DJM mixing machines are equipped with one or two mixing shafts, and are able to install columns to a maximum depth of 16–33 m (DJM Assoc., 2002). A dual mixing shaft is the current standard outfit, while a single shaft may be used in narrow working areas or for sites with headroom restrictions. The driving unit of the mixing shafts is located at the foot of the tower to improve machine stability while the shafts are kept together with a transverse steel bar, allowing for interlocking or tangential positioning of the mixing blades. The bar, and sometimes additional freely

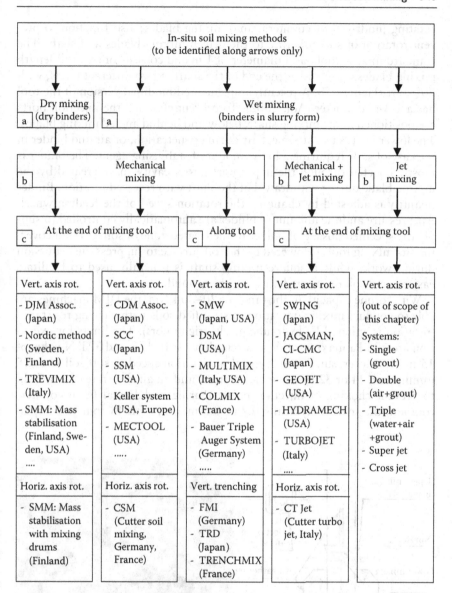

Figure 9.6 General classification of *in-situ* soil mixing based on (a) binder form, (b) mixing principle, and (c) location of mixing action, with allocation of selected fully operational methods developed in various countries, split with respect to rotation characteristic of the cutting/mixing tool.

rotating (undriven or counteracting) mixing blades, also function to prevent rotation of soil adhering to the driven mixing blades and shaft. The standard mixing tool has a diameter of 1 m and consists of two full-length mixing blades, mounted at the end of the shaft at two different levels, with 90° shift (Figure 9.7). A recently implemented modified version of the tool has a 1.3-m diameter (Aoi, 2002). To prevent choking, the injection ports are positioned at the mixing shaft, below and behind angled mixing blades. The lower port is used to inject air during penetration, or air and binder in the case of soils requiring a high amount of stabilising agent. The binder is mostly injected through the upper port into a cavity space created by the mixing blade during withdrawal of the shaft with reversed rotation. Binder quantity is adjusted by changing the rotation speed of the feeding wheel. Air pressure and the amount of binder are automatically controlled to supply the specified dosage of binder to the treated zone of soil. A hood covering the mixing tools is lowered to the soil surface to suppress dust emission during work, while a square mixing shaft is generally used to facilitate easier expulsion of injected air from the ground.

With torque capacity in the range of 20–30 kNm, the DJM machines are able to conduct mixing operation in stratified soils with varying resistance (rotation is agitated by hydraulic or electric motors). The limits for execution are 70 kPa maximum shear strength for stiff clays and SPT N-value of 15 in sands (Terashi, 2003). Typical penetration speed in soft soil is 1–1.5 m/min with 24 or 32 rpm (electric motors) and an air flow rate of 2 m³/min to prevent choking of injection ports. During withdrawal (with counter-rotation) the speed is typically 0.7 or 0.9 m/min, with 48 rpm or 64 rpm,

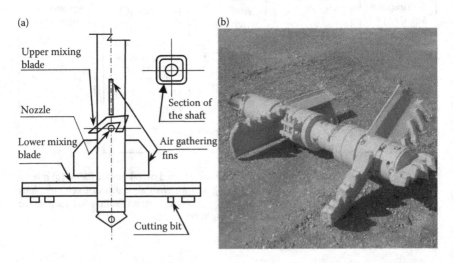

Figure 9.7 Mixing tools of the DJM method. (a) Construction scheme. (Redrawn from DJM Association, 2002.) (b) Recently used single mixing tool of 1.0 m diameter.

respectively (electric motors), and air flow rate of 5 to 3 m³/min at shallow depths. Consumed air volume may vary between 2–9 m³/min, requiring heavy-duty compressors with a capacity of 10.5–17 m³/min/shaft. The volume of the binder tank is usually 2–3.5 m³ per one mixing shaft (DJM Assoc., 2002).

The mixing machines developed in Sweden and Finland are lighter than the Japanese rigs and are equipped with one mixing shaft. They are constructed to work mainly in soft to very soft soils with undrained shear strength below approximately 25 kPa (maximum 50 kPa). The torque capacity at 180 rpm is typically about 7 kNm, and increases to 30–40 kNm at 20–30 rpm (some machines have two engines driving the Kelly rotation). This allows for the installation of 0.6–0.8 m diameter columns to a depth of about 25 m, 1–5 m from the edge of the base unit. The columns can be also inclined up to about 1:4; maximum 1:1.

The equipment on site usually consists of a drill rig and a separate self-driven mobile shuttle, hosting pressurised binder material tank(s), an air dryer and a compressor. In addition to the plant items working on the construction site, there is usually a requirement for a premix station including a filter unit, especially when delivery of ready-to-use binder is too expensive. The binder shuttle moves between the premix station and the drill rig, which is normally working several hundred metres away from the premix station. During production, the shuttle is connected to the drill rig by an umbilical through which the binder passes (via compressed air), along with monitoring information on the binder mixing and supply rate. For shallow penetration depths, combined mixing machines with on-board installations are available. The amount of discharged binder is controlled with a cell feeder mechanism, located at the bottom of the supply tank.

The air containing the binder is transported through the hollow Kelly bar to an exchangeable mixing tool, mounted at the end. Typical mixing tools consist of horizontal and curved or angled cutting/mixing blades, as shown in Figure 9.8. The injection outlet is located at the central shaft, close to the upper horizontal mixing blade. After the required depth is reached, the mixing tool is lifted and simultaneously rotated in reverse, while the binder material is horizontally injected to the soil. Typical withdrawal speed is 15–25 mm per rotation, with about 150–180 rotations per minute.

A summary of mixing conditions for selected dry DM methods is presented in Table 9.1.

9.3.2 Dry method shallow mixing

Shallow dry method mixing offers a cost-effective solution for ground improvement works or site remediation when dealing with substantial volumes of very weak or contaminated superficial soils with high water content, such as deposits of dredged sediments, wet organic soils, or waste

Figure 9.8 Selected mixing tools. (a) SD 600 mm, (b) modified SD 600 mm, (c) PB3 600 mm, (d) peat bore 800 mm. Note: changed location of binder outflow hole in relation to the horizontal mixing blade in standard (a) and modified (b) tool. (Courtesy of LCM.)

sludges. In-situ mixing of the encountered soil mass with dry reagents to the depth of a few metres can be economically carried out with large-diameter single-axis augers, or with recently developed mass mixing tools implemented in Finland and Sweden. In such applications it is also quite common for the topsoil to be too weak to provide safe support for heavy mixing machines. Therefore, it is best to use execution methods that employ mixing tools suspended from the crane or mounted on elongated cantilever arms, as they usually offer more flexible operation in the field.

The shallow soil mixing (SSM) method, modified for accommodation of dry binders, utilises a crane mounted single auger tool 1.8–3.7 m in

Table 9.1 Mixing conditions for selected dry deep mixing methods

	Selected dry DM methods		
Technical specification	DJM	Nordic method	Trevimix
Number of mixing shafts	2 (standard), 1	1	2 (more common), 1
Diameter of mixing tool [m]	1.0 (standard) 1.3 (modified)	0.5–1.0 possible 0.6, 0.8 standard	0.8–1.0 (standard)
Realistic maximum penetration depth [m]	33	25 (30)	30
Penetration/Retrieval velocity [m/min]	0.5–3 (4), 7 (1 shaft) typically: P: 1.5, R: 0.7, 0.9 (R: 15 mm/rev.)	P: 2–15 R: 2–6 (R: 15–30 mm/ rev.)	P: 0.4 R: 0.6
Penetration/Retrieval rotation speed [rpm]	P: 24, 32 (Electr.) R: 48, 64 (Electr.) P/R 21–64 (Hydr.)	R: 100–220 (150–180 typically)	10–40 P: 20 typically R: 30 typically
Injection during Penetration/Retrieval	R (P used: air/binder)	R (P possible)	R (P used: air/binder)
Footprint area of the mixing tool (max.) [m²]	0.78 : 1 × 1.0 m 1.56 : 2 × 1.0 m 2.65 : 2 × 1.3 m	0.28, 0.5 (0.78)	0.78 : 1 × 1.0 m 1.56 : 2 × 1.0 m
Amount of injected dry binder [kg/m³]	100–400 Cem.:sands 200–600 Cem.: peat 50–300 Lime: clay	70–150 150–250 organic soils	150–300 250 typically
Binder supply capacity per shaft [kg/min]	25–120 standard, up to 200 mod. version	40–230	around 100
Injection pressure [kPa]	P: 100–600 R: 600–100	400–800	600–1,000
Productivity [m³/shift]	300–700	150–300	150–220

diameter (Jasperse and Ryan, 1992). The driver for the tool is a drilling system. It can be a conventional hydraulic drill or a high-torque dual motor turntable. The auger tool itself is specially designed to break up the soil and/or sludge and mix it with dry reagent without bringing the material to the surface. To suppress emissions from the mixing process and/or for environmental applications, the mixing tool can be enclosed in a hood or bottom-opened cylinder to control dust and airborne contaminants (Figure 9.9). Further components of environmental control may also include a low pressure blower or vacuum pump to keep negative pressure inside the hood during operation, a dust collector, a fume incinerator, or an activated carbon scrubber, depending on site-specific conditions and contaminants (Aldridge and Naguib, 1992).

Figure 9.9 SSM mixing tool diameter 3.7 m for the dry method. (Courtesy of Geo-Con Inc.)

Treatment reagents are transferred pneumatically to the mixing unit. The delivery system consists of bulk storage tanks, several pneumatic pumps, portable booster stations, and material receivers. The final application of the reagent to the treated soil is made with the hood lowered by dropping the reagent into the hood through a calculated rotary valve located at the bottom of a material receiver. Various cementitious, chemical, or even biological reagents can be added to soil or waste with this method.

The shallow mixing machines developed in Finland and Sweden are essentially different from the column stabilisation machines. The mass mixing tools are typically attached to the arm of a crawler mounted excavator to enable vertical and horizontal movements of the tool through the soil to complete mixing (Figure 9.10). The binder is fed from a separate unit

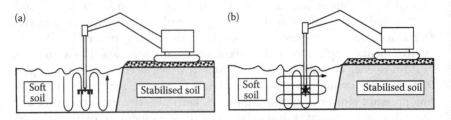

Figure 9.10 Mass stabilisation using dry binders. (a) Mainly vertical mixing. (b) Vertical and horizontal mixing.

which houses the pressurised binder container, compressor, air dryer, and supply control unit. The operator injects the binder into the soil in such a manner that it is equally distributed and mixed.

Different mixing tools have been tried in the past to treat very soft and organic soils. The tools currently in use comprise cutting/mixing heads equipped with blades rotating about a vertical axis, as shown in Figure 9.11, as well as very efficient cutting/mixing drums developed by Allu Finland Ltd. (Powermixer systems PM and PMX) and suitable for large mass stabilisation projects (Figure 9.12). The diameter of the mixing tool rotating about a vertical axis is normally 600–800 mm, and the rotation speed lies between 80–100 rpm. This method can be applied for soft clays and organic soils with shear strength below 25 kPa. The more powerful PM/PMX mixing drums, with maximum torque of 7 and 23 kNm for two drums, respectively, can be fitted with different types of exchangeable blades, teeth, or wings enabling enhanced mixing or cutting/breaking action in the treated soil. A typical mixing drum has plan width of 1.5–1.6 m and outside tooth to tooth diameter of 0.85–0.95 m. The working depth is 0–5 m. The novel aspect of the PMX system consists of drums mounted on two inclined axes of rotation. This arrangement reduces the distance between both drums and improves the homogeneity of mixing by eliminating potential zones of unmixed soil which may appear behind the mounting arm in case of the PM tools, especially when mixing soils with higher shear resistance.

The mixing pattern of mass stabilisation is planned taking into account site-specific conditions and capabilities of the mixing machine and the

(a)   (b)

Figure 9.11 Mass stabilisation of organic soil in Sweden. (a) Equipment. (b) Mass mixing tool diameter 1.0 m. (Courtesy of LCM.)

(a)

(b)

Figure 9.12 Mass mixing tools developed by Allu Finland Ltd. (a) Powermixer system PM 500, with cutting/mixing drums diameter 0.95 m, mounted on a horizontal axis of rotation. (b) Powermixer system PMX 500, with cutting/mixing drums diameter 0.85 m, mounted on two inclined axes of rotation.

mixing tool. Usual practice is to stabilise in one sequence a block of soil within the operational range of the machine (e.g., 4 × 4 m in plan and 5 m deep). When the prescribed amount of binder is mixed into the volume treated, remoulding is continued in order to obtain a homogeneous soil-binder mixture. The productivity rate is approximately 200–400 m³/shift of stabilised soil, depending on the system/tool used and type of soil. In heavier soils the production rate decreases. The amount of binder used is typically in the range of 100–250 kg/m³. In Scandinavia the objective for shear strength in peat is usually 50 kPa (Jelisic and Leppänen, 2003).

9.3.3 Wet method mechanical deep mixing

The wet DM methods applied for ground treatment on land in Japan, US, and Europe are generally developed to produce similar quality columns or panels/walls of stabilised soil, with unconfined compressive strength in the order of 0.5–5 MPa, or even more in granular soils, while the machines, mixing tools, execution procedures and productivity differ considerably.

Typical wet-method DM construction equipment consists of a batch mixing plant to supply proprietary slurry, and of a mixing machine for injection and mixing of slurry into the ground. The plant generally includes silos, water tank, batching system, temporary storage tank, slurry pumps (equipped with flow metres), and power supply unit. Cutter soil mixing and vertical trenching processes can be also supported by compressed air and a desanding plant. The batching system can be varied from manual or computer-controlled colloidal shear mixer to a very fast inline jet mixing system. The storage tanks have paddle agitators to keep the component materials from settling out of the slurry. Delivery pumps are duplex or triplex reciprocating piston pumps, or variable speed progressive cavity pumps. Pumping rates typically range from 0.08–0.4 m³/min, but can reach up to 1 m³/min for high-capacity mixing tools. Any changes in the slurry are made by adjusting the weight of each ingredient. Since fluid volume is being introduced into the ground, spoils must come to the surface.

The operational systems involving wet method mechanical deep mixing can be subdivided into three groups, taking into account rotation of the mixing tool about vertical or horizontal axes, or around the whole cutting arm (cf. Figure 9.6).

9.3.3.1 Mixing tools rotating about vertical axes

The machines that are used for on-land applications usually have 1–4 shafts mounted on fixed or hanging leads, and are equipped with specially designed mixing tools. A multi-axis gearbox distributes the torque from a rotary drive unit to each shaft for penetration to the intended depth. The penetration speed is typically in the range of 0.5–1.5 metres per minute and is usually increased during withdrawal. The mixing tools are kept in parallel by joint bands mounted at vertical intervals along the drive shafts. With some machines the spacing between individual shafts can be adjusted within prescribed limits to produce overlapped columns, which is beneficial when forming continuous panels or blocks of stabilised soil in a single-stroke operation. A summary of mixing conditions for selected wet DM methods used for on-land applications and utilising mechanical mixing is presented in Table 9.2.

For marine applications, large execution vessels equipped with the mixing machine, batching plant, storage tanks, and a control room are usually used for rapid treatment of considerable soil volumes. The area of treatment in single-stroke operation with 2–8 mixing shafts ranges from 1.5–9.5 m², and the productivity rates are in excess of 1000 m³ per day.

The mixing tools for the wet method are designed for various improvement purposes and are configured to soil type and available turning equipment. Since there is no one tool that can successfully treats all soils, field adjustments are typical. The mixing tools can be broadly classified into blade-based and auger-based constructions (cf. Porbaha et al., 2001).

The tools of the first group have an assortment of flighting and mixing blades of full or near-full diameter and different orientation to efficiently break down the soil structure. Steel hard-facing and an arrangement of purposely located teeth serve to aid penetration and reduce maintenance. A small diameter lead auger/drilling bit usually extends below the cutting blades to centre and control penetration and verticality. The mixing process is mainly conducted at or within a short distance from the tip of the drill shaft(s). Injection nozzles are strategically located on the tool to uniformly distribute the slurry into the soil. They are usually found near the shaft tip, but can also be located along and above the mixing blades. Sophisticated single-axis systems with double cutting/mixing blades, spaced apart and rotating in opposite directions, have also been developed in Japan (Horpibiulsuk et al., 2002) and in Germany (Bauer Machines, 2012). The counter-rotating components provide an exceptionally high degree of shearing, especially in cohesive soils, and uniformity of soil-mix. Examples of single- and multiple-shaft mixing tools are shown in Figures 9.13 and 9.15, and a single-column mixing tool for double rotary drives is shown in Figure 9.14. The latter tool is designed in such a way that the outer cutting frame, driven by a rotary head with a larger torque, is used to loosen the soil whereas the faster rotating inner blades are used to enhance the mixing process.

The second group employs discontinuous or continuous helical augers for drilling and mixing or several levels of inclined paddles located above the cutter head of the mixing shaft. In these systems interlocking or closely spaced multiple shaft arrangements are typically used and the mixing operation is enhanced by counter-rotating action of adjacent shafts. The mixing process occurs along all or a significant portion of the drill shaft(s). The direction of rotation is usually reversed during withdrawal. In most systems the slurry is fed through nozzles located at the bottom of each shaft. Examples of mixing tools are shown in Figures 9.16 through 9.18.

Table 9.2 Mixing conditions for selected wet DM methods used for on-land applications

Technical specification	Selected wet DM methods for mechanical mixing about vertical axis(es)							
	CDM (Stand. and MEGA)	CDM Land4	SCC	HB-Keller USA/Europe	Bauer	SMW	DSM	COLMIX
Number of mixing shafts	2 / 1:older syst.	4	1 possible 2	1, 2, 3	3 possible 1	1-3,5 usually 3	1-6 usually 4	2, 3, 4
Diameter of mixing tool [m] (shaft spacing)	1.0: S / 1.2/1.3: M (variable)	1.0/1.2 (variable)	0.6-1.5 / 1.2 (2 shafts)	0.5-2.4 :U / 0.6-1.2 :E	3 × 0.37 / 3 × 0.55 / 3 × 0.88	0.55-1.5 usually 0.9 (variable)	0.8-1 usually 0.9	0.23-0.75 :2 / 0.36-0.50 :3 / 0.50-0.75 :4
Realistic maximum penetration depth [m]	50 (55) / 30: M 1.2 / 20: M 1.3	25	20	20	0.37: 10.5 / 0.55: 15.7 / 0.88: 25	35 (50)	35	20
Penetration/Retrieval velocity [m/min]	P: (0.3) 0.5-1 / R: 0.7-1 (2)	P: 0.7-1 / R: 1.0	P: 1.0 / R: 1.0	P: 0.3-1 / R: 1-2	P: 0.2-1 / R: 0.7-1 (5)	P: 0.5-1 / R: 1.5-2	P: 0.6-1 / R: 1-2	P: 0.8 / R: 1.0
Penetration/Retrieval rotation speed [rpm]	P: 20 / R: 40	P: 20 / R: 40	30-60	P: 20-25 / R: 40-60	20-40 (80)	P: 14-20 / R: higher	15-25	8-30
Injection during Penetration and/or Retrieval	P and/or R, restroking at the bottom	P and/or R, restroking at the bottom	usually P, restroking at the bottom	P (+R) :U / P + R with restroking:E	P and/or R, P (30%-50%) restroking	P and R, restroking common	P (+R), restroking at the bottom	P (+R) / ev. restroking in clays
Water/Cement ratio	0.6-1.3 av. 1.0	0.6-1.3 av. 1.0	0.6-0.8 clays, 1.0-1.2 sands	1-1.5: U / 0.6-1.2: E	0.6-2.5	0.7-2.5	1.2-1.75 av. 1.5	0.7-2.5
Footprint area of the mixing tool [m²]	1.5: 2 shafts / 0.8: 1 shaft / 2.17/2.56:M	2.83-3.14 or 4.21-4.52	0.3-1.75 / 2.25 two shafts	usually / 1.1-4.5: U / 0.5-1.5: E	0.44: 3 × 0.37 / 0.94: 3 × 0.55 / 2.35: 3 × 0.88	0.7: 3 × .55 / 1.7: 3 × 0.9 / 4.7: 3 × 1.5 (4 shafts, tangential)	2.5 (4 shafts, tangential)	0.08-0.95 :2 / 0.29-0.57 :3 / 0.76-1.60 :4
Amount of added (dry) binder [kg/m³]	70-300 av. 140-200	70-300 av. 140-200	150-400	150-275: U / 250-450: E	80-500	200-750	120-400	100-550
Productivity per shift (one rig)	100-200 m³	500-700 m³	100 m² wall 400 m col.	250-750 m³:U / 75-200 m³:E	30-300 m³	100-200 m³	200-300 m² wall	100-300 linear m

(a) (b)

Figure 9.13 Single-shaft mixing tools. (a) Diameter 0.8 m. (b) Diameter 2.4 m, with a free-blade system. (Courtesy of Keller-Hayward Baker.)

Figure 9.14 Single-column mixing tool for double rotary drives (SCM-DH); column diameter 1.8 to 2.4 m, max. depth 23.5 m. (From Bauer Maschinen GmbH. (2010). *Cutter Soil Mixing: Process and Equipment.* Brochure No. 905.656.2.)

(a)

(b)

(c)

(d)

Figure 9.15 Multiple shaft mixing tools. (a) Standard CDM 2 × 1.0 m. (From Cement
Deep Mixing Association. (2002). *CDM Cement Deep Mixing Bulletin as of
2002.*) (b) CDM Land4, 4 × 1.0 m. (From Cement Deep Mixing Association.
(2002). *CDM Cement Deep Mixing Bulletin as of 2002.*) (c) SMW mixing paddles 3 × 1.5 m. (d) Cutter head. (Courtesy of R. Jakiel.)

In addition to the information presented in Table 9.2, another CDM
machine for on-land applications deserves more specific attention. The
CDM-Column21 machine uses two shafts with large mixing heads of 1.5 m
(1.6 m) diameter consisting of an upper and lower mixing unit, both of
which are equipped with inner and outer mixing blades that rotate in
opposite directions (Figure 9.19). The unique counter-rotating action of the
blades accentuates the shearing mixing effect and ensures uniform mixing

Figure 9.16 DSM mixing tools. (a) Four blade-based mixing shafts. (b) Four discontinuous, interlocking augers, diameter 0.9 m. (Courtesy of Geo-Con Inc.)

of the cement slurry with the soil. The unit is capable of treating harder ground formations sandwiched between softer layers. The area of treatment is 3.5 m² and the required capacity of slurry supply is up to 1.0 m³/min. The maximum depth of treatment is 40 m. This modern system not only reduces the unit cost of soil treatment due to its very high productivity,

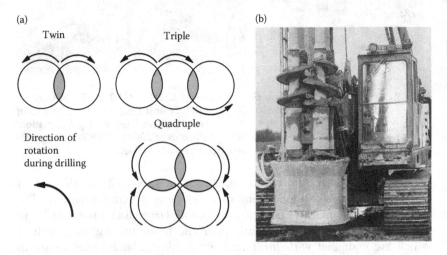

Figure 9.17 COLMIX mixing tools: (a) Possible arrangements. (b) Four discontinuous, interlocking augers diameter 0.5 m. (Courtesy of Bachy-Soletanche.)

(a)   (b)

Figure 9.18 Bauer mixing tool with three closely spaced, continuous augers, diameter 0.55 m. (Courtesy of Bauer Spezialtiefbau.)

but also offers higher-quality soil improvement through increased mixing operation efficiency (CDM Assoc., 2002).

Except for special situations and projects executed very close to sensitive objects, wet method deep soil mixing has a very low impact on nearby structures. To avoid net volume increase and corresponding lateral stress in the ground caused by penetration of the mixing tool

Figure 9.19 CDM-Column 21 mixing tool. (Courtesy of CDM Association, Japan.)

and injection of cement slurry, a dedicated method called CDM-LODIC (low displacement and control) has been developed and modified since 1985 (Sugiyama, 2002). In this system the upper part of the mixing shafts are equipped with auger screws to forcibly expel equivalent soil volume during penetration and withdrawal stages of the mixing tool (see Figure 9.2). The screws have standardised dimensions (diameter and pitch), and can be changed to best suit the ground conditions. It has been demonstrated that the installation of CDM and CDM-LODIC columns in soft clay 1.5 m from an inclinometer installed in a vertical borehole causes at the depth of 17 m maximum horizontal displacements of 16.11 and 1.01 cm, respectively, confirming efficiency of the LODIC method (Horikiri et al., after CDIT, 2002). In addition to the normal quality control system used for the conventional CDM method, an automatic system has recently been developed to display the volume of extracted soil. Since the cement slurry is normally injected during withdrawal through the nozzles located above the mixing blades, the extracted soil is free of cement and can be deposited or reused without any restrictions, if not contaminated.

9.3.3.2 Mixing tools rotating about horizontal axes

The cutter soil mixing (CSM) method utilises special cutting and mixing heads derived from diaphragm walling cutter technique. They are attached to round or rectangular shaped Kelly bars, reaching penetrations of up to about 20–40 m, respectively, or are suspended from ropes to reach depths up to about 50 m. The standard head consists of two sets of counter-rotating, vertically mounted, cutting/mixing wheels, as shown in Figure 9.20a. The wheels run on independent drives and can be driven in both directions. They are equipped with cutting teeth capable of drilling and mixing even in stiff ground and keying into bedrock. The in-situ soil is broken up, while at the same time a specific slurry is pumped to the nozzles centrally positioned between the wheels. The rotating wheels and cutting teeth push the loosened soil through vertically mounted shear plates that have the effect of a compulsory mixer to form homogeneous soil-cement panels, usually 2.4 or 2.8 m long and 0.55 to 1.2 m wide. Compressed air can be also used to assist cutting and mixing operation during down stroke phase. Typical penetration speed of the cutting/mixing head is 25–30 cm/min in sand and gravel and 5 cm/min in cobbles. Withdrawal speed is usually 50 cm/min.

The CSM tool cuts vertical rectangular panels resulting in fewer vertical joints when compared to multiple shaft systems and making this system ideally suited for linear in-situ structures such as cut-off barriers, retaining walls and liquefaction mitigation cells. Reinforcement in the form of steel sections can be used to provide additional structural strength where needed. An advantage of the method is that both fresh in fresh and fresh to hard

(a)
(b)

Figure 9.20 The cutter soil mixing heads. (a) Standard BCM 5, type 3-2 (three tooth holders in one row of teeth). (b) QuatroCutter. (Courtesy of Bauer Spezialtiefbau GmbH.)

panel construction joints can be facilitated. For bigger depth, two standard cutting and mixing heads attached at the bottom and at the top of a special frame are used (Figure 9.20b). This system, known as QuatroCutter (Bauer Maschinen, 2010), ensures intensive and homogeneous mixing as well as high directional accuracy of CSM walls up to 60 m deep.

In relatively uniform soils, or for retaining structures up to about 20 m deep, mixing is conducted during cutting (penetration) and withdrawal. The backflow of soil and slurry is collected in a pre-excavated trench. For deeper cut-off walls and for less uniform soils, a two-phase approach is adopted for CSM using bentonite for temporary trench support during cutting. As with conventional diaphragm wall construction, the bentonite slurry is recirculated and cleaned by passing through desanding equipment. Grout is injected during the withdrawal phase and mixed with remaining soils. The speed of extraction and flow of grout are adjusted to ensure that the desired quantity of cement is blended with the soil.

The CSM system allows complete instrumentation inside the cutter gearbox support frame to read and control in real time the coordinates of the cutting head. This inclinometer system, coupled with the advantage of a steerable tool, provides assurance of complete overlap between panels. In favourable soil conditions the net productivity can reach about 40 m² of a CSM wall/hour. However, a daily output is usually 100–200 m² (based on 70–90-tonne machines with power outputs of 260–300 kW) due to high maintenance on the rig and cutter head. Wear rates are different for each type of soil. For instance, in compacted sand and gravel re-welding of the head was needed every 500–1,500 m² and the wear was 0.1–0.2 teeth/m². Changing of the head takes one day.

9.3.3.3 Vertical trenching

A further facet of soil mixing has been provided by the development of equipment which enables wet mixing while cutting trench structures in the ground (e.g., FMI, TRD, and TRENCHMIX). Key advantages of these single-phase walling methods, given the right conditions, are a reduction in the number of joints over competing methods and less waste for any required wall thickness as overlap is minimised, leading to reduced costs and improved quality.

TRD equipment consists of a large machine about 100 tonnes and 7 m tall and effectively enables cutting and mixing by means of a chainsaw concept (Figure 9.3a and b). An initial starter trench is formed into which a post is assembled and lowered into the ground to the required depth. The post holds the cutting chain and injection proceeds as the machine crawls forward cutting a full-depth face and providing a uniform mix in place material devoid of the original soil stratification. Joints in the soil-mixed trench material only occur if production is stopped, for example if only day shifts are being used. In this way joints are few and the system is particularly efficient for long cut-off barriers (cf. Evans and Garbin, 2009). Depth is limited to roughly 60 m. Wall thicknesses of 550–800 mm are possible with presently available equipment. For earth retention applications, steel beams are inserted in the freshly constructed wall to provide the required lateral strength. Productivity is highly influenced by depth, width, soils (rock), excavation support versus cut-off, and length of segments. For trenches about 20 m deep and 700 mm wide, average productivity is about 250–300 m²/shift. The teeth can need changed once a week to once a shift if working in hard soil/rock.

Quality control of TRD walls during construction includes monitoring of the grout components and specific gravity (SG) testing of the neat grout in real time using a mass flow sensor. The same sensor also measures and records the flow rate, volume, and temperature of the grout being pumped through the system. It is typical to also verify SG several times each shift using a mud balance as a check test for the instrumentation. Additionally, the wet soil-mix material from the trench is sampled and subjected to flow table testing in order to assess the mix viscosity. Maintaining the flow table value within experimentally established range ensures proper material flow around the cutter post, which is essential for uniform full-depth mixing (Garbin et al., 2010). Wall verticality is controlled by the operator and monitored in real time using inclinometers installed inside the TRD cutter post at various elevations. Additionally, the position of the cutter post can be tracked using a differential GPS (Global Positioning System), as well as with routine surveys using a total station.

TRENCHMIX uses a modified Mastenbroek ditchdigger (Figure 9.21a). The trench cutting chain is reversed to enable breaking up and mixing of

(a)

(b)

Figure 9.21 TRENCHMIX equipment. (a) Rig in operation on a river dike. (b) Close-up of cutting/mixing teeth. (Courtesy of Bachy Soletanche and Mastenbroek.)

the soil with slurry injected at a controlled rate along the boom. Effective and thorough mixing is ensured by the specially designed teeth (Figure 9.21b) and high energy mixing process, controlled by a purposely designed QC/QA system. The fact that the soil-mix material is drawn to the surface allows also a good visual inspection of the efficiency of mixing. Vertical mixed elements of about 0.4 m width to depths of typically 4–10 m (max 13 m) can be constructed. These elements can be used to form cut-off walls to control pollution or groundwater in the ground, or as improved ground foundation bearing elements when spaced close together.

The trencher, despite its long boom, is highly manoeuvrable and capable of operating in narrow and limited headroom spaces. This technique is faster than sheet piling, produces significantly less spoil than conventionally dug walls and has no issues with noise and vibration. In favourable soil conditions, the driving speed may reach 30–40 m/hour at the depth of 10 m and width of 0.4 m.

Mixing in the trench using dry binders is also possible given the right conditions. The binder is placed in the shallow pre-trench and water is added during mixing to achieve the required workability.

9.3.4 Wet method mechanical shallow mixing

Wet method mechanical shallow mixing can be used to improve substantial areas of soft or loose superficial soils in ground engineering applications, as well as for stabilisation and fixation of contaminated soils.

The SSM method uses specially designed single augers of 1.8–3.7 m diameter, attached to a hollow-stemmed Kelly rod suspended from a crane. Similar systems offer rigid attachment of the mixing tool to the base unit, and can therefore incorporate down-pressure capability. The Kelly transfers the torque and feed pressure to the mixing tool, while the swivel mounted at the top of the rod seals the connection for delivery of the binder during rotation. The binder is usually injected during penetration, in slurry form, through several ports mounted at the bottom of the mixing augers. The pitch on the auger flights and the centrifugal force help to distribute the binder to all parts of the column during rotation. Cycling up and down with reduced binder delivery rates is often performed to improve mixing efficiency. An overlapping pattern of primary and secondary columns is normally used to ensure that the entire volume of treated soil is thoroughly mixed. A high-torque driver in the range of 400–600 kNm and high-capacity batching plant are generally required since the treatment area may reach about 10 m^3 of soil per meter of penetration. Examples of large-diameter mixing tools are shown in Figure 9.22.

Wet method mass stabilisation can be also carried out with specially designed mixing tools that are similar to those presented in Section 9.3.2.

(a)   (b)

Figure 9.22 Crane-mounted SSM tools for the wet method. (a) Diameter 2.4 m. (b) Diameter 3.7 m. (Courtesy of Geo-Con Inc.)

Druss (2002) describes a major project conducted at the Fort Point Channel site in Boston, where very soft organic sand and organic silt deposits were shallow mixed prior to the execution of DM in underlying marine blue clay. The works were mostly performed underwater, in areas initially dredged to remove obstructions and timber piles. The objective of stabilisation was to construct a temporary support for a drill bench required for land-based DM operations. Shallow mixing was performed using a sectional barge, an excavator with extended reach, and a shallow mixing bucket containing blades rotating about a horizontal axis (Figure 9.23). Jet nozzles delivering the fly ash and cement grout were located inside the bucket and were directed towards the mixing blades. The bucket mixed horizontal trenches 1.2 m wide and about 10 m long in 1-m vertical lifts, moving from the surface of soft sediments to the top of clay or finishing at partial depth.

A similar application has also been mentioned by Terashi (2002a). The original ground was an artificial landmass in Imari City, Japan, reclaimed by dredged sea-bottom clay with undrained shear strength on the order of 1 kPa. A 2-m thick block of treated soil was used to provide a working platform and/or temporary access road floating on the extremely soft soil deposit. In this particular case, a special floater equipped with four mixing shafts was placed directly on the soft soil and dragged horizontally by winches while the mixing tools were moved up and down vertically. Similar shallow mixing tools as used in Finland and Sweden are also available in Japan.

Figure 9.23 Shallow mixing equipment used at Fort Point Channel Site in Boston. (Courtesy of R. Jakiel.)

9.3.5 Wet method hybrid deep mixing

In addition to mechanical mixing, these methods employ high-velocity jet grouting in order to reduce penetration resistance and improve mixing operation and/or to increase the diameter of the improved ground.

The SWING (spreadable wing) method, initially developed as a mechanical mixing system, uses a retractable mixing blade mounted at the end of a single drilling shaft. The position of the blade in the ground can be changed from a vertical to horizontal alignment and vice versa, as shown in Figure 9.24a and b. A combination of mechanical and jet mixing with cement slurry enables columns of up to 3 m diameter to be constructed, and the addition of compressed air allows columns greater than 3 m diameter. During penetration of the ground, the soil is broken down by rotation of

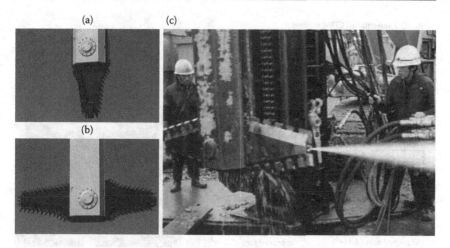

Figure 9.24 Spreadable wing (SWING) method. (a) Blade position during penetration. (b) Blade expanded. (c) Demonstration of jetting action. (Courtesy of SWING Assoc.)

the blade and jetting action of water. Cement slurry is injected during withdrawal, with the jetting energy supplemented by air pressure. Air is used when the larger diameters are required or when the soils under treatment are too stiff. This method also enables the installation of inclined or even horizontal columns and therefore allows soil mixing in areas of difficult access.

The JACSMAN system consists of two 10-bladed mixing tools, each combined with a pair of jet grouting nozzles aligned for Cross Jet (XJET) to ensure that over-cutting does not occur. As compared with the conventional CDM method, JACSMAN offers significant improvements that contribute to a more economical, high-quality product. Due to XJET cutting with air-enveloped, high-velocity cement slurry during withdrawal, the treatment area of single-stroke operation increases considerably and equals 6.4 m² for type A arrangement, with a 75% share of jet mixing, and 7.2 m² for type B arrangement, with a 63% share of jet mixing, as shown in Figure 9.25. Moreover, the diameter of the soil-cement column can be controlled and changed over the column's length through stopping and starting XJET action, not affecting the surrounding soil due to the dissipation of jet energy at the cross point (Figure 9.26). This allows for the adjustment of the column's diameter to soil stratification, as well as for controlled mixing operations close to structures or excavation walls. The main operating parameters of JACSMAN are as follows: jetting pressure 30 MPa, jetting slurry flow rate 4 × 150 l/min, air pressure 0.7 MPa, grout flow rate 2 × 200 l/min, grout pressure 5 MPa, and withdrawal speed of 0.5 and 1 m/min for type A and B arrangement, respectively (Kawanabe and Nozu, 2002).

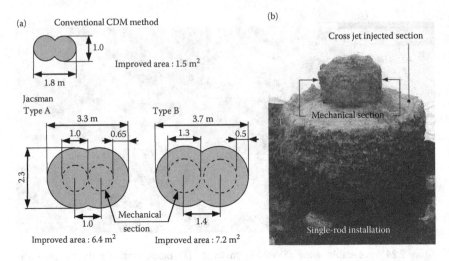

Figure 9.25 The JACSMAN method. (a) Comparison of treatment areas. (b) Exposed single column. (Redrawn from Kawanabe, S. and Nozu, M. (2002). Combination mixing method of jet grout and deep mixing, *Proceedings from the Deep Mixing Workshop 2002*, Port and Airport Research Institute & Coastal Development Institute of Technology, Tokyo, Japan.)

HYDRAMECH utilises mechanical mixing with a single shaft, fitted with 1.2-m diameter paddles and a 0.9-m diameter auger, in combination with high-velocity grout injection at 40 MPa through eight 2-mm 'hydra nozzles' on the outer edges of the mixing tool. HYDRAMECH is capable of creating soil-cement columns with diameters of 2 m. Mechanical mixing

Figure 9.26 JACSMAN mixing tool. (a) Twin head assembly (note grout nozzle in front of the tool). (b) XJET demonstration with increased pressure. (Courtesy of R. Essler.)

occurs smoothly in the centre of the column, and chunks of material are forced to the perimeter, where they are disaggregated by the jets. Treatment with jets can be switched on and off throughout the column length to create plugs of treated soil. Realistic maximum penetration depth is 20 m. Penetration/retrieval velocity is 1–3 m/min, with 5–20 rpm during penetration and 10–30 rpm during retrieval (additional mixing). Industrial productivities are in the range of 250–500 m³/shift. The main objective for developing this method was to improve on current jet grouting technologies that can create subsurface problems with the use of compressed air. HYDRAMECH can create an extended diameter soil-cement column without the injection of compressed air and still provide the continuous overlap that is a very positive aspect of jet grouting systems, particularly when installing horizontal barriers.

TURBOJET (GEOJET in the US) combines mechanical mixing with single fluid jet grouting technology. Jetting is used during insertion of the tool to increase penetration velocity while extraction is carried out solely with mechanical mixing. A specially designed mixing tool (or processor), fitted at the end of a tubular Kelly bar, consists of two levels of inclined blades and is furnished with several 4–8 mm diameter high-pressure nozzles mounted along the shaft and tip (Figure 9.27). The exact nature and composition of the processor can be varied, depending on soil conditions. Grout can be pumped with a discharge rate of 450 l/min at 30 MPa, although lower flows rates and pressure (15 MPa) are the norm. Tool diameters range from 0.6–1.5 m, usually 0.9–1.2 m, and the practical available depth of treatment is 25 m (Lebon, 2002). Instantaneous rates include 2–12 m/min (6 m/min typical) during penetration and 15 m/min during withdrawal. Computer control of the equipment during column formation is therefore required. The computer analyses the rate of tool rotation and penetration, slurry pressure, torque, crowd force, and soil-mix volume and density as a function of depth. The system also reacts to changing parameters and automatically adjusts to maintain specified soil-cement properties, even in varying subsurface soils. Because of the additional mixing energy supplied, restroking is not required. Industrial production rates in excess of 150 m/h and 1100 m/shift are possible. The system produces low waste volumes (20%–30% of ground treated).

9.3.6 Installation process (mixing about vertical axis)

The typical installation process consists of positioning the mixing shaft(s) above the planned location, penetration of the mixing tool, verification and improvement of the bottom soil layer, withdrawal, and movement to a new location if necessary. The details of execution depend on the type of method applied (dry or wet), technical features of the equipment, and the site-specific and functional requirements. Frequently used execution procedures are shown schematically in Figure 9.28.

Figure 9.27 TURBOJET deep soil mixing equipment and processor. (Courtesy of M. Siepi.)

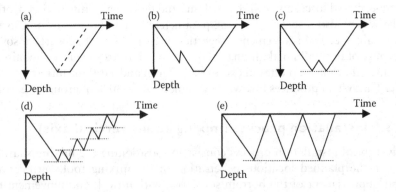

Figure 9.28 Typical execution procedures of deep soil mixing. (a) Without or with bottoming. (b) With reversal(s) during penetration. (c) With bottom restroking. (d) With stepped restroking during withdrawal. (e) With full-depth restroking.

The position and verticality of the shaft is checked first, and zero adjustments of the logging system are conducted. For on-land application optical survey devices are normally used, whereas for marine operations the use of the GPS has become common. The GPS is also advantageous in the case of large on-land projects, especially those involving treatment of very weak superficial soils.

During penetration the mixing tool is delivered to the required depth. In this phase compressed air (dry method), or slurry (wet mechanical mixing), or high-velocity jetting with slurry or water and air (hybrid mixing) is used to support mechanical drilling. Mechanical penetration may be difficult when the tool hits a hard layer or when the improvement depth is relatively deep, leading to possible damage or deadlock of the tool in the ground. This danger may be reduced with partial restroking to minimise rotation resistance along the shaft (Figure 9.28b), or by means of pre-boring with an auger machine.

After the basal treatment depth is reached, the mixing tool remains on the bottom rotating about 0.5–2 minutes for complete mixing. This phase is often called 'bottoming' and serves to ensure sufficient contact of the column(s) with the bearing subsoil. Penetration into the bearing layer should be confirmed by a rapid change of penetration velocity of the tool, required torque, and rotation speed. At this stage the tool can be raised about 0.5–1 m and lowered again to treat more effectively the transition zone between soft and bearing soils (Figure 9.28c). Withdrawal may be conducted as a continuous upstroke, but can include stepped or even full restroking if needed (Figure 9.28d and e). Full restroking is beneficial in the case of interchanged soft/stiff layers and stratified soils, leading to more uniform properties of stabilised soil across the depth of treatment.

The accompanying delivery of the stabilising agent to the subsoil is operator/computer controlled and linked to the energy of mixing in the specific layers of treated soil. In general, injection of the stabilising agent can take place during penetration, withdrawal, and restroking; however, two main injection methods are distinguished—the penetration injection method, which is a top-bottom process, and the withdrawal injection method, which is a bottom-top process. Penetration injection is typically used for on-land applications of the wet method because the slurry helps to lubricate the mixing tool and assists in breaking up the soil into smaller pieces. Normally, 80%–100% of the total slurry volume is used in this stage. This method is also beneficial to the homogeneity and strength of the manufactured column because the native soil is mixed twice with the binder.

Withdrawal injection is typically used for the dry method, usually with the whole amount of binder delivered to the soil during this phase. However, if very high binder concentration is needed to reach the design strength, part of the stabiliser may be injected during penetration phase and the rest during withdrawal of the mixing tool. Withdrawal injection also prevails for marine operations with the CDM method, but not for all. For on-land applications with the wet method, withdrawal injection is also possible, but usually at a reduced flow rate to minimise the volume of spoil (except of the CDM-LODIC method where withdrawal injection is a standard installation process).

The sequence of mixing operations will need to be adjusted to suit each site's specific conditions, but in general the most efficient sequence is to work the stabilisation machine within its radius of operation as much as possible before it is moved.

9.3.7 Details of construction and execution (mixing about vertical axis)

9.3.7.1 Number of shafts

Models and field observations indicate that multiple-shaft arrangements generally provide better homogeneity of DM columns than those produced with single-shaft mixing tools furnished with fixed cutting/mixing blades rotating in one direction. This has been especially observed in clays, which may tend to stick to the mixing blades and hence rotate with the mixing shaft, resulting in poor mixing. As a countermeasure, nonrotating vanes have been mounted on single auger shafts, close to cutting blades. An example is the free blade, developed in Japan (CEDIT, 2002), which extends beyond the reach of the mixing blades and is therefore supposed to stay in the ground to provide sharing capability (cf. Figure 9.13b). The 'entrained rotation' phenomenon is significantly reduced with closely spaced augers when the neighbouring shafts rotate in opposite directions, or are eliminated with overlapping augers or interlocking mixing blades providing greater soil shearing and particle milling (the same applies for sophisticated single-shaft tools with counter-rotating blades). Transverse steel bars used to keep multiple shafts in position have a similar function as the free blade.

Besides contributing to interactive mixing and increased productivity rates, multiple shaft arrangements also minimise the countermovement against shaft rotation as compared with single-shaft rigs, and improve stability of the machine. This further contributes to more precise control of shaft verticality during penetration. Linear arrangements of mixing shafts, applied for the construction of retaining and cut-off walls, enable easier and safer connection of individual wall panels using the intercut principle. Furthermore, limited adjustments of deviations occurring during penetration can be made by altering the rotation of coupled shafts.

Multiple shaft arrangements are, however, more demanding in terms of constructional requirements, generally leading to more complicated mechanical systems and larger/heavier machines. This may result in reduced flexibility in some applications, as well as increased mobilisation and operational costs, as compared with single-shaft machines.

9.3.7.2 Shape and orientation of mixing blades

The function of the mixing tool is to disaggregate the soil during penetration and to facilitate binder injection and immixing with the native soil. In the case of purely mechanical interaction, the mixing tool should also create the appropriate column diameter. A wide variety of different mixing tools have been tried so far, ranging from very simple to quite complicated constructions, with the obvious outcome that no single construction method can equally serve all soils. Nevertheless, some general indications can be formulated, keeping in mind that when compared to the wet method, the dry method requires more vigorous mixing to achieve the same level of homogeneity of soil-binder mix.

During downward movement of the shaft the mixing tool has to loosen the soil, while during withdrawal the soil should be thoroughly mixed with the binder and recompacted as much as possible to reduce excess spoil and to ensure maximum mixed soil density (recompaction does not matter in saturated conditions but is rather important for the dry method). This generally occurs when the inclination of the blades to the rotating direction produces mixing movements from the outside inward and from above downwards, opposite to the lifting movement of the tool. The degree of mixing increases when the soil is finely divided into horizontal, inclined, and vertical directions during tool rotation. This explains why in the case of single-axis shafts the window-type mixing tools, such as shown in Figure 9.8a and b, may perform slightly better than the tools with several separated horizontal blades, as also corroborated by the investigation conducted by Abe et al. (CDIT, 2002). On the other hand, soils like peat require more shearing action to be thoroughly mixed, and this is usually better achieved with multiple horizontal blades. Furthermore, window-type tools cannot overlap when used in multi-shaft arrangements. In the case of continuous or discontinuous augers, counter-rotation against the direction of auger pitch permits the soil-binder mix to be recompacted during withdrawal. Counter-rotation and/or shear bars also generally reduce the mixing energy required.

When designing a mixing tool, it is usually necessary to find a balance between a good ability to penetrate stiff or compacted layers of soil and a good mixing performance in soft soils. The same applies to the speed of rotation and associated higher wear of the mixing blades, as well as the possibility of easy repair and quick replacement of the mixing equipment. Consequently, the goal of designing a mixing device that creates sufficient movement in the soil without a great mixing effort or long mixing time is difficult to achieve.

9.3.7.3 Position of injection nozzles

To ensure optimum mixing efficiency, the position of injection nozzle(s) is different for various methods and installation processes. For the penetration injection method the outlet is normally placed close to the bottom end of the mixing tool, while for the withdrawal injection method it is above the mixing blades or at the level of the upper blade. Besides these two standard outfits some mixing tools have nozzles at both levels to also facilitate a combined penetration/withdrawal injection (e.g., CDM and DJM methods). During the penetration phase, the lower port is used and the upper one is closed. When withdrawing, the opposite combination is applied.

With the wet method and single-axis mixing tools there is usually one centre injection nozzle located close to the shaft tip, while large-diameter tools have several injection nozzles located along the blades at specified distances from the central shaft. In multiple-axis tools the grout is usually fed independently to each shaft, with the outlet port placed at the shaft tip. In some linear arrangements grout can be also supplied through the central shaft, incorporating 1 or 2 nozzles at the bottom, depending on the auger diameter, as done for the Bauer triple method. The direction of grout injection is generally horizontal.

As for the mixing tools of the hybrid method, the high-velocity jet nozzles are purposely located on the outer ends of the mixing blades to increase the range of mixing, but they can be also located at the tip or along the Kelly bar if jetting is primarily used to increase the rate of mixing tool penetration. The direction of the jet stream may be horizontal or inclined, depending on the nozzle orientation.

9.3.7.4 Degree of mixing

The efficiency of in-situ soil mixing with a stabilising agent is one of the key factors affecting column homogeneity and strength. The degree of mixing depends on the mixing time, type of mixer, characteristics of the native soil, and the form of applied binder (slurry or powder) and the energy of injection (low or high output velocity). The overall mixing process is rather complex, especially for the dry method (cf. Larsson, 1999), and difficult to quantify. Therefore, in an attempt to specify a criterion for the required mixing work, which could be controlled and altered on site during execution, a simplified index named 'blade rotation number' has been introduced in Japan (e.g., CDIT, 2002). The blade rotation number, T, is defined as the total number of mixing blades passing during 1 m of single shaft movement through the soil, and is expressed as follows, considering:

(a) complete injection during penetration and outlet located below the blades:

$$T = \Sigma M \times (R_p/V_p + R_w/V_w), \tag{9.1}$$

(b) complete injection during withdrawal and outlet located above the blades:

$$T = \Sigma M \times (R_w/V_w), \text{ and} \tag{9.2}$$

(c) partial injection during penetration and main injection during withdrawal, with the lower outlet active only during penetration and the upper outlet active during withdrawal:

$$T = \Sigma M \times (R_p/V_p \times W_p/W + R_w/V_w) \tag{9.3}$$

where: T = blade rotation number [rev/m], ΣM = total number of mixing blades, R_p = rotational speed of the mixing tool during penetration [rev/min], V_p = penetration velocity [m/min], R_w = rotational speed of the mixing tool during withdrawal [rev/min], V_w = withdrawal velocity [m/min], W_p = amount of binder injected during penetration [kg/m³], W = total amount of injected binder [kg/m³].

The total number of mixing blades, ΣM, is assessed by counting all cutting/mixing blades that are effective in the mixing process, taking into consideration the method of injection and position of the injection outlet(s) in relation to the blades. A full-diameter blade is counted as two blades. For example, when the outlet port is located beneath two levels of blades and when injection is carried out during penetration, as is common for the wet methods, the total number of mixing blades is $\Sigma M = 4$ and Equation 9.1 is used to evaluate T. In case of the withdrawal injection method and the outlet port located above all blades, as is common for the dry methods, ΣM is also four but only the withdrawal stage is considered (Equation 9.2). This example demonstrates that higher rotational speeds are required for the withdrawal injection method to obtain a comparable degree of mixing. In case multiple restroking along the whole depth is used, as shown in Figure. 9.27e, the resulting blade rotation number is a sum of T values calculated for each penetration/withdrawal cycle of the mixing tool.

The blade rotation number is used for mechanical mixing only, and the soil conditions are included indirectly (i.e., through selection of appropriate input values), taking into account accumulated experience and technical specifications of the equipment. Based on field data obtained in loose sands (Mizuno et al.) and clays, the blade rotation number of 360 has been recommended in Japan for the wet method to ensure reasonably low value of the coefficient of variation, v, of the unconfined compressive strength (CDIT, 2002). Field tests using wet mixing in silty/sandy clay in Poland revealed $T = 430$ to satisfy $v \leq 0.3$ (Topolnicki, 2009). For the dry mixing methods the blade rotation number is typically 274 or 284 for the DJM and 200 to 400 for the Nordic method, noting that dry binders are injected mainly only during withdrawal. For special mixing tools using cutting/mixing blades that rotate in opposite directions on a single shaft, like for instance the Bauer SCM-DH and the CDM Column 21 methods, there is a need to conceive a new guideline for the quality of mixing.

9.3.7.5 Control of binder supply

The amount of binder injected in a certain soil volume is easier to control for the wet method than for the dry method, where the binder is fed into a stream of compressed air.

The wet mixing process blends the materials with water to form a slurry at the design water to binder ratio. The quantity of binder components needed for each batch is weighed and added to the measured water volume in the mixer. Alternatively, ready-batched or preweighted bagged materials can be used to simplify this process. The binder slurry is then transferred to a temporary storage tank that continually agitates the slurry to ensure that the constituents of the mix do not separate. The slurry is then pumped at a specified flow rate to the mixing tool. In order to obtain the required amount of binder per soil volume, the penetration and withdrawal velocities of the mixing tool and the applied flow rates have to be simultaneously adjusted, taking into account the number of restroking passes with slurry injection. The flow rate of binder slurry is controlled at the delivery pump and monitored with a flow meter.

With the dry method, the weight loss of the binder storage tank, continuously measured by means of load transducers and averaged in such a manner that acceleration components are cancelled out, is used as an indicator of the amount of used binder. This information is combined with the corresponding geometry of stabilised soil to evaluate the binder output in kg/m of column, or in kg/m^3 of treated soil. To reach the predetermined target value it is necessary to control the feeding rate of binder into the air stream until the specified rate of loss of the material is obtained. This is mainly accomplished by adjusting the rotation speed of an impeller provided at the bottom outlet of the binder storage tank (Figure 9.29). The feeding mechanism must be manufactured with a very high precision since the distance between the rotating blades and the cylinder walls is on the order of 1/100 mm. However, the throughput of the impeller is not a linear function of the rotating speed and also depends on: (1) wear of impeller, blades, and wall; (2) pressure and amount of binder in the tank; (3) air and material flow below the impeller; and (4) flow properties of the binder, making binder output control more sophisticated (Bredenberg, 1999). The downward movement of binder in the tank towards the impeller is facilitated by 'fluidisation' of the stored material, caused by blowing compressed air from the tank bottom. To ensure high productivity with this system, it is important that the air blown into the binder storage tank is sufficiently dry and that the binder material is free of particles able to cause blocking

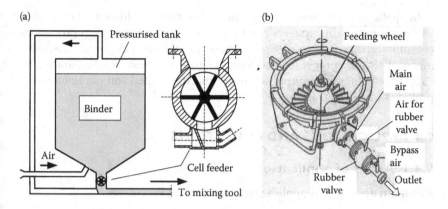

Figure 9.29 Binder feeding systems. (a) Cell feeder used in the Nordic method (courtesy of LCM). (b) Impeller used in the DJM method. (Redrawn from Aoi, M. (2002). *Execution procedure of Japanese dry method (DJM), Proceedings of the Deep Mixing Workshop 2002*, Port and Airport Research Institute & Coastal Development Institute of Technology, Tokyo, Japan.)

or damage to the feeding control system. Moreover, suitable hose diameter for the equipment used must be carefully selected to ensure smooth binder flow to the mixing tool.

9.3.7.6 Control during construction

Soil mixing, like other ground improvement technologies, uses indirect control measures to ensure the quality of work and product during execution. The main objective of a control system is to ensure delivery of a correct amount of binder and mixing energy along the installed element. The extent of in-situ mixing operation monitoring is closely associated with the type of project and the required level of quality control.

For a typical production process of DM, the following should be documented: element identification and/or position, mixing tool details, working grade, mixing depth, start time, time at bottom, finish time, mixing duration, agent specification, injection flow rate and pressure, total amount of agent used, tool rpm on penetration, tool rpm on withdrawal, and torque of the shaft. From this information the mixing energy and binder content can be calculated to match laboratory and/or test columns. The standard criteria to ensure quality of tip bearing are the penetration velocity and the applied torque. Centralised control systems are usually available to digitally record all parameters and display information at the control panel to facilitate real-time adjustments (e.g., Yano et al., 1996, Bredenberg, 1999, Burke, 2002, Hioki, 2002). They also simplify the task of preparing daily reports by recording the daily performed activities of soil mixing works.

In applications requiring automatic and/or more sophisticated control, a variety of measuring systems can be used to control the mixing process or column verticality or to observe horizontal and vertical ground displacements. The computer reacts to changing ground conditions and automatically adjusts injection output to ensure specific treated soil parameters are provided for each stratum.

9.4 APPLICATIONS AND LIMITATIONS

9.4.1 Areas of application

The main areas of SM applications are as follows:

(1) Foundation support
(2) Retention systems
(3) Ground treatment
(4) Liquefaction mitigation
(5) Hydraulic cut-off walls
(6) Environmental remediation

Case histories relating to each application group may be found in the cited bibliography as well as in the proceedings of specialty international conferences on deep mixing, held in Tokyo (1996), Stockholm (1999), Helsinki (2000), New Orleans (2003, 2012), Stockholm (2005) and Osaka (2009). It should be pointed out, however, that in many cases SM works are conducted to fulfil combined functions. Consequently, certain projects fall into more than one general category of application.

Selected case histories are included in Section 9.7 and in Chapter 10.

9.4.2 Patterns of deep soil mixing installations

Soil mixing can be done to a replacement ratio of 100% wherein all the soil inside a particular block is treated, as is usually the case for shallow mixing applications, or to a selected lower ratio, which is often practised with deep mixing. The chosen ratio reflects, of course, the mechanical capabilities and characteristics of the applied method. Depending on the purpose of deep mixing works, specific conditions of the site, stability calculations and costs of treatment, different patterns of column installations are used to achieve the desired result by utilising spaced or overlapping and single or combined columns. Typical patterns are presented in Figure 9.30.

Square or triangular grid patterns of single or combined columns are usually applied when the purpose of SM is reduction of settlement and, in some cases, improvement of stability. Common examples are road and railway embankments. Walls are used for excavation control, to stabilise

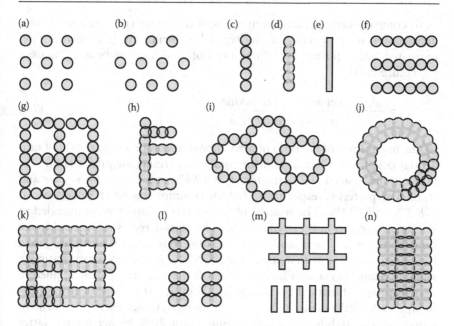

(a)   (b)   (c)   (d)   (e)   (f)

(g)   (h)   (i)   (j)

(k)   (l)   (m)   (n)

Figure 9.30 Examples of deep soil mixing patterns. (a), (b) Column-type (square and triangular arrangement). (c) Tangent wall. (d) Overlapped wall. (e) Trench/CSM wall. (f) Tangent walls. (g) Tangent grid. (h) Overlapped wall with buttresses. (i) Tangent cells. (j) Ring. (k) Lattice. (l) Group columns. (m) Multiple trenches/CSM walls. (n) Block.

open cuts and protect structures with shallow foundations surrounding the excavation, and as a measure against seepage. They are also constructed to increase the bearing capacity of improved soil against horizontal or sliding forces, with column rows installed in the direction of horizontal loading or perpendicular to the expected surface of failure. Walls can be constructed with tangential or overlapping columns, joint panels, or as trench structures. Overlapping is particularly important when executing cut-off walls or environmental barriers. In the case of DM machines equipped with linearly arranged multiple shafts, walls are usually executed using intersecting primary and secondary panels, with partial or even full-column diameter overlap. Groups of columns can be utilised to support embankments and foundations in order to reduce settlements and/or increase the bearing capacity. Various combinations of columns or panels are also used to build grid, U-formed, cellular, or circular installations with tangential or overlapping elements to improve the interaction with the untreated soil. Lattice-type improvements are considered an intermediate, cost-effective system between the wall-type and the block-type improvement. Full blocks are used to create large, highly stable volumes of stabilised soil, which act as gravity structures.

To compare various patterns in terms of the treatment area and to evaluate composite properties of the treated elements and the surrounding untreated soil, a purposely defined ratio of area improvement, a_p, is used (cf. Figure 9.31):

$$a_p = \frac{A_t}{A} = \frac{\text{net area of soil mixing}}{\text{respective total area}}. \tag{9.4}$$

The upper limit of the ratio of area improvement for a square grid of tangential columns is 78.5%, and for equilateral triangular grid it is 90.7%. For columns spaced at 2 diameters a_p is 19.6% and 22.7% for square and triangular patterns, respectively; and for columns spaced at 3 diameters it is 8.7% and 10.1%. The spacing of 3 diameters, usually recommended to minimise interaction between piles, can be considered as a practical lower limit of the area improvement ratio. Numerous embankments in Japan have been stabilised with a_p usually 30%–50% (due to seismic excitations), while in Scandinavia area ratios 10%–30% have been typically applied in case histories. Statistical evaluation of about 2,770 embankment projects in Japan revealed a distinct difference between a_p values used for settlement reduction and stability problems, being about 20% higher for the latter cases (Terashi et al., 2009).

Column/panel installation patterns may not only vary in plan view but also with respect to the depth of treatment. In the wall-type improvement, short and long walls can be alternately installed in the soft soil to reduce the costs of soil mixing (Figure 9.32a). The long walls transfer the loads exerted by the superstructure and external excitations to the bearing stratum, while the intermediate short walls provide connection between the long walls, increasing the rigidity of the total improved soil mass. This type of improvement has been commonly applied in port and harbour constructions in Japan (e.g., Kansai Airport man-made island; CDIT, 2002). Another example is the variation of column/panel lengths in transition and/or purposely determined zones of soil treatment, as shown in Figure

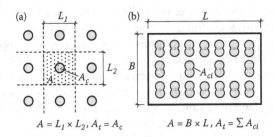

$$A = L_1 \times L_2, A_t = A_c \qquad A = B \times L, A_t = \sum A_{ci}$$

Figure 9.31 Evaluation of the ratio of area improvement. (a) Regular grid of columns. (b) Foundation slab.

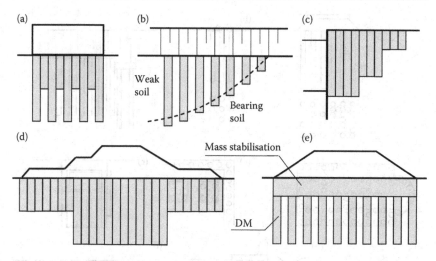

Figure 9.32 Examples of deep soil mixing with varying column/panel lengths (schematic). (a) Combined short and long walls. (b) Transition zone. (c) Stepped elements. (d) Embankment with berms. (e) Combined mass stabilisation and DM.

9.32b through d. Furthermore, a combination of different soil mixing techniques may be applied to treat specific soil depths, such as with a combined application of deep mixing and mass stabilisation resulting in a column-supported raft structure (Figure 9.32e), as practised in Scandinavia (e.g., Rogbeck et al., 1999, Jelisic and Leppänen, 2003).

9.4.3 Foundation support

The purpose of using DMM is mainly the reduction of settlement and the increase of bearing capacity of weak foundation soil, as well as prevention of sliding failure (Figure 9.33). For on-land projects the applications usually comprise road and railway embankments, buildings, industrial halls, tanks, bridge abutments, retaining walls and underground facilities. For waterfront and marine applications they can include quay walls, wharfs, revetments and breakwaters. Novel applications include also foundation of wind turbines (Topolnicki and Sołtys, 2012).

The installation patterns typically employ single or combined columns/panels with variable spacing for settlement reduction applications, while combined walls, lattices, and blocks are used when dealing with high loads and/or horizontal forces. An increasing tendency to apply economical low values of the area improvement ratio can be observed in recent times, depending on the adopted DM method and the available column strength. Design of such patterns requires rigorous analysis of the interaction

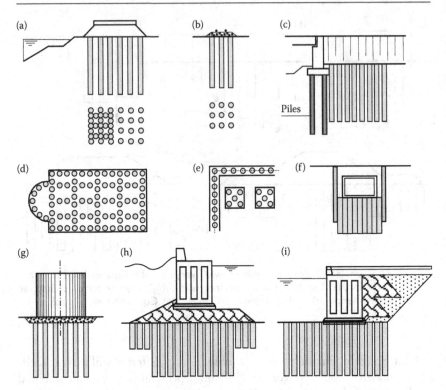

Figure 9.33 Examples of DM application for foundation support (schematic). (a) Road embankment. (b) Railway embankment. (c) Bridge approach zone. (d) Slab foundation. (e) Strip and pad foundations. (f) Culvert. (g) Tank. (h) Breakwater. (i) Quay wall.

between treated and untreated soil. The strength of DM elements may differ significantly within the range determined by low-capacity lime/cement columns, with say 0.15 MPa shear strength, and high-capacity structural elements having unconfined compressive strength on the order of 5 MPa, which act like piles or caissons. The external loads are usually transferred down to the bearing layer, resulting in a fixed-type improvement, but can be also partly or wholly transferred to the foundation soil when a more interactive or even a floating type of improvement is desired. The choice of the required strength and of the load transfer system is dictated by the purpose of the DM application, and reflects the mechanical capabilities and characteristics of the particular method used.

When deep soil mixing is applied under embankments or foundation slabs to reduce differential settlement and increase bearing capacity of the foundation soil, it can be noticed that individual column/panel quality is less important and that it is the overall performance taking into account soil to column/panel interaction that matters most. Such a concept of soil/structure interaction, practised for instance in Scandinavia using the Nordic method and often combined with preloading and drainage function of the columns to accelerate settlement, has proved to be efficient and cost-effective compared to other methods. On the other hand, when DM is performed to support high embankments or heavily loaded foundations, and where horizontal loadings or shear forces are significant, the quality of load-bearing columns/panels is essential to prevent progressive failure mechanisms. The same applies for low values of the ratio of area improvement.

In bridge construction the DM elements can be used to act as the pier foundation for the abutment, or to prevent lateral thrust and sliding by reducing the earth pressure behind the abutment. They can also reduce settlement of the bridge approach zone. In the case of buildings, DM is an alternative solution to conventional deep foundation methods, particularly in seismic-prone areas. Since the DM columns/panels can be closely spaced, the foundation dimensions in plan remain relatively small, which contributes to the overall cost-effectiveness of this foundation solution.

Waterfront or marine gravity-type structures are subjected to large horizontal forces caused by earth pressure or wave loading. Therefore, DM patterns typically comprise blocks, lattices, or combined walls created by overlapping columns or joint panels.

9.4.4 Retention systems

Retention systems comprise applications associated with restraining the earth pressure mobilised during deep excavations and vertical cuts in soft ground, with protection of structures surrounding excavations, measures against base heave, and prevention of landslides and slope failure (Figure 9.34). In these applications, wall- and grid-type column/panel patterns are mainly used, while the soil-binder mix is typically engineered to have high strength and stiffness. Wall construction is especially effective using CSM and TRD methods. To overcome soil and water lateral pressures, the DM elements should have adequate internal shear resistance. Other key requirements for successful construction are a high degree of homogeneity and maintaining verticality tolerance to achieve the minimum required designed thickness, especially in the case of columns effectively in continuous contact. It is also important that early strength gain is sufficiently retarded to prevent problems when constructing secondary intercut columns or panels.

Steel reinforcement can be installed in DM elements executed with the wet method to increase the bending resistance and create a structural wall for excavation support (Figure 9.34b). Elongated mixing time and/or full restroking are usually applied to ensure easier installation of soldier elements immediately after mixing. Panels of mixed soil between H-beam reinforcement are designed to work in arching, as in a 'Berlin'-type wall (e.g., Außenlechner et al., 2003). Concrete facing, tieback anchors, or stage

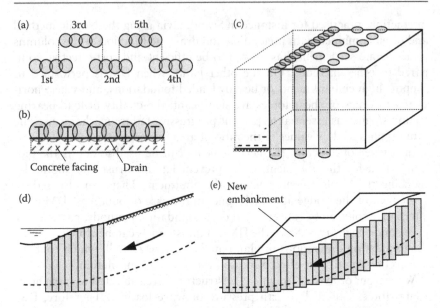

(a)   3rd   5th

1st   2nd   4th

(b)

Concrete facing   Drain

(d)

(c)

(e) New
    embankment

Figure 9.34 Examples of DM applications for retention systems (schematic). (a) Typical DM wall. (b) DM wall with concrete facing. (c) Composite gravity wall (relieving platform not shown). (d) Landslide protection. (e) Slope stabilisation.

struts are typically used in combination with the DM walls. Drainage media may be required behind the wall to prevent build-up of excess hydrostatic pressures. Deep circular shafts can be constructed using two or three concentric rings of overlapping DM columns or with joint panels, acting together in hoop compression.

Another innovative concept to support vertical excavations is to construct composite gravity structures, which do not require anchors or braces (Andromalos et al., 2000). This vertical earth reinforcement technology (VERT) top-down retaining wall system typically consists of a continuous front row of DM columns and one or more rows of isolated columns or staggered panels (Figure 9.34c). The back rows of DM columns are sized and spaced to ensure composite action between the wall elements and to provide external stability to the wall in conjunction with the relieving platform constructed from spoil at the top of the wall. The column edge-to-edge spacing should not exceed 1.2–1.5 times a column diameter. The ratio of area treatment is typically 30%–40%. The cemented-soil relieving platform is used to tie the DM columns together to transfer the load to the bottom of the vertical columns. This external stability requirement implies the need of a site-specific minimum tensile strength and absolute continuity for the cemented-soil relieving platform (HITEC, 2002). Light steel reinforcement may be used in some of the front face columns to provide anchorage for permanent cast-in-place facing. High-quality DM and rigorous analyses are required for such retaining systems. Limitations may also result from high water level, freeze-thaw durability, surcharges or structures behind the wall, and acceptable horizontal displacements. Terashi (2003) mentioned, for instance, that some cut slopes improved by a group of columns have suffered from excessive horizontal deformation, although not documented.

Measures against base heave are comprised of DM columns installed within an excavation site to act like dowels penetrating through potential sliding planes. In some cases the sides of the excavation are stabilised to increase the passive earth pressure and to reduce the penetration length of sheet piling or diaphragm walls.

SM is also applied to stabilise landslides and critical slopes. With suitable column arrangements, typically in the form of walls, grids, cells and blocks which intersect a potential failure surface, the combined shear strength of soil is improved and the factor of safety is increased (Figure 9.34d and e).

There are also applications comprising soil nailing and installation of special anchors using DM. An example is the RADISH (rational dilated short) anchor 40 cm in diameter, which has been used to modify existing embankments to steep slopes (Tateyama et al., 1996). Special anchors can be also installed with the Nordic method.

9.4.5 Ground treatment

Ground treatment works usually involve substantial volumes of unob-structed soft soils and fills to be improved on-land, at waterfront areas, and offshore with relatively high area improvement ratios. Typical examples are large developing projects including the construction of roads and tunnels on soft soils, stabilisation of reclaimed areas (Figure 9.35a) or river banks, and the strengthening of sea-bottom sediments. The purpose of improvement is mainly the reduction of settlement and an increase of bearing capacity, as well as prevention of sliding failure. Other applications include

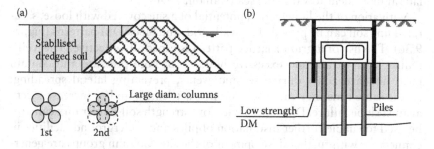

Figure 9.35 Examples of ground treatment with SM (schematic). (a) Stabilisation of reclaimed area. (b) Low strength DM.

the installation of wave-impeding DM blocks of high rigidity beneath or near the foundation to reduce adverse effects caused by vibration on surrounding structures, as well as DM rings around the pile foundation of a vibrating machine (Takemiya et al., 1996). Depending on the project requirements, deep and shallow soil mixing methods can be applied, including mass stabilisation.

Ground treatment works also comprise of dry and wet method soil stabilisation to a low strength, in the order of 0.2–0.5 MPa UCS using a reduced amount of cement and cheaper supplementary binders, like fly ash and gypsum. For the wet method, this allows increasing the amount of slurry injected into the soil, hence improving the uniformity of mixing as compared to standard DM applications using cement grout (e.g., Azuma et al., 2002). High initial moisture content of the treated soil may have an adverse effect on the available compressive strength and/or hardening process after treatment, as observed in soft Finnish clay in the Old City Bay area in Helsinki (Vähäaho, 2002). As a consequence, dry mixing may be the better option for very wet soils.

Underground blocks of low-strength DM may be used to increase passive resistance and minimise heave at the bottom of excavation, allowing at the same time easy driving of sheet pilling elements or piles directly into or through the improved ground (Figure 9.35b). Moderate strength DM can also be used to improve soft soil to allow steady digging by the shield tunnel machine, as applied for instance during construction of the Trans-Tokyo Bay highway project (design UCS 1 MPa, CDIT, 2002).

9.4.6 Liquefaction mitigation

The effectiveness of DMM to prevent liquefaction was confirmed during the magnitude 7.2 earthquake in Kobe in 1995. A hotel under construction on a reclaimed sand area, supported on drilled piers, actually survived because the piers have been protected with a DM grid against liquefaction and the accompanying lateral flow, while the nearby seawall suffered large lateral movement towards the sea (Kamon, 1996).

Mitigation of the liquefaction potential of a site covered with loose, saturated fine soil can be provided by wall, grid, and block DM patterns (Figure 9.36). The use of a grid or lattice pattern is especially effective. The 'cells' reduce shear strain and excessive build-up of pore pressure and contain local liquefied zones during seismic events, preventing lateral spreading. At the same time they can also minimise settlement and/or increase safety against slope failure. DM blocks with low-strength soil-binder mix can also be used to enable further installation of piles and underground facilities in connection with further development of the site. Column groups are generally not recommended because they may suffer from stress concentrations and bending failure.

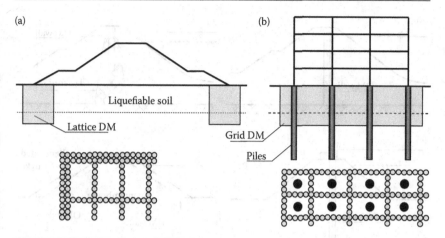

Figure 9.36 Examples of DM application for liquefaction mitigation (schematic). (a) Protection of a river dyke. (b) Improvement of the lateral resistance of piles.

Seismic prevention by DM to existing structures also comprises perimeter walls installed to isolate and contain liquefiable soils under the structure. The groundwater within the enclosed zone is then permanently lowered to provide nonliquefiable conditions. This solution is used where other more conventional remedial measures are not viable.

9.4.7 Hydraulic cut-off walls

Hydraulic cut-offs are constructed by installing DM walls to intercept the seepage flow path. The columns/panels or trench walls are typically installed through the permeable strata to some cut-off level, usually penetrating 0.5–1 m into a clay layer or finishing at the top of the bedrock. The soils treated are generally highly permeable coarse deposits, or interbedded strata of fine- and coarse-grained soils.

The applications mainly involve rehabilitation and/or upgrading of older water-retaining structures to meet new regulations for safe operation. Typical examples are earth-fill dams, dyke embankments and river banks (Figure 9.37). In the case of excavations, the supporting DM walls may additionally serve to prevent seepage of groundwater towards the pit. When a conventional elevation of a river dyke crest is not possible, steel H-beams can be installed in DM columns or panels to support concrete superstructures or light dismountable protection walls on the crest to prevent overtopping (e.g., Topolnicki, 2003).

Since the hydraulic conductivity and continuity of the cut-off wall are of primary importance, careful design of slurry mixes tailored to soil conditions, and adequate control of overlapping zones and verticality are

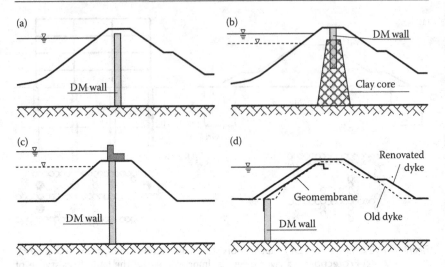

Figure 9.37 Examples of DM applications for cut-off walls (schematic). (a) Dam sealing. (b) Extension of the clay core. (c) River dyke with superstructure on the crest. (d) Seepage protection.

required, especially when cut-off walls are executed to a large depth with single-shaft mixing equipment. For DM walls the unconfined compressive strength is typically in the range of 0.7–3 MPa, and higher if steel reinforcement is installed, while the permeability is normally 10^{-8} to 10^{-9} m/s. When bentonite and/or clayey stone dust and/or fly ash are added to the slurry mix, the permeability can be reduced to 10^{-9} to 10^{-10} m/s, with associated decrease of the unconfined compressive strength usually below 1 MPa.

Related case histories may be found in Yang and Takeshima (1994), Walker (1994), Nagata et al. (1994), Schwarz and Seidel (2003), Evans and Garbin (2009).

9.4.8 Environmental remediation

Environmental applications mainly involve installation of containment barriers and solidification/stabilisation of contaminated soils and sludges. Fixation is much harder to achieve, as it requires contact of the chemical reagent with the contaminant. This is easier in sandy soils but very tough in clayey soils. At an experimental level, soil mixing has also been used to introduce microorganism–based grout for bioremediation purposes, acid/base reagents for neutralisation, and oxidation reagents for chemical reaction.

Soil mixing containment systems include passive and active barriers constructed around a part or a whole periphery of the contaminated site. Passive barriers resemble hydraulic cut-off walls described above and are installed to prevent migration of polluted leachates out of the contaminated site. Active barriers have permeability comparable to the native soil. They are typically constructed as 'gates' in passive barriers to reduce significant effects of the containment on the existing groundwater regime (Figure 9.38). With appropriate soil-mixed materials, such as modified alumina silicates and adsorbance capacities, gates act as microchemical sieves, removing contaminants from groundwater as it passes through and allowing, in principle, only clean water to emerge on the other side. The DM containment barriers are suitable for existing waste disposal dumps and new landfill facilities. However, grout composition and binder reactions with the contaminants in the short and long term perspective are key factors in the success of such applications.

Solidification/stabilisation of contaminated soils and sludges containing metals, semi-volatile organic compounds, and low-level radioactive materials using wet and dry method soil mixing started to be recognised as a favoured remediation option. Advantages of this option include reduced health and safety risks, elimination of off-site disposal, low cost, and speed of implementation.

By selecting appropriate equipment and procedures, the reagents can be uniformly injected at depth, and efficiently and reliably mixed with the soil or sludge present. In the case of soil contaminated with volatile compounds, negative pressure is kept under a hood placed over the mixing tool to pull any vapours or dust into the vapour treatment system.

Related case histories can be found in Walker (1992), Aldridge and Naguib (1992), Jasperse and Ryan (1992), Hidetoshi et al. (1996), and Lebon (2002). Recent innovations in soil mixing technology for remediation of contaminated land have been summarised by Al-Tabbaa et al. (2009).

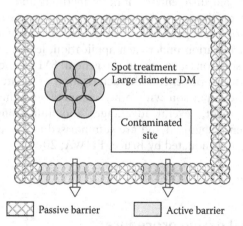

Figure 9.38 Containment system consisting of passive and active DM barriers (schematic). Barriers can be created with overlapping columns or panels, or as vertical trenching structures.

Table 9.3 Main advantages and limitations of the use of deep mixing method

Main advantages of DMM	Main limitations of DMM
High productivity usually possible (hence, economical) for large-scale projects	Depth limitations (depending on the method applied)
Can be potentially used in all types of soils and fills (without obstructions)	Not rational in soils that are very dense, very stiff, or that may have boulders
Spacing and patterns of DM elements highly flexible, arrangements tailored to specific needs	Limited or no ability to install inclined elements (depending on the equipment applied)
Engineering properties of treated soil can be closely designed	Uniformity and quality of mixed soil may vary considerably in certain conditions
Causes minimal lateral or vertical stress that could potentially damage adjacent structures	Elements cannot be installed in close proximity to existing structures (except hybrid mixing)
No vibration, medium-low noise	Freeze/thaw degradation may occur
Very low spoil (dry method, trenchers)	Significant spoil produced with the wet method
Can be used for on-land, waterfront and marine projects	Weight of the equipment may be problematic for weak soils (depending on the method)
Quality of treatment verifiable during construction	Air pressure or grout injection pressure may cause heave
Minimum environmental impact	Limited ability to treat isolated strata at depth

9.4.9 Advantages and limitations

When evaluating possible advantages and limitations of the use of soil mixing for each of the six application areas mentioned above, it should be remembered that different soil mixing methods and machines provide different types of treated soil geometries and treated soil parameters. Therefore, one or a few particular soil mixing variants may be practically feasible for consideration under each application, leading to a stimulating competition of solutions also within the general SM technology.

However, in terms of general characterisation of in-situ soil mixing, as well as relative comparison with other competitive technologies of deep ground improvement, several advantages and limitations of the use of DMM have been recognised and are summarised in Table 9.3 based on a similar evaluation conducted by Bruce (FHWA, 2000).

9.5 DESIGN

9.5.1 General design procedure

The current design procedures for DM works, mainly developed in Japan for wet and dry mixing and in Sweden for the Nordic method, are to a large extent based on accumulated experience. They are still being modified in accordance with new findings from numerous research studies and developments of the execution methods. In general terms, however, the planning of DM application involves the following main steps:

(1) Selection of a suitable DM process (i.e., wet or dry) and the construction method
(2) Selection of the strength of stabilised soil in specific ground conditions (mix design)
(3) Selection of the installation pattern and dimensions of improved ground (geotechnical design)

Clearly, all three steps are closely interrelated and iterative procedure is usually needed to achieve the full benefit of the planned DM application, cf. Porbaha (2000a), Filz (2009). Laboratory tests, model investigations and even field trials are often conducted to assist this selection process.

In the first step, the most suitable treatment process (dry or wet) and binder should be considered taking into account actual soil conditions, site constraints, and functional requirements of the structure to be built as well as related economical aspects of the design. At this stage, possible advantages and disadvantages of the application of DMM in comparison to other competitive technologies should also be examined.

The expected 'minimum' strength of stabilised soil is selected next in relation to the treatment process, type and amount of binder(s), and working specifications such as the rate of penetration and withdrawal, rotation speed, injection method, mixing tool, water/binder ratio, etc. At this stage a good understanding of the intricacies of soil-binder physics, chemistry, and mechanics is required in view of the variability of soils and stabilisation agents. In all cases where the DM elements are much stiffer than the surrounding soil there is a stress concentration on these elements and the internal stability of soil-binder mix starts to be a dominating aspect of the design. The required strength of stabilised soil must therefore be carefully selected, taking into account functional requirements and inevitable and significant scatter of the field strength. It should be also remembered that the quality of soil-binder mix directly impacts the cost of treatment as it will govern the mix energy required as well as the binder content and type necessary. Specifying a quality higher than actually needed for acceptable performance can significantly impact the cost of ground improvement.

The geotechnical design determines the installation pattern and the dimensions of the improved ground to satisfy internal and external stability conditions. In the current practice, the approach based on the allowable-stress concept is generally used at this stage. Furthermore, displacements of the superstructure are examined. On the other hand, the limit state design (LSD) approach is being progressively adopted in civil and geotechnical engineering. To satisfy ultimate limit states (ULS) requirements, the design of stabilised ground must be such that there is a sufficiently low possibility of collapse of the supported structure with respect to all potential failure modes, including internal and external stability and progressive failure. To satisfy serviceability limit states (SLS) requirements, soil mixing shall be designed in such a way that the total and the differential displacements, including long-term creep movements, are below accepted limit values. Although the failure modes and design formulas remain the same, this approach can be used in practice when respective partial safety factors (requiring lengthy calibration work) are established.

9.5.2 The choice between dry and wet process of soil mixing

As it is with other ground improvement technologies or foundation methods, selection of the most suitable process of deep soil mixing is a matter of somewhat arbitrary engineering judgement. In such an evaluation various technical and economical aspects of both processes should be considered in relation to the type of structure and soil for which DM will be used, including all site-specific conditions and availability of the equipment. Although in practice the decisions in favour of the dry or the wet process usually have a rational background, it is rather difficult to detect distinct application limits and to provide widely valid criteria for selection purposes. Instead, few characteristic features of both processes may be appointed to enable easier selection of the most suitable DM process in the initial stage of design.

From the point of view of engineering properties of soils stabilised with the same type of binder applied in dry or slurry form, there is no substantial difference between both mixing processes, as also underlined by Terashi (2003). This has been demonstrated by numerous investigations conducted on laboratory mixed soils, but is difficult to repeat in field conditions. For example, when a sufficient but not too high amount of cement in powder or slurry form is thoroughly mixed with a soft soil, higher compressive strength is expected for the dry than for the wet process due to lower value of the resulting water/cement ratio. To obtain a comparable strength with the wet process the amount of immixed cement must be increased, which is usually easily managed at the construction site although more spoil is inevitably produced. However, when still-higher strength is requested, the amount of cement has to be increased for both methods. In such cases, the mixing conditions in situ start to play a dominant role in the stirring process since it is much easier to dispense and to mix cement slurry with the soil than dry cement, especially when the natural water content of the soil is low and the shear strength is high. Furthermore, problems with complete recovery of the increased amount of air from the ground may also appear in the dry process. Decreasing efficiency of dry mixing will directly affect the homogeneity of DM columns, increasing strength and stiffness variation of stabilised soil. This explains why in the current design practice and for 'normal' soft soils, lower strengths of stabilised soil are adopted for the dry rather than for the wet process. In the case of the wet process, the amount of cement injected is usually 100–500 kg/m³ of soil to be treated and the strength target is typically 0.3–2 MPa, based on the four-week unconfined compressive strength. For the dry process applied in soils with 40% to more than 200% natural moisture content, the amount of cement used is typically 100–300 kg/m³, with a compressive strength target of about 0.5 MPa for the Japanese DJM method using mainly cement or cement-based binders, and 0.15 MPa shear strength for the Nordic method using mostly combined cement/lime binders. On the other hand, for very soft soils with very high moisture content, reasonable strength gain is easier and more effective using dry binders. This illustrates that the initial moisture content of the soil to be treated, besides the penetration resistance of the ground and the purpose of ground improvement, is a major factor affecting the choice between the dry and the wet processes.

A summary of the distinct characteristics of both mixing processes that are relevant for selection purposes is presented in Table 9.4. Related case histories where the applicability of the dry or the wet process was actually investigated by means of field trials are not yet very common. However, three case studies are excellent examples of such an approach. Huiden (1999) describes field trials for the Botlek Railway tunnel in The Netherlands, partly embedded in a mass of stabilised soil. The subsoil was very heterogeneous, consisting of peat, clay, silt, and sand. Shiells et al. (2003) report trial installations for the project at the Woodrow Wilson Bridge in Virginia, comprising ca. 135,000 m³ of deep soil mixing at the I-95/Route 1 Interchange. The subsoil is built of highly compressible organic silts and clays of 3–12 m thickness. The organic content of the alluvial clay soils commonly varies between 5% and 15%, with maximum organic content approaching 50% at isolated locations. In both cases the wet process was chosen. Vähäaho (2002) describes field trials of DM in the Old City Bay area in Helsinki. The area is part of an old sea bed, reclaimed in the 1960s and 1980s. In this case wet and dry mixing were tested in order to stabilise very weak deposits of organic clay and clay, with about 2–3 MPa cone tip resistance. A very high shear strength requirement, set to 750 kPa in clay at 10–19 m depth, was fulfilled only with the dry method (1,300 kPa on average), although a clear asymmetry in the distribution of binder in column cross-section was noticed. The shear strength obtained with the wet method averaged 230–340 kPa. It should be noted that in all reported cases the final choice was dictated by technical requirements.

In Japan the market share of the wet and dry methods in on-land projects is almost equal. The total volume of soil treated with the wet method from

Table 9.4 Characteristic features affecting the choice of dry or wet soil mixing

Item of concern	Expectations
Initial water content of the soil to be treated	Cohesive soils with moisture content $w = 60\%$–200% are best suited for the dry process (lower limit $w > 20\%$, note that water content below plastic limit is not fully available for hydration); for soils with very high w dry process is more effective than wet
Quality of mixing	Wet process usually provides better homogeneity of stabilised soil because of longer mixing time, prehydration of cement, and easier distribution of slurry across the column area
Compressive strength of soil-binder mix	Higher strength is more reliably obtained with the wet process, except for very wet soils
Ability to penetrate through hard soil layers	Much higher for the wet process due to the 'lubrication' effect of the injected slurry and due to higher torque capacity of mixing shafts
Stratified soils	Wet mixing provides better homogeneity and more uniform strength in the vertical profile, especially when using trenchers and cutter/mixing tools; quality control more difficult for the dry process
Spoil	Dry mixing creates very little or no spoil
Use of combined binders and industrial by-products	Quite frequent in dry mixing; slag cement often used in wet mixing, other binders and by-products very rare in wet mixing
Air temperature below 0°C	Dry process is significantly less affected by low temperatures since compressed air is used to transport the binder
Reinforcement	Possible in combination with the wet process
Targeted treatment at depth	Dry mixing allows for targeted treatment at depth while leaving the overlying soils untreated, since columns are formed during withdrawal of the tool

1976–2007 is about 36 million m³ (3,000 projects). For the dry method, since 1981, the total volume exceeded 30 million m³ (4,700 projects), as reported by Terashi et al. (2009).

9.5.3 Engineering properties of stabilised soil

The most common engineering parameters of the stabilised soil that are measured and/or inferred on DM projects and used in the design analyses comprise compressive, shear, and tensile strengths, modulus of elasticity, unit density and permeability (hydraulic conductivity). For ground improvement applications with low-strength DM, the coefficient of consolidation and the coefficient of volume compressibility of treated soil also may be important for settlement prediction. In more typical cases, however, the working stresses acting on the columns are far below the consolidation yield pressure, which for cement-treated soils is typically about 30%

higher than the unconfined compressive strength. In a sort of overconsolidated state, the coefficient of consolidation of soil-cement is improved at least 5–10 times and the coefficient of volume compressibility is improved to 1/10 or less (CDIT, 2002). Therefore, the consolidation settlement of cement-treated ground is small and usually can be neglected. For dynamic analyses, including seismic, cyclic, and repeated loading excitations, assessment of dynamic properties is required, as for instance the shear modulus and damping at different strain levels. More information on dynamic properties can be found in the state-of-the-art paper of Porbaha et al. (2000).

Low hydraulic conductivities are of primary concern for cut-off walls and retention systems involving groundwater control, as well as for most environmental applications. Hydraulic conductivity is also important when dealing with long-term behaviour of DM columns installed in aggressive soil-water environments. In other areas of DM application the actual permeability of soil-binder mix should be compared to the permeability of soil before treatment to inspect whether the installed columns are likely to act in the ground as vertical drains or as semipermeable or (almost) impermeable soldier elements. For instance, significantly increased permeabilities are usually observed after treatment when quick lime is used as stabilising agent in soft clayey soils, with a tendency to increase further in time. In this case, a combined action of the DM columns in the ground must be considered, including strengthening and drainage function. Contrary to that, stabilisation with cement usually effectively reduces initial permeabilities and the treated soils are practically impermeable. Therefore, such columns are not expected to function as drainage elements.

In the case of combined lime/cement treatment, the permeabilities are generally also low, with a tendency to decrease in time and with increasing confining pressure, while the actual permeability of soil-binder mix will depend greatly on the site-specific conditions. It is thus evident that the resulting permeability of stabilised soil will significantly affect the interaction pattern between the treated and untreated ground, and consequently the design approach.

The initial wet density of the soil may slightly increase or decrease after treatment. Sample Japanese data presented in CDIT (2002) and referring to field investigations indicate that soil density after treatment may change within ±5% in case of stabilisation with quick lime, or increase 3%–15% if cement in dry form is used (in peat, an even higher increase was observed).

In the case of the wet process, the density was found to be almost unchanged after treatment, irrespective of the mass of cement admixed per 1 m³ of soil within 50–250 kg/m³ range. Similar observations are known from numerous jet grouting works. However, lower densities after treatment are also occasionally observed, especially on wet grab samples (e.g., O'Rourke et al., 1997, cited after FHWA, 2001). Since the expected changes of soil density are generally small, it is often assumed for design purposes that the wet soil density is not affected by the in-situ treatment. More critical evaluation is recommended if the soil weight plays a significant role in the design, as for instance in the uplift stability analysis.

Three other basic parameters, the shear and tensile strengths and the modulus of elasticity, can be correlated with reasonable accuracy to the compressive strength of stabilised soil. Consequently, the compressive strength, and in practice, the unconfined compressive strength (UCS) due to the simplicity and cost-effectiveness of testing, is the key parameter for the current design practice. In Scandinavia, the shear strength is equivalently used.

At present, the UCS of soil-binder mix cannot be reliably forecasted on the basis of properties of the native soil and the type and amount of admixed binder(s). Therefore, it is generally recommended that advance appropriate trial tests be conducted on stabilised soils to obtain more adequate data regarding UCS for each project. These pilot investigations typically include testing of laboratory mixed samples, but may also involve full-scale trials in the cases of more challenging designs. At this stage, it is a common practice to inspect the UCS in relation to the binder factor, α (kg/m^3), expressed as the weight of injected dry binder divided by the volume of ground to be treated. The weights can refer to the weight of binder used in dry methods, or the weight of binder used in the slurry in wet methods. It should be kept in mind, however, that in field situations the injected binder quantities may not necessarily be those that actually remain in place. Therefore, the established correlations should be critically evaluated before being used for optimisation purposes. Likely field strength can be also estimated on the basis of accumulated experience from previous projects, and by exercising engineering judgement.

The expected field strengths and permeabilities for ranges of cement factors and different soils are listed in Table 9.5. The corresponding volume ratios, defined as the ratio of the volume of slurry injected to the volume of ground to be treated, vary greatly and reflect the type of DM technique used, but are in the range of 15%–50% (25%–40% in most cases). Lower-volume ratios can be applied when the efficiency of mixing is high due to higher rotational speed or jet assistance. The data presented in Table 9.5 are useful for an early assessment of technical and economical aspects of the DM design. With additional cement, the strengths generally increase. However, laboratory tests have also indicated that in some clays additional cement dosage will not increase UCS values. In such soils, blast-furnace slag has proven to be very effective. Generally, with increased cement dosage permeabilities decrease, but not to the order of magnitude required for effective cut-off barriers. For this purpose, bentonite or other proprietary reagents should be used to lower the hydraulic conductivity in a more effective way. The amount of spoil increases as the volume ratio increases.

Table 9.5 Typical field strength and permeability for ranges of cement factors and soil types (data based on soils stabilised with the wet process[a])

Soil type	Cement factor[b], α [kg/m³]	UCS 28 days, q_{uf} [MPa]	Permeability, k [m/s]
Sludge	250–400	0.1–0.4	1×10^{-8}
Peat, organic silts/clays	150–350	0.2–1.2	5×10^{-9}
Soft clays	150–300	0.5–1.7	5×10^{-9}
Medium/hard clays	120–300	0.7–2.5	5×10^{-9}
Silts and silty sands	120–300	1.0–3.0	1×10^{-8}
Fine-medium sands	120–300	1.5–5.0	5×10^{-8}
Coarse sands and gravels	120–250	3.0–7.0	1×10^{-7}

[a] Data compiled from Geo-Con, Inc. (1998), FHWA (2001), and Keller Group (unpublished data).
[b] Binder factor, or Cement Factor in the table, represents the weight of injected dry binder divided by the volume of treated ground, whereas binder factor in place represents the weight of injected dry binder divided by the total volume of treated ground and injected slurry.

There are numerous factors that can affect the strength increase of in-situ treated soil, as well as the quality and reliability of data collected and/or reported on soil-binder mix strength. The most important factors are summarised in Table 9.6.

A great deal of valuable published information is available on the relative importance of the factors indicated in Table 9.6, and on the mechanical behaviour of stabilised soils, based on dedicated laboratory and field studies and performance observations (extensive bibliographies can be found in the state-of-the-art reports of Porbaha et al., 2000, FHWA, 2001, and CDIT, 2002). However, the majority of these studies have been conducted

Table 9.6 Main factors affecting the observed strength of stabilised soil

Source	Specific items
Physical and chemical properties of the soil to be treated	Grain size distribution, mineralogy, natural water content, Atteberg limits, organic matter content, reactivity, and pH of pore water
Binder, additives, and process water	Type and quality of hardening agent(s), binder composition, quantity of binder and other additives, quality of mixing water
Installation technique and mixing conditions	Tool geometry, installation process, water/binder ratio, energy of mixing (speed and period), time lag between overlaps and working shifts
Curing conditions, time	Curing time, temperature (heat of hydration in relation to treated volume), humidity, wetting/drying and freezing/thawing cycles, long-term strength gain, and/or deterioration
Testing and sampling	Choice of testing method, type of test, sampling technique, sample size, testing conditions (stress path and drainage conditions, confining pressure, strain rate, and method of strain measurement)

on laboratory mixed soils, actually violating in-situ mixing and curing conditions except for the amount of binder and the curing time. Field investigations have been mostly carried out to solve site-specific problems and/or to provide quality evidence of the executed works, and often have inherent limitations. Data on long-time performance of DM columns are still insufficient, although generally show promising results (see Section 9.5.4). It is therefore rather difficult and challenging to compare extremely detailed experimental data and to assess the real mechanical behaviour of soil-binder mix, especially in view of changing conditions of mixing in-situ and a variety of native soils. On the other hand, it is believed that a good understanding of a generalised behaviour of stabilised soils is needed to meet static and functional requirements of any DM design. Therefore, a synthesised overview of the most pronounced relationships between selected 'input factors' and expected 'responses' of cement-treated soils, as revealed by published experimental evidence, is presented in Table 9.7. Obviously, for any specific project these qualitative relationships must be validated and carefully quantified before being used during design or construction stage.

Bearing in mind that the UCS of stabilised soil is a result of many variables, including construction variability itself, useful relationships and data for practical design have been compiled in Table 9.8. In general, laboratory and field investigations show reasonable correlation between early UCS of 4 or 7 days, and strengths observed at longer cure times. On this basis, relatively quick assessment of the expected strength after 28 or 56 days can be made to confirm initial assumptions. Prediction of soil-binder strength prior to construction is very often based on laboratory mixed samples and correlations established between the UCS of laboratory and in-situ mixed soils. Due to inherent limitations of laboratory-prepared samples, discussed above, such strength data must be applied with appropriate safety margins and considered rather as an index of the actual field strength than a precise prediction.

The information on the secant stiffness modulus and on the axial strain at failure, presented in Table 9.8 for the wet process, gives only a crude engineering estimation of very complex stress-strain behaviour of stabilised soil. For dry mixing with cement the ratio E_{50}/UCS is somewhat lower and in a tighter range than for the wet process, being roughly 25–50 for compressive strength less than 0.3 MPa and 50–250 for strength of 0.3–2 MPa. As the cement factor increases, soil-cement becomes stiffer and more brittle. It should be noted, however, that only in unconfined compression does soil-cement lose nearly all its strength at strains beyond peak strength. In field conditions, when surrounded by untreated soil or a large mass of treated ground, and when loaded axially, it will rather exhibit a ductile behaviour such that sudden breakage or failure does not occur. In undrained triaxial compression, for example, a clear shear band develops and most of the soil-cement strength is actually retained at larger strains, depending on the confining stress (e.g., Yu et al., 1997).

Table 9.7 Summary of generally observed relationships for cement-treated soils

Factor of influence	Expected reaction on stabilised soil
Granular soils	Increase strength and allow reduction of the cement factor, shorten curing time to reach the design strength, simplify distribution of cement throughout the soil, impede very low permeabilities
Clayey soils	Reduce strength and require higher cement factors than sands, slow the rate of initial strength gain compared to sands, involve pozzolanic reaction and strength growth over elongated time (with different rates), impede uniform distribution of agents throughout the soil, enhance low permeabilities of treated soil
Fine soil fractions	The smallest 25% of particle size controls strength; silty sands have noticeably lower strength than clean sands
Natural water content	Compressive strength decreases almost linearly with increasing water content; flow of groundwater may cause cement washout
Organic matter, low pH	Significant negative impact on strength, soils with organic contents over 6% and having pH <5 are difficult to improve
Cement factor, α (typical range 100–400 kg/m³)	In silts and clays: almost linear strength gain with increasing α, in sands and gravels: overproportional strength increase with increasing α, especially for higher cement factors; higher α improve durability and decrease permeability
Cement type	In clays, long-term strength gain is higher for blast-furnace slag cement than for Portland cement
Water/cement ratio (typical range 0.8–1.2, band extends from 0.5–2.5)	Increasing W/C ratios more directly decrease strength than α, higher W/C ratios slow the rate of hydration-related strength gain and lower long-term strength gain beyond 28 days; low W/C ratios minimise extra water, higher promote mixing
Additives (e.g., dispersants, retarders, anti-washout agents, etc.)	Change fluid and set properties of slurry mix; prevent binder washout in dynamic water situations
Substitutes for cement (used alone or with cement):	Significantly influence all properties, but rare in wet mixing (except for cut-off walls and environmental applications)
• Bentonite, clay	• Improve stability of high W/C slurries, reduce permeability
• Slag	• Improves chemical stability and durability, retards strength gain
• Kiln dust	• Used in environmental applications
• Fly ash	• Increases chemical durability, reduces heat of hydration
• Lime, gypsum	• Used in low-strength DM
• Silicates, polymers, etc.	• Used in special environmental applications
Air entrainment (in jet mixing)	Lowers strength, may increase freeze-thaw resistance
Mixing operation	Mixing efficiency improves with higher rotary speeds, is usually easier with thinner blades; UCS improves and variation of strength decreases with increasing blade rotation number
Installation process	High volume ratios cause high volume of spoil; weaker strength observed in overlapping zones when separated in construction by considerable time
Sampling	Good quality core samples have often higher strength than wet grab samples; small samples usually yield higher strength than bigger samples

Table 9.8 Typical correlations and data for cement-treated soils using the wet process[a]

Selected parameters	Expected values/ratios or relationships
UCS: rate of strength gain	28 days UCS = ca. 2 × 4 days strength 28 days UCS = 1.4–1.5 × 7 days strength (silts, clays) 28 days UCS = 1.5–2 × 7 days strength (sands) 56 days UCS = 1.4–1.5 × 28 days strength (clays, silts) long-term strength increase typically observed
Generalised relationship for all soils (after Filz, 2009)	$UCS(t) = (0.187 \ln(t) + 0.375) \cdot UCS_{28days}$ t = curing time (days);
UCS: coefficient of variation, v (COV)	0.2–0.6 (typically 0.35–0.5), v is lower for laboratory mixed samples than for field samples
UCS - relative strength ratios: - core samples to laboratory mixed samples, λ, - core samples to wet grab samples	0.5–1, lower values for clays higher for sands (1.0 - for offshore works in Japan) 1–1.5
Shear strength (direct shear, no normal stress)	0.4–0.5 × UCS, for UCS < 1 MPa 0.3–0.35 × UCS, for 1 < UCS < 4 MPa 0.2 × UCS, for UCS > 4 MPa
Tensile strength	0.08–0.15 × UCS, but not higher than 200 kPa, indirect splitting tests yield lower values than direct uniaxial tests
Secant stiffness modulus, E_{50}, at 50% peak strength	50–300 × UCS, for UCS < 2 MPa 300–1,000 × UCS, for UCS > 2 MPa (ratio increases with increasing UCS)
Axial strain at failure, ε_f : - unconfined compression test (crushing failure) - confined compression tests (plastic shear failure)	0.5%–1% for UCS > 1 MPa 1%–3% for UCS < 1 MPa, 2%–5% (undrained triaxial test)
Poisson's ratio	0.25–0.45, typically 0.3–0.4
Unit density	No noticeable relationship with UCS

[a] Data compiled from CDM (1996), BCJ (1997), Porbaha et al. (2000), FHWA (2001), CDIT (2002), Matuso (2002), Filz (2009).

The stiffness modulus of stabilised soil is also highly dependent on the strain level. When a DM column is stressed in the field only to a low portion of the peak strength, the secant modulus may be unrealistically small, leading to overprediction of deformations. In such cases it may be more appropriate to use the initial stiffness modulus in design analyses.

9.5.4 Long-term strength gain and deterioration of stabilised soil

For a rigorous design of DM it is necessary to collect data on long-term behaviour of stabilised soils. Terashi (2002b) has summarised Japanese field investigations concerning two aspects of long-term characteristics of stabilised ground with respect to lime and cement-treated soils using both wet and dry processes.

One aspect is the strength gain in the long term. The data presented in Table 9.9 reveal that stabilised soils exhibit significant strength increases, although the rate of strength increase varies strongly with different case records. Roughly, a strength increase of two to three times may be expected during 10–20 years after treatment. Stronger increases are observed for the dry process, while increasing W/C ratios may impede long-term strength gain. No substantial change with time of the unit density and moisture content of treated soil was noticed. In the current design practice, possible long-term gain of the compressive strength, beyond a 90-day period, has not been yet accounted for.

The second aspect is the long-term deterioration of stabilised soil exposed to different environmental conditions, like untreated soil, fresh water, salt water, polluted ground, etc. The possibility of deterioration was first addressed by Terashi et al. in 1983. They found that a slow deterioration process starts from the exposed boundary of treated soil and progresses inwards. Strength reduction is associated with leaching of Ca dissolved from hydrated cement component into pore solution of treated soil, and then migrating to the surrounding environment. Saitoh (1988) observed that the rate of deterioration depends mostly on the strength of treated soil and partly on the type of soil and binder. Existing experimental evidence from laboratory and field observations suggests that the rate of deterioration is almost linear with logarithm time. The depth of deterioration is smaller for soil with a higher cement factor. For the treated soil at Daikoku Wharf, the depth of deterioration is 30–50 mm over the past 20 years (Terashi, 2002b).

In view of the above investigations it may be tacitly assumed for current design practice and unpolluted soils, as long as new information are available, that both effects compensate.

9.5.5 Design strength of stabilised soil

A common feature of all DM applications is a large scatter of the field strength of stabilised soil. A rational selection of the design compressive strength of soil-binder mix should therefore be based on a statistical approach. According to accumulated experimental evidence, it can be assumed, with reasonable accuracy, that the UCS data fits a normal distribution curve (e.g., BCJ, 1997, Matuso, 2002). Bearing in mind the limitations of the unconfined compression test to represent the actual field strength of stabilised soil mentioned in Section 9.5.3, the design compressive strength, f_c, may be related to the mean, $\bar{q}_{uf}$, and the standard

Table 9.9 Long-term strength gain based on Japanese field data[a]

Mixing process	Soil type	Binder	UCS and standard deviation, s_d		UCS ratio
			short term	long term	
Dry deep mixing	Reclaimed clay 60% clay, 39% silt, 1% sand	Quick lime, 12.5% dry weight	64 days: 1.02 MPa mean s_d = 0.2 MPa	11 years: 3.5 MPa mean s_d = 0.78 MPa	3.4
Dry deep mixing	Upper volcanic ash, underlined by peat, clay, silty fine sand and silt	Blast-furnace slag cement type B, 290 kg/m³ upper 3 m, 130 kg/m³ lower 5 m	28 days: 0.2–0.5 MPa at 0–1 m depth, 0.2–0.5 MPa at 4–6m depth	17 years: 1.5 MPa mean s_d = 0.98 MPa 0.53 MPa mean s_d = 0.35 MPa	>3 1.5
Dry deep mixing	4.5 m peat layer w = 300%–500%, 5.5 m organic clay w = 150%–200%	Cement-type agent, 265 kg/m³	28 days: 0.58 MPa mean	14 years: 3.5 MPa mean	6.0
Wet shallow mixing	Reclaimed clay 30% clay 70% silt	Cement-type agent W/C = 1.5, 5% wet weight	21 days: 74 kPa mean	15 years: 220 kPa mean s_d = 139 kPa	3.0
Wet deep mixing	Deep marine clay (offshore)	Cement	93 days: 6.1 MPa mean s_d = 2.0 MPa	20 years: 13.2 MPa mean s_d = 5.19 MPa	2.2

[a] Data compiled from Terashi (2002b), based on the investigations of Terashi and Kitazume (2002), Yoshida et al. (1992), Hayashi et al. (2003), Inagaki et al. (2002), Ikegami et al. (2002).

deviation, s_d, of the strength of field samples using the following equation (see Figure 9.39):

$$f_c = \bar{q}_{uf} - m s_d \qquad (9.5)$$

The m-value determines the confidence that any measured $q_{uf} \geq f_c$ (f_c is sometimes referred to as guaranteed strength). For a relatively high confidence level of 95%, as is usually applied in the case of a structural concrete, m is equal to 1.64. For wet soil mixing using cement, $m = 1.3$ has been recommended in Japan by BCJ (1997), corresponding to 90% of confidence.

Introducing the coefficient of variation (COV), $v = s_d / \bar{q}_{uf}$, Equation 9.5 can be rewritten as:

$$f_c = (1 - m v)\bar{q}_{uf} = \eta_1 \bar{q}_{uf} \qquad (9.6)$$

where η_1 is a measure of the scattered strength. Based on BCJ (1997) data, Taki (2003) reported that the average ratio between compressive strengths of DM columns, evaluated from 26 case histories of strength tests performed on full-scale columns of 0.6–1 m diameter, and the mean UCS of the cores obtained from these columns was $\eta_1 = 0.62$ for cohesive soils and $\eta_1 = 0.80$ for cohesionless soils (mean for all soils was 0.69, CDIT 2002). Both values are relatively high due to rather low COV equal to 0.26 and 0.18 respectively. For a more conservative estimation, Taki proposes to use $\eta_1 = 0.5$ for cohesive soils and $\eta_1 = 0.64$ for cohesionless soils, which corresponds well with the design recommendation of CDIT (2002) quoting $\eta_1 = 0.5$–0.6. Applying the latter range of η_1 and assuming that $m = 1.3$ is a reasonable choice for the acceptance criterion, it can be back-calculated that

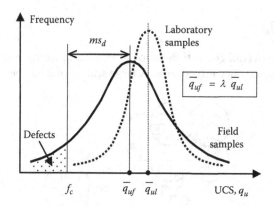

Figure 9.39 Normal distribution curves for field and laboratory strength data and assessment of the design strength f_c.

the corresponding coefficients of variation should not exceed $v = 0.38$ for cohesive soils and $v = 0.31$ for cohesionless soils. Although both variation coefficients fall into the observed range specified in Table 9.8, a relatively high degree of mixing must be assured to fulfil these criteria. It should be noted that the reported field data of BCJ are average values of many different mixing methods. For a specific project, the level of confidence, and especially the expected coefficient of variation, should be carefully assigned in relation to the type of support and the selected method of treatment to obtain a reliable estimate of the field design strength.

Equation 9.6 can be used directly in the case of non-overlapping execution of DM columns. If intersecting columns are installed, the strength in the overlapped area may be lower than in other parts of the columns, depending on the time interval until overlapping, execution capacity of the DM machine and the type of binder used. The corresponding reliability coefficient of overlapping, η_2, is typically set in Japan in the range of 0.8–0.9 (CDIT, 2002). Furthermore, when the improved ground is composed of serial overlapping columns, small areas of untreated soil remain enclosed between the joint columns (Figure 9.40). In this case a correction factor for the effective width of treated column, η_3, is used to compensate for the untreated part. Typical values of η_3 are within 0.7–0.9 (CDIT, 2002), depending on the applied installation pattern.

With the additional correction factors, Equation 9.6 reads:

$$f_c = \eta \, \bar{q}_{uf} \tag{9.7}$$

where $\eta = \eta_1 \, \eta_2 \, \eta_3$.

If the mean UCS of laboratory mixed specimens, $\bar{q}_{ul}$, is used in Equation 9.7 instead of $\bar{q}_{uf}$, then the following practical relationship for an early estimation of the design field strength is obtained (cf. Figure 9.39):

$$f_c = \eta \lambda \, \bar{q}_{ul} \tag{9.8}$$

where λ is the overall correction factor representing strength discrepancy of the field and laboratory mixed soils. A large scatter of λ has been observed in

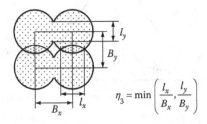

$$\eta_3 = \min\left(\frac{l_x}{B_x}, \frac{l_y}{B_y}\right)$$

Figure 9.40 Correction factor for the effective width according to CDIT (2002).

a number of investigations, with values ranging roughly from 0.3–1.5 for the wet method and about 0.5–2 for the dry method (based on CDM and DJM experience). For design purposes λ is usually assumed in the range of 0.5–1 because lower mixing efficiency *in situ* is generally expected (see Table 9.8).

The allowable compressive strength, f_{ca}, of stabilised soil is subsequently calculated as follows:

$$f_{ca} = \frac{f_c}{F_s} \tag{9.9}$$

where F_s is the adopted safety factor. Since f_{ca} is based on UCS in which no account for creep and cyclic loading is incorporated, relatively high global safety factors of 2.5–3 for static conditions and 2 for dynamic (earthquake) conditions are typically used. They also incorporate the importance of the structure, the type of loads, and the design method (CDIT, 2002). The corresponding allowable shear and tensile strengths of stabilised soil can be calculated from the allowable compressive strength f_{ca} using empirical relationships specified in Table 9.8. If more advanced analyses with respect to shear deformation are required, direct data from drained or undrained triaxial compression tests may be used to account for actual confining pressure and drainage conditions in the field, while the residual shear strength rather than the peak shear strength should be used in the design.

If the limit state design philosophy is followed, then F_s can be replaced with appropriate partial safety factors applied to various elements of the design according to the reliability with which they are known or can be calculated. The obtained factored design compressive strength, $f_{cd} < f_c$, is then compared with the maximum factored design normal stress acting on the column. The same applies for the shear and tensile strengths.

In general, the overall procedure aiming for selection of the design strength of stabilised soil should be more rigorous for all cases where failure of a single column may be critical to the performance of stabilised ground. For DM applications with a sufficiently high ratio of area improvement and appropriate installation patterns, individual column quality is usually less important and lower safety margins can also be accepted. Evidently, the final judgement regarding the field design strength should be based on local experience in terms of the improvement effect on the soils found in the region, the properties of the stabiliser, data from preconstruction trial tests, the sensitivity of the project, the experience of the contractor (i.e., the COV), and the expected level of quality control and quality assurance, as pointed out by Porbaha et al. (2000).

9.5.6 Geotechnical design

The purpose of geotechnical design is to determine the final installation pattern and dimensions of improved ground on the basis of appropriate stability analyses to satisfy functional requirements of the supported structure. Depending primarily on the adopted arrangement of DM columns/panels and on the selected design strength of stabilised soil, which in general may represent hard to semi-hard material, the improved ground can be considered as a rigid body or as a geocomposite system.

If a stabilised soil is likely to behave as a rigid structural member embedded in the ground, its external stability can be evaluated under modes of failure typical for gravity-type structures, including horizontal sliding, overturning, bearing capacity and rotational sliding. Related DM patterns which can be analysed with this approach comprise mainly block-type improvement and, with certain simplifications, also 'blocks' composed of long and short walls (cf. Figure 9.32a), as it has been practised in Japan for various port facilities (cf. CDIT, 2002). In the latter case, however, it is also necessary to examine the extrusion failure mode of untreated soil remaining between the long walls of stabilised soil when subjected to unbalanced active and passive earth pressure, generated for instance by an earthquake (Figure 9.41). One method to inspect this failure mode is to assume that the soft soil in between the long walls moves as a rigid rectangular prism, while the height of the prism is varied (cf. Terashi et al., 1983). Any safe design requires also that the stresses inside the stabilised soil body do not exceed the capacity of soil-binder material. The earth pressures applied to the internal stability analysis should be carefully assigned in relation to the margin of safety adopted for the external stability to maintain compatibility of displacements. Sequence and method of construction may also affect the internal stability of the soil-binder product as cold joints may form at the intersections of primary and secondary panels or overlapping columns. After the final pattern of treatment is determined by the above analyses, the immediate and long-term displacements of the stabilised ground should be calculated. Since the deformation of the stabilised soil itself is usually small, the displacement of the improved ground results from the deformation of the soft layers surrounding or underlying the treated soil mass.

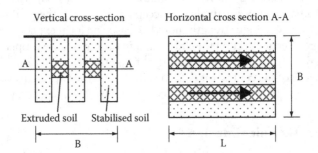

Figure 9.41 Extrusion failure of combined wall-type DM.

For a majority of on-land applications, there is a general tendency to use more economical installation patterns than block treatment, especially for settlement reduction and improvement of stability of embankments and for foundation of structures. Consequently, column-, group column-, lattice- and grid-type solutions are frequently designed with area improvement ratios between 15% and 50%, depending on application and sensitivity to earthquake excitations. A common feature of this type of solution is that untreated soil is surrounding an individual column or a column group, or is left within enclosed spaces formed by stabilised soil (cf. Figure 9.30). As a result, a geocomposite system is created and the interaction between the stabilised and the native soil must be carefully considered and understood at the stage of design. The mode of deformation and/or the mode of failure of composite ground are primarily dependent on the relative stiffness of stabilised soil, on the selected installation pattern and on external loading conditions (cf. Filz, 2009, Wehr et al., 2012).

Corresponding stability analyses of composite ground usually begin with initial assumption of the installation pattern (i.e., with selection of area improvement ratio, a_p) and initial evaluation of necessary compressive strength of stabilised soil with respect to acting loading (internal stability). Examination of sliding stability is carried out considering equilibrium of horizontal forces acting on the boundary of improved ground assumed to behave as a unit block. Rotational sliding is usually checked by the slip circle analysis, taking into account the average shear strength of composite ground, $\bar{c}$, calculated as a weighted mean of the strength of the columns and the strength of the unstabilised soil, also noting the difference in the strain levels, so that

$$\bar{c} = f_t \, a_p + r \, c_0 \left(1 - a_p\right) \tag{9.10}$$

where f_t is shear strength of stabilised soil, c_0 is shear strength of untreated soil and r represents reduction factor for soil strength due to limited strain level. Centrifuge tests conducted by Kitazume et al. (2000, 2009) revealed, however, that the group of columns might fail by several failure modes governed not only by shear failure but also by tilting and bending failure of columns, as shown in Figure 9.42. Consequently, the design procedure based only on simple slip circle analysis with averaged shear strength may overestimate the external stability of group columns, especially if horizontal forces in excess are encountered, leading to progressive bending failure, which actually happened in a couple of unreported cases as mentioned by Terashi (2003). Unexpected deformations of column-stabilised embankments have also occurred, despite the fact that the undrained shear strength of the columns was much higher than the design shear strength adopted for stability calculations (Kivelö, 1998). To avoid too-risky designs with low a_p values and high strength elements, it has been generally recommended that

(a)

(b)

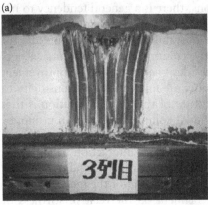
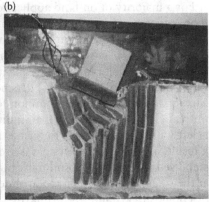

Figure 9.42 Typical modes of failure observed in centrifuge tests. (a) Vertical loading— shear failure of columns just beneath the foundation, and bending failure of the outer columns. (b) Inclined loading—all columns collapsed by bending failure. (From Kitazume, M., Okano, K., and Miyajima, S. (2000). *Soils and Foundations* 40(4):43–55.)

the width of the improved ground should be larger than half of the thickness of the soft soil (CDIT, 2002).

Deformation and stability analyses for composite ground are generally very complex, except for vertical or nearly vertical loading (horizontal forces less than 3%) and regular patterns of separated columns. In such cases, classical calculation methods based on stress concentration and uniform settlement of the stabilised and untreated soil may be applied. For more complicated installation patterns, complex loadings, and difficult soil conditions, sophisticated 2D or 3D FEM calculations as well as model or even field tests are indispensable (cf. Topolnicki, 2009, and Topolnicki and Sołtys, 2012). In general, numerical analyses and parametric studies are an exceptionally good mean of assessing how alternative column patterns and combinations of soil and column stiffness will affect the behaviour of the structure, especially when the properties of stabilised soil are not well known.

A consistent design procedure for DM support of embankments and levees has been developed in the US (and applied for the large levee rehabilitation project in New Orleans). The design shear strength of treated ground is evaluated taking into account residual strength, curing time, and variability of soil-mix material (Filz, 2009). Modified analysis and design procedures for external and internal stability of embankments and levees are also presented.

The design of reinforced DM columns/panels for excavation support is similar to the traditional approach, while ensuring safety against failure in bending and shear in the soil-mix is an important addition to the design of the structural resistance of the wall system (Rutherford et al., 2007). Simplified beam-column methods are recommended for a preliminary design purposes. Finite element calculations allow for better modelling of stage construction phases and for more realistic deformation, deflection, and bending moment predictions. Current design practice does not account, in general, for any contribution of the soil-mix to the bending stiffness of the wall.

When considering the geometry of the in-situ treatment volume, one should also consider the treatment costs. These costs can vary widely depending on the specialist contractor's equipment and procedures. In general, the larger the treatment area per penetration, the lower the treatment cost per unit volume. Be on guard to this as a design with the least amount of treatment volume may not be the least costly to satisfy the support or resistance requirements. Performance specifications for the quantities and quality offer the best solution for lowest cost construction.

The design for composite ground DM is further illustrated with the selected case histories in Section 9.7.

9.6 QUALITY ASSESSMENT OF DEEP MIXING

Quality assessment (QA) of the finished DM product is regarded as one of the pressing issues in the implementation of soil mixing (FHWA, 2001, Porbaha, 2002, Terashi and Kitazume, 2009). Quality assessment is obtained from the installation records of the columns/panels and from the results of appropriate laboratory and field verification tests. A summary of current testing methods for QA is presented in Table 9.10. Traditionally, the most frequently used testing methods are coring (Japan), core and wet grab samples (US), and probe testing (Scandinavia). The selection of suitable verification methods should depend on careful analysis of their relevance, accuracy, applicability, and cost in relation to the purpose and pattern of soil treatment, strength of stabilised soil, and the applied process of soil mixing (dry or wet).

Wet grab samples taken from different depths shortly after construction of wet-mixed columns are used to make cubes or cylinders for later laboratory testing. Typical sampling tools consist of a hollow rod suspending a sample bucket or a tube at its tip, with inside dimensions sufficient to hold the required quantity of treated soil. The sampling device must be able to reach a prescribed depth in order to take a wet sample from a representative soil layer, and must allow it to be retrieved without contamination. Wet grab sampling is relatively easy in operation and permits a large number of specimens to be obtained at low cost. It is common to take wet samples from each day's work (e.g., from two columns per shift at different sampling levels) and to check the strength at 3, 7, 14, 28, and 56 days of curing, all depending on project specifications. An early estimate of the

Table 9.10 Summary of verification methods for quality assessment of deep mixing

Laboratory testing	(1) Wet grab samples	Only for the wet method: samples taken during construction from different depth along the treated columns
	(2) Core samples	Dry and wet method, typically 50–100 mm diameter, taken after construction
	(3) Block samples from extracted or exposed columns, or cut column sections	Block sample dimensions ranging from standard cubes 150×150 mm to full-diameter columns, taken after construction
In-situ testing	(1) Penetration methods	Static cone penetration test (CPT), standard penetration test (SPT), rotary penetration test (RPT), pressuremeter test (PMT), conventional column penetrometer (CCP) test, reverse column penetrometer test (RCP, FOPS), column vane penetrometer (CVP) test, dynamic cone penetration (DCP) test, static-dynamic penetration (SDP) test
	(2) Loading methods	Single-column loading test, group-column loading test, plate loading test, screw-plate test, post-construction monitoring
	(3) Geophysical methods	Seismic methods (inhole, downhole, crosshole), electrical resistivity
	(4) Nondestructive methods	Sonic integrity test
	(5) Drilled shafts or excavations	For visual observation and testing (and sampling)

field strength is also used for optimisation of the mix design. In the case of less efficient mixing, however, the presence of soil clods (i.e., unmixed soil masses) may prevent the sampler from functioning correctly and/or from obtaining a wet sample whose composition is truly representative of the overall mixed volume. While preparing small specimens for testing, soil clods greater than 10% of the mould diameter should be screened off.

Core drilling is frequently used to obtain field specimens for testing and to inspect continuity and uniformity of DM columns (to a less extent for the Nordic method). In the case of major Japanese DM applications, it is typical to core one hole per every 10,000 m³ of treated soil for marine projects and one per 3,000 m³ for on-land projects (Okumura, 1996). At Boston's Central Artery/Tunnel Project, involving over 500,000 m³ of soil-cement, one corehole was drilled per ca. 2,250 m³. It is important to bear in mind that retrieval of representative core samples from a stabilised soil is difficult and sensitive, both to the sampling device and to the sampling technique. Cracks or microcracks may occur in the cores during sampling for a variety of reasons, such as bend in the borehole, rigidity of the sampler, wobble of the drill steel, locking of the sampler, and rotation of the sampling tube with the outer barrel. Therefore, special drilling methods and lengthy field adjustments are often required to improve the quality of core recovery through treated soils of widely varying strength and composition. As an illustration, the coring activities at the CA/T project in Boston, comprising 225 coreholes and more than 7,000 lin. m of core, may be cited after Lambrechts and Nagel (2003). At the start of the core drilling programme a double-tube core barrel equipped with a standard carbide tipped bit was employed but the core recovery rate in the variable composition soil-cement material was poor (mean 71.5%, standard deviation 16.7%). A significant improvement in core recovery (mean 87%, standard deviation 17.7%) occurred after a number of field trials as a result of the following modifications to coring technique: (1) switch to a triple-tube core barrel, (2) replacement of the carbide bit with a fine diamond step-bit with side discharge waterways to minimise sample washout, and (3) use of synthetic drilling mud and additives. In general, cores with diameters greater than 76 mm are recommended (e.g., 86 mm, 100 mm, or more) since they provide less disturbed core specimens. Moreover, core samples should not be taken exclusively from the middle of the column but rather from all across the radius, including overlapped zones, and also from the weakest soil layers to ensure collection of representative specimen.

The core runs should be visually examined for continuity and uniformity of the soil-binder material. Continuity is evaluated by determining the core recovery rate, as measured similar to a rock quality designation (RQD) value. The recovery rate is defined as the total length of full-diameter core obtained per coring depth, expressed in percentages. In Japan the core RQD should exceed 90% for clayey soils and 95% for sandy soils, while for each individual core run the RQD may be 5% less than specified above (Futaki and Tamura, 2002). These requirements are rather high and may be difficult to fulfil even with sophisticated coring techniques, especially in medium to stiff clay. At Boston's CA/T Project, for example, only 38% of the borings drilled with the improved equipment achieved better than 90% overall recovery, while with the double-tube barrel equipment used initially only 1 core boring out of 25 achieved 90% RQD. Although much of the recovered core failed to meet the minimum required compressive strength and RQD criteria, the overall DM mass performed adequately during excavation and tunnel construction (Lambrechts and Nagel, 2003). Uniformity is dependent upon the quality of mixing. An assessment of the uniformity with respect to the distribution of binder should preferably be based on the concepts of scale and intensity of segregation or some other form of defined mixing indices to avoid subjective visual judgement, as pointed out by Larsson (2001). However, in practice it is difficult to draw any definite conclusions about the homogeneity of soil-binder mix from a limited number of samples, especially if the degree of mixing is low and the sample size is small. Therefore mixing indices and statistical methods for assessing uniformity have been used only sporadically, mainly in connection with R&D projects in Japan and Sweden.

Controversial opinions exist about the true relationship of UCS data from cores to those from wet grab samples (cf. Bruce et al., 2002). It may be expected, however, that core strength is higher than that of wet grab sample, providing good quality specimens are obtained from careful drilling operations. For instance, Taki and Yang (1991) published data for various soils which show that the core strengths were about twice those obtained by samples made from wet grabs. The scatter of core sample results is, however, typically larger.

The opportunity to expose the treated ground permits block samples to be taken with different shapes, sizes, and orientations. Moreover, it is also possible to verify column position and to examine column shape, homogeneity, integrity, nature of column, overlap, and so on. Single columns can even be fully extracted, while multiple columns can be constructed in a ring or box arrangement to allow a self-supporting excavation to be completed. Drilled observation shafts may serve the same purpose. This kind of QA inspection is often limited due to cost, time, and site logistics constraints, but may be very useful to resolve any apparent anomalies identified by coring or penetration testing (e.g., Burke et al, 2001).

Evaluation of the results obtained from testing of wet grab, core and block samples should be based on a statistical approach taking into account the actual number of tested specimen, especially if the number of data is low. Considering, for example, compression testing, the mean value of field UCS, $\bar{q}_{uN}$, evaluated for N randomly distributed sampling points within a specific layer of stabilised soil treated in the same manner, is calculated as follows:

$$\bar{q}_{uN} = \frac{\sum_{i=1}^{N} q_{ui}}{N} \tag{9.11}$$

where q_{ui} represent the average UCS of a series of three tests conducted for each sampling point so that

$$q_{ui} = (q_{u1}^{i} + q_{u2}^{i} + q_{u3}^{i})/3, \text{ for } i = 1 \text{ to } N \tag{9.12}$$

The corresponding QA criterion can be derived from Equations 9.6 and 9.9 and reads:

$$\bar{q}_{uN} \geq \frac{f_{ca} F_s}{1 - k_a v_d} \tag{9.13}$$

where k_a is the coefficient of acceptance depending on the number of sampling points N, and v_d represents the designed value of the coefficient of strength variation for a specific DM method. The k_a-values used in Japan are shown in Table 9.11. Assuming for example $v_d = 0.35$ and $F_s = 3$ it can be calculated that for two sampling points ($N = 2$) the mean field UCS should be at least 7.4 times larger than the allowable design compressive stress f_{ca}, while for $N \geq 9$ the ratio drops to 5.5 (assuming $\eta_2 = \eta_3 = 1$, see Equation 9.7).

In the US, the practise of DM also simplified evaluation criteria for QA used in the case of a large number of wet grab samples and for relatively high area improvement ratios. For instance, the following acceptance criteria have been developed (Burke, 2002):

- The strength of stabilised soil must achieve an average UCS equal or greater than the minimum design stress value necessary multiplied by a safety factor of 2.5.
- No greater than 5% of the test results shall be less than the minimum design stress value.
- A ceiling value of twice the average required strength shall be used for individual UCS values in calculating the average strength achieved in the field.

Laboratory tests provide verification data only in discrete points of DM columns and may not localise weak zones along the entire depth of treatment. Accordingly, various in-situ techniques have been developed mainly in Japan, Scandinavia and the United States, usually adapting existing geotechnical and geophysical testing methods. Current state-of-the-art summaries of available in-situ verification techniques have been published by FHWA (2001), Porbaha (2002), Axelsson and Larsson (2003), Porbaha and Puppala (2003), and Terashi and Kitazume (2009).

Penetration testing is feasible in low-strength DM or young soil-mix columns designed for higher strength. For the Nordic method, all routine testing on installed columns, comprising usually 0.5%–2% of columns, is carried out by some form of penetrometer testing, including specially developed reverse testing methods (e.g., FOPS, limited to column's shear strength of ca. 600 kPa) to avoid the problem of the cone's tendency to steer out of the column. CPT is most effective when the length of the column is less than 10 m and the shear strength is below 1 MPa. Stepwise

Table 9.11 Number of sampling points, N, and coefficient of acceptance, k_a, used in Japan

$N =$	1	2	3	4–6	7–8	≥ 9
$k_a =$	1.8	1.7	1.6	1.5	1.4	1.3

Source: Data from Futaki, M. and Tamura, M. (2002). The quality control in deep mixing method for the building foundation ground in Japan, *Proceedings of the Deep Mixing Workshop 2002*, Port and Airport Research Institute & Coastal Development Institute of Technology, Tokyo, Japan.

pre-boring may be used to keep the cone inside long columns. SPT and DCP tests are also limited to UCS of about 1.5 MPa, while uncertainty exists as to the reliability of the correlation between UCS and the number of blow counts. For hard-treated columns with shear strength lower than 2 MPa, static-dynamic (SPD) test may be used, combining the mechanical CPT and dynamic probing. In Japan, rotary penetration tests (RPT) has been developed to allow column testing to a large depth in stratified ground. The cutting resistance during rotary penetration is measured using sensors installed at a special drilling bit. Pressuremeter tests (PMT) were also conducted in Sweden and the US inside a drilled hole at the centre of the treated column to estimate the in-situ strength and compression modulus (e.g., for the Nordic method).

Various loading tests carried out on the ground surface or at depth in test pits or inside the columns (screw-plate test) provide reliable information about the strength and deformation of the treated soil. They include small diameter plate loading tests as well as full-scale load tests of individual columns or a group of columns. Examples of full-scale tests are presented in Section 9.7. Post-construction monitoring of settlements provides most objective data of the treated ground's overall performance, and is practised on many projects.

Modified geophysical methods are undergoing extensive testing, especially in Japan and Sweden, as a means of assessing column strength, integrity, and homogeneity. Types include: seismic methods (inhole P-S logging, downhole logging, crosshole logging), borehole electrical resistivity profiling and low strain sonic integrity testing. Broadly, each can be described as 'promising', having provided reasonable correlation with cores, but it does not seem that any geophysical method is routinely used (FHWA, 2001). According to Japanese experience it is possible to obtain reliable information from integrity tests if the column compressive strength exceeds 1 MPa and the column length exceeds 4 m (Futaki and Tamura, 2002).

Experiments have also been made in Finland and Japan with, respectively, 'measurement while drilling' (MWD) and 'factor of drilling energy' tests, which relate the records of various drilling parameters to the strength properties of the treated soil (Bruce et al., 2002).

Finally, it should be emphasised that the QA programmes and the adopted control criteria should be dictated by the main purpose of soil treatment and by careful evaluation of associated design limit states. Even with close controls, significant field variability of the properties of in-situ treated soil is most probable. This should be understood as an inherent characteristic of SM technology. Therefore, QA control programmes need sufficient flexibility to respond to variable characteristics of soil-mix, avoiding too restrictive criteria for occasional low strength or existence of soil clods inside the treated ground if the overall performance of stabilised soil is satisfied.

9.7 SELECTED CASE HISTORIES

As discussed in Section 9.4, the applications of SM are numerous and varied. The case histories presented in this chapter are typical for some of these applications and methods used but they also leave many applications untouched. Examples of dry soil mixing executed with the Nordic method are included in Chapter 10.

9.7.1 Metropolitan intercity expressway (dry DM)

Source	:	Technical site visit (courtesy of DJM Association), also Ohdaira, H., Hashimoto, H., Gotoh, K. and Nozu, M. (2002)
Location	:	Metropolitan highway at Kawashima, Japan
Construction site/ description	:	ca. 100 m long, 47.6–57.4 m wide, case A: settlement reduction for a box-culvert, case B: protection against slope failure (Figure 9.43)
Soils	:	Soft clay, 5–7 m thick, N = 0, c_u = 18–26 kPa, Sand, 3.3–5.4 m thick, N = 10–20, Clay, 9–13.5 m thick, N = 4, c_u = 50–70 kPa
Embankment height and load	:	Average 8 m

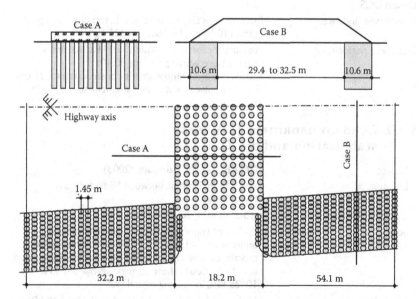

Figure 9.43 Applied column patterns at the Kawashima site. (Redrawn from DJM Association (2002).)

(a)   (b)

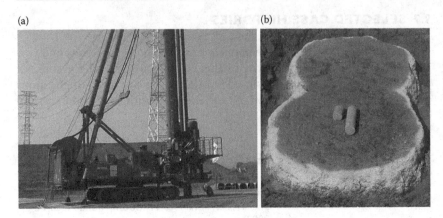

Figure 9.44 (a) DJM machine at the construction site. (b) Exposed twin columns diameter 1 m.

Design requirements   :   Residual settlement 30 cm
Applied DM method   :   Dry jet mixing (DJM), two mixing shafts (Figure 9.44)
Column data   :   Diam. 1.0 m, length 20–22 m, overall 34,200 lin. m
Column pattern   :   Case A: detached single columns, spacing 1.5 m × 1.7 m,
                              a_p = 31%
                              Case B: overlapped columns 0.8 m c/c, wall spacing 1.45 m
                              c/c, a_p = 61%
Design UCS   :   230 kPa
Binder type and factor   :   Blast-furnace slag cement type B, top clay 150 kg/m³,
                              sand 110 kg/m³, bottom clay 125 kg/m³
Observed performance   :   Pressure increase on column 130 kPa, maximum
                              excess pore water pressure 80 kPa
                              Settlement after embankment construction ca. 29 cm,
                              lateral displacement at embankment toe ca. 2 cm

9.7.2 Road embankment (dry mass stabilisation and dry DM)

Source   :   LCM AB, a Keller Company (2003)
Location   :   Road 255 in Sweden, between Södertälje and
                              Nynäshamn
Construction site   :   ca. 500 m long, 20–25 m wide
Soils   :   0.5–3 m of superficial organic soil (peat/gyttja),
                              underneath 3–15 m of soft silty clay, laying upon
                              moraine; water content: peat up to w = 1200%, gyttja
                              w = 300%–500%, shear strength: gyttja 3–7 kPa, clay
                              10–25 kPa, increasing with depth
Embankment height and   :   Height h = 1.4–5.6 m, equivalent traffic load 20 kPa
load   (typical cross-section shown in Figure 9.45)

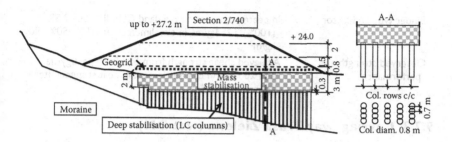

Figure 9.45 Typical cross-section of the road embankment and treated zone.

Design requirements	:	Max. settlement 30 cm in 12 months of operation, safety factor for slope failure 1.5 (long term)
Applied methods	:	The Nordic method (lime cement column) for deep stabilisation of clay (Figure 9.46a), LCM system for mass stabilisation of organic soils (Figure 9.46b)
Execution of LC columns (first phase)	:	Detached columns diam. 0.8 m, square grid of 1.8 m c/c for $h < 4$ m, for $h > 4$ m rows of overlapping columns at c/c 0.7 m spaced 2 m, $a_p =$ 15.4 to 22%, column length 2–10 m, mean length approx. 7 m, total column length ca. 57,000 m
Execution of mass stabilisation (second phase)	:	Grid pattern of soil blocks 3 × 4.5 m, 0.3 m vertical overlap with respect to column heads, total volume of stabilised soil 34,000 m³
Embankment construction (third phase)		Geotextile placed on stabilised soil, crushed fill 0.3–0.5 m thick + geogrid + 0.3–0.5 m crushed fill placed over the stabilised block to preload the peat/gyttja layer, subsequent embankment construction after ca. 1 month
Design shear strength	:	50 kPa in peat/gyttja, 100 kPa in clay

(a)   (b)

Figure 9.46 (a) First phase: installation of LC columns. (b) Second phase: mass stabilisation. (Courtesy of LCM AB.)

| Binder type and factor | : | In peat/gyttja: cement Portland CEM II/A-LL 42.5R (100%), 175 kg/m³, in clay: lime/cement (50%/50%), 80 kg/m³ |
| Observed strength and performance | : | Achieved shear strength from 50 kPa in peat/gyttja and 200 kPa in clay, settlement in peat stopped after approximately 4 months |

9.7.3 Carriageway Trasa Zielona (wet DM)

Source	:	Own project (design and build)
Location	:	Lublin, Poland
Construction site/ description	:	235 m long, 44 m wide, dual three-line urban carriageway
Soils	:	Weak soils down to 3–8 m, loose anthropogenic fill, underlined by peat (w = 400%) and organic clay (w = 35%), organic soils 1–4 m thick
Embankment height and load	:	1.3–2.5 m, equivalent traffic load 30 kPa
Design requirements	:	Max. differential settlement 0.2%
Design compressive stress	:	480–676 kPa, assuming reduced column diameter to 0.7 m
Applied DM method	:	Wet DM, single-shaft equipment (Figure 9.47)
Column data	:	2402 No., diameter 0.8 m, length 3–8.5 m, mean 6.5 m, overall 15,538 lin. m

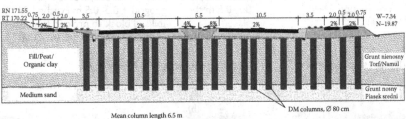

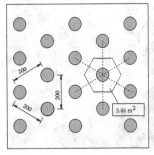

Figure 9.47 Column-type design for a road embankment (single-axis mixing tool, diameter 0.8 m).

Column pattern	:	Detached columns, equilateral triangle, side 2 m c/c, a_p = 14.5%
Design UCS strength	:	1.5 MPa (with 90% confidence)
Cement type and factor	:	Slag cement CEM III/A, 350 kg/m³
Embankment reinforcement	:	Two layers of TENSAR geogrid, 20 and 30 kN/m, separated by 30 cm crushed aggregate

9.7.4 Deepwater bulkhead in soft soils (wet DM)

Source	:	Burke, Lyle, Sehn, and Ross (2001)
Location	:	Ham Marine Inc. Facility, Pascagoula, Mississippi, USA
Construction site	:	Ca. 610 m long, ca. 15.5 m wide, outboard soil dredged to −10.7 m upon completion of DM works and the anchored wall
Soils	:	Loose silty sand, followed by medium stiff to very soft clay (slightly organic), w = 34%–56%
Client's request	:	UCS of 96 kPa (1,000 psf) at 14 days of cure
Applied DM method	:	Wet DM, large-diameter single shaft equipment (Figure 9.49a)
Column data	:	Diam. 2.13 m (7 ft), length 3.5–14.5 m, overall 72,633 m³
Column pattern	:	Cellular grid created with overlapping columns, a_p = ca. 85%, stepped arrangement to minimise the quantity of DM works (Figure 9.48)

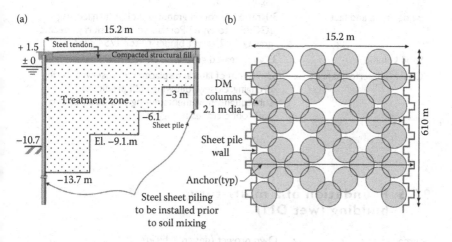

Figure 9.48 Bulkhead design: (a) Cross-section. (b) Cellular DM column pattern. (Redrawn from Burke, G.K., Lyle, D.L., Sehn, A.L., and Ross, T.E. (2001). Soil mixing supports a deepwater bulkhead in soft soils, *Proceedings of the American Society of Civil Engineers Ports Conference*, April 29–May 2, 2001, Norfolk, Virginia, United States.)

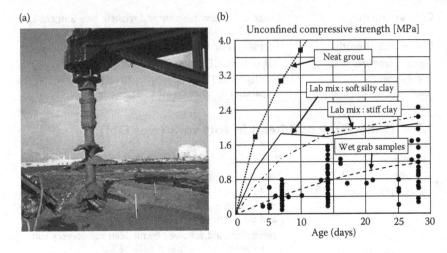

(a)   (b)

Figure 9.49 (a) Single axis mixing tool, diameter 2.1 m. (b) UCS of laboratory and field sampling (Redrawn from Burke, G.K., Lyle, D.L., Sehn, A.L., and Ross, T.E. (2001). Soil mixing supports a deepwater bulkhead in soft soils, *Proceedings of the American Society of Civil Engineers Ports Conference*, April 29–May 2, 2001, Norfolk, Virginia, United States.)

Design UCS	:	Average UCS of 689 kPa (100 psi) (obtained laboratory and field strengths depicted in Figure 9.49b)
Binder type and factor	:	3:1 ratio of ground granulated blast furnace slag (GGBFS) to type I Portland cement, slurry specific gravity 1.42, target binder factor 175 kg/m³
Spoil volume	:	33% of treated ground
Observed performance	:	63–76 mm of lateral displacement observed during dredging, closely following FEM prediction, subsequent movements during operation of two 250-tonne cranes negligible

9.7.5 Foundation of a multistory building (wet DM)

Source	:	Own project (design & build)
Location	:	Kielce, Poland
Construction site/ description	:	ca. 35 × 55 m, foundation slab 1497 m², slab thickness 45 cm (60 cm along the edges)

Soils	:	Heterogeneous soft soils, extending 3–7.5 m below the slab level, including silt, organic clay, fine sand, peat inclusions 0.5–0.8 m thick, constrained compression modulus 2.1–5.4 MPa
Mean loading pressure	:	112 kPa (under the foundation slab)
Expected settlement	:	Without ground improvement: 70–500 mm
Design requirement	:	Max. settlement less than 30 mm
Applied DM method	:	Wet DM, single-shaft equipment
Column data	:	461 No., diameter 0.8 m, length 5–9.2 m (from working level), mean 7.3 m, overall 3,370 lin. m
Column pattern	:	Detached single columns, arrangement adjusted to load distribution and slab-soil-columns interaction, average a_p = 15.4% (Figure 9.50)
Cement type and factor	:	Cement CEM III/A 32.5 NA, 340–380 kg/m³, slurry specific gravity 1.7–1.75
Design compressive stress	:	Max 0.86 MPa (factored value), on a single DM column
Design UCS	:	1.9 MPa (with 90% confidence)
Measured UCS strength	:	32 specimens 15 × 15 cm (wet grab), mean UCS at 28 days is 5.72 MPa, standard deviation 2.14 MPa, COV 0.38
Observed performance	:	Below 10 mm

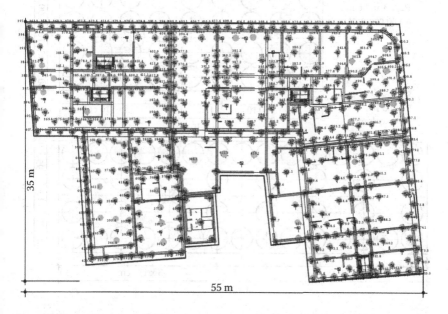

Figure 9.50 Arrangement of DM columns diameter 0.8 m under the foundation slab.

9.7.6 Foundation of highway bridge supports (wet DM)

Source : Own project (design & build)

Location : A2 highway n/Poznań, Poland, bridge WD-23 with five supports (i.e., P1 and P5 are bridge heads, P2 and P4 are intermediate supports (supports in plan view shown in Figure 9.51)

Construction site/ : The whole project included construction of 39 new
description bridges across and along A2 highway, with 2 to 5 foundation supports each, DM applied for reduction of total and differential settlements

Soils : Boulder sandy clays (CPT log shown Figure 9.53a)

Loading pressures : P3 (central support): $\sigma_{mean} = 251$ kPa, $\sigma_{max} = 406$ kPa, $\sigma_{min} = 96$ kPa
P5 (bridge head): $\sigma_{mean} = 138$ kPa, $\sigma_{max} = 248$ kPa, $\sigma_{min} = 27$ kPa

Maximum column load : P3: 399 kN, P5: 418 kN

Design requirements : Settlement < 2 cm, settlement difference between supports <1 cm

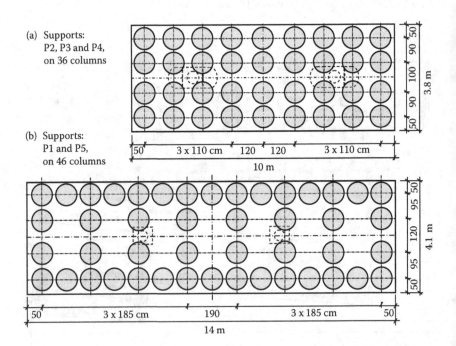

(a) Supports: P2, P3 and P4, on 36 columns

(b) Supports: P1 and P5, on 46 columns

Figure 9.51 Arrangement of DM columns diameter 0.8 m under bridge supports (bridge WD-23).

Applied DM Method	:	Wet DM, single-shaft equipment, diam. 0.8 m (Figure 9.52)
Column length	:	For P3 at elevation of +87.0 m: l = 3.75 m, for P5 at +89.78 m: l = 6.3 m, overall for five supports: 200 columns and 880 lin. m.
Column pattern	:	Detached columns, P3: a_p = 47%, P5: a_p = 40% (Figure 9.51)
Cement type and factor	:	Cement CEM III/A 32.5 NA, 320 kg/m³, slurry spec. gravity 1.65
Design compressive stress	:	Max 0.83 MPa (characteristic value) on a single DM column
Design UCS	:	2.5 MPa (with 90% confidence), global safety factor 3.0

Figure 9.52 (a) Single axis mixing tool diameter 0.8 m. (b) Exposed columns (WD-23, P5).

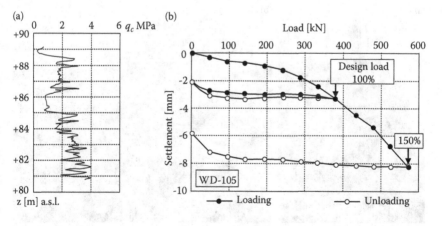

Figure 9.53 (a) Soil conditions (CPT log). (b) Typical result of column static loading test.

Measured UCS strength   :   30 specimens 15 × 15 cm (wet grab), mean UCS at 28 days is 8.73 MPa, standard deviation 1.60 MPa, COV 0.18

Observed settlement   :   P1: 12 mm, P2: 7 mm, P3: 9 mm, P4: 6 mm, P5: 9 mm, typical result of a single column static loading test is shown in Figure 9.53b for a similar bridge (WD-105)

9.7.7 Excavation protection (wet DM, reinforced columns)

Source   :   Own project (design & build)

Location   :   Royal Castle courtyard in Warsaw, Poland

Construction site/ description   :   Temporary excavation pit, 10 m deep (Figure 9.54b) Structural wall constructed with overlapping DM columns; steel H-beams installed every second column. DM-wall supported with a system of two and three rows of pre-stressed ground anchors, 17–18 m long

Ground conditions   :   At the ground surface mixed anthropogenic fill, 2–8 m thick, underlain by medium compacted fine sands and stiff Pliocene clay. Groundwater level 2–6 m below surface

Applied DM method   :   Wet DM, single-shaft equipment; elongated mixing time and full restroking required to ensure installation of soldier elements immediately after mixing

DM columns   :   0.7 m diameter, in 0.55 m spacing. Primary (reinforced) columns embedded 4.5 m below excavation bottom, secondary columns finished 1 m below excavation bottom and designed to work in arching

(a)   (b)

Figure 9.54 (a) Installation of H-beams in DM columns. (b) Completed excavation pit.

Column reinforcement	:	H-beams: IPE 360 and IPE 400 (Figure 9.54a)
Cement type	:	Portland cement CEM I 32.5 R
Cement slurry	:	Slurry specific gravity 1.60, avg. consumption: 0.336 m³/m
Design UCS	:	2.5 MPa (with 90% confidence), global safety factor 2.5
Observed wall performance	:	Lateral displacement of adjacent palace's walls below 4 mm

9.7.8 Cut-off block to install a crosswise connection between two tunnels (CSM, wet DM)

Source	:	BAUER Spezialtiefbau Switzerland AG
Location	:	Biel, Switzerland
Construction site/ description	:	Watertight cut-off block 13 × 65 m, up to 24 m deep, volume ca. 19,300 m³, constructed between and across two planned TBM tunnels, enabling later execution of the connection tunnel with small excavators. Block composed of primary and secondary rows, oriented perpendicular to tunnel axis. Each row, 13 m long, build of three primary (P) panels and two interlocking secondary (S) panels, installed with 25 cm overlap on both sides (Figure 9.55a)
Soils	:	Down to 14 metres, mainly sand and gravel with various cohesive components, followed by medium-dense to dense silty sand and sandy silt up to the final depth. Many small stones up to huge boulders encountered between 10–15 metres depth.
Applied methods	:	Primary panels in primary rows partly constructed using the slurry wall (SW) system (single-phase cut-off wall, width 1.5 m) and partly with the cutter soil mixing (CSM) system (Figure 9.55b). All secondary rows constructed with CSM. Intercut of secondary rows on primary rows was only 15 cm on each side owing to precise verticality control of CSM
CSM panels	:	$w \times l = 1.0 \times 2.8$ m². CSM panels executed in pilgrim step method, all secondary panels installed 'fresh in hard'.
Cement type and factor	:	Cement CEM I 42.5 and bentonite, 440 kg/m³ treated soil, slurry specific gravity about 1.7
Productivity (CSM)	:	210–240 min/panel, including moving rig, changing teeth, etc.
Required properties	:	UCS > 5 MPa at 28 days; permeability $k \leq 1 \times 10^{-7}$ m/s
Measured UCS strength	:	Wet grab specimen, cylinders 100 × 100 mm; mean UCS at 28 days about 5.5 MPa
Additional controls	:	Diagonal borings across cut-off block to confirm water tightness

(a)

Plan view

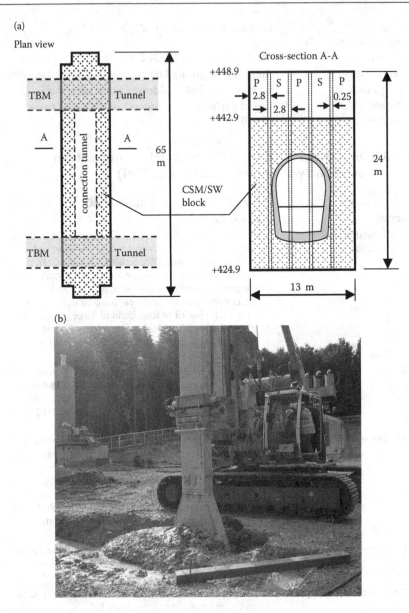

Figure 9.55 (a) Plan view and the cross-section of watertight cut-off block. (b) CSM rig at work. (Courtesy of BAUER Spezialtiefbau GmbH.)

9.7.9 Excavation control (hybrid wet mixing)

Source	:	Technical site visit (courtesy of Chemical Grouting Co., Japan)
Location	:	Tokyo, Japan
Construction site/ description	:	Four-storey basement, 17.2 m deep, part of the site close (ca. 6 m) to existing Japan Railway tunnels at depths from 13.5–22 m (Figure 9.56)
Purpose of DM works		Reduction of the risk of potential movements resulting from the excavation (movements limited to 9 cm)
Soils	:	17 m of very soft silts/clays overlying very dense silts and sands
Applied DM method	:	Wet DM, JACSMAN method, double shaft equipment (combined XJET and mechanical mixing)
Column data	:	Two columns diam. 2.3 m, spaced 1.4 m (type B), improving area 7.2 m²

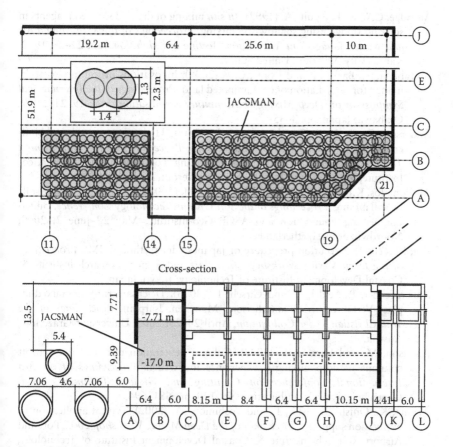

Figure 9.56 Hybrid deep mixing for excavation control (based on project design).

Column pattern	:	Block pattern ca. 12.5 × 56 m, area improvement ratio a_p = 80%, column length 9.3 m (from 7.7–17 m), total 6423 m³, 95 No.
Cement factor	:	α = 160 kg/m³, slurry specific gravity 1.50
Design UCS	:	0.98 MPa (applied factor of safety 3)
Penetration phase	:	1 m/min, utilising 4 × 75 l/min of water through the jetting system and 2 × 50 l/min of water through the mixing system
Withdrawal phase	:	0.7 m/min, grout through both XJET nozzles at 4 × 150 l/min and through the mixing system at 2 × 200 l/min,
Spoil	:	Minimal spoil production

REFERENCES

Aldridge, C.W. and Naguib, A. (1992). *In situ* mixing of dry and slurried reagents in soil and sludge using shallow soil mixing, *Proceedings of the 85th Annual Air and Waste Management Association Meeting & Exhibition*, June 21–26, 1992, Kansas City, Missouri, United States.

Al-Tabbaa, A., Barker, P. and Evans, C.W. (2009). Innovations in soil mixing technology for remediation of contaminated land. *Proceedings of the International Symposium on Deep Mixing & Admixture Stabilisation*, May 19–21, 2009, Okinawa, Japan, pp. 3–35.

Andersson, R., Carlsson, T. and Leppänen, M. (2000). Hydraulic cement based binders for mass stabilization of organic soils, *Proceedings of the Soft Ground Technology Conference*, United Engineering Foundation and ASCE Geo-Institute, May 28–June 2, 2000, Noordwijkerout, Netherlands.

Andromalos, K.B., Hegazy, Y.A. and Jasperse, B.H. (2000). Stabilization of soft soils by soil mixing, *Proceedings of the Soft Ground Technology Conference*, United Engineering Foundation and ASCE Geo-Institute, May 28–June 2, 2000, Noordwijkerout, Netherlands.

Aoi, M. (2002). Execution procedure of Japanese dry method (DJM), *Proceedings of the Deep Mixing Workshop 2002*, Port and Airport Research Institute & Coastal Development Institute of Technology, Tokyo, Japan.

Außenlechner, P., Seidel, A. and Girsch, E. (2003). Deichsanierung im mixed-in-place-Verfahren (MIP) am Beispiel München, Isarplan, *Proceedings of the 18th Christian Veder Colloquium*, April 24–25, 2003, Universität Graz,Graz, Austria.

Axelsson, M. and Larsson, S. (2003). Column penetration tests for lime-cement columns in deep mixing – experience in Sweden, *Proceedings of the 3rd International Conference on Grouting and Ground Treatment*, ASCE Geotechnical Special Publication No. 120, pp. 681–694.

Azuma, K., Ohishi, K., Ishii, T., and Yoshimoto, Y. (2002). Typical application to excavation support, *Proceedings of the Deep Mixing Workshop 2002*, Port and Airport Research Institute & Coastal Development Institute of Technology, Tokyo, Japan.

Babasaki, R., Terashi, M., Suzuki, T., Maekawa, A., Kawamura, M., and Fukazawa, E. (1996). Japanese Geotechnical Society Technical Committee report – Factors influencing the strength of improved soil. In: Yonekura R., Terashi M., and Shibazaki M. (eds.), *Grouting and Deep Mixing: Proceedings of the 2nd International Conference on Ground Improvement Geosystems*, May 14–17, 1996, Tokyo, Japan: Balkema Publishers, pp. 913–918.

Bauer Maschinen GmbH. (2012). *SCM-DH Single Column Soil Mixing Double Head Mixing Tool*. RTG Rammtechnik GmbH. Brochure No. 902.497.1+2.

Bauer Maschinen GmbH. (2010). *Cutter Soil Mixing: Process and Equipment*. Brochure No. 905.656.2.

Blackwell, J. (1994). A case history of soil stabilisation using the mix-in-place technique for the construction of deep manhole shafts at Rochdale. In: Bell A.L. (ed.), *'Grouting in the Ground': Proceedings of the Conference Organized by the Institution of Civil Engineers and Held in London on 25–26 November 1992*, London, England: Thomas Telford, pp. 497–510.

Bredenberg, H. (1999). Keynote lecture: Equipment for deep soil mixing with dry jet mix method, *Proceedings from the International Conference for Dry Mix Methods for Deep Soil Stabilization*, October 13–15, 1999, Stockholm, Sweden, pp. 323–331.

Bruce, D.A., Bruce, M.E. and DiMillio, A.F. (2002). Deep mixing: QA/QC and verification methods, *Proceedings of the Deep Mixing Workshop 2002*, Port and Airport Research Institute & Coastal Development Institute of Technology, Tokyo, Japan.

Building Center of Japan (BCJ). (1997). *Guidelines on Ground Improvement Design and Quality Control for Building Construction* (in Japanese). Tokyo, Japan: Building Center of Japan.

Burke, G.K. (2002). North American single-axis wet method of deep mixing, *Proceedings of the Deep Mixing Workshop 2002*, Port and Airport Research Institute & Coastal Development Institute of Technology, Tokyo, Japan.

Burke, G.K., Lyle, D.L., Sehn, A.L., and Ross, T.E. (2001). Soil mixing supports a deepwater bulkhead in soft soils, *Proceedings of the American Society of Civil Engineers Ports Conference*, April 29–May 2, 2001, Norfolk, Virginia, United States.

Cement Deep Mixing Association. (1996). Promotional information.

Cement Deep Mixing Association. (2002). *CDM Cement Deep Mixing Bulletin as of 2002*.

Coastal Development Institute of Technology (CDIT), Japan. (2002). *The Deep Mixing Method: Principle, Design and Construction*, Rotterdam, The Netherlands: A.A. Balkema.

Druss, D. (2002). Case history: Deep soil-cement mixing at the I-90/I-93 NB Interchange on the Central Artery/Tunnel Project, Contract C09A7 at Fort Point Channel Site, Boston, MA, *Proceedings of the Deep Mixing Workshop 2002*, Port and Airport Research Institute & Coastal Development Institute of Technology, Tokyo, Japan.

Dry Jet Mixing Association. (2002). *DJM-Dry Jet Mixing Method Bulletin as of 2002*.

European Standard EN 14679. (2005). *Execution of Special Geotechnical Works – Deep Mixing*.

EuroSoilStab. (2002). *Development of Design and Construction Methods to Stabilise Soft Organic Soils*. Design Guide Soft Soil Stabilisation, CT97-035I, European Commission Project BE 96-3177.

Evans, J.C. and Garbin, E.J. (2009). The TRD method for *in situ* mixed vertical barriers. *Proceedings of the US-China Workshop on Ground Improvement Technologies IFCWEEF09*, March 14, 2009, Orlando, Florida.

FHWA-RD-99-138 (2000) An Introduction to the Deep Soil Mixing Methods as Used in Geotechnical Applications, 143 p., Prepared by Geosystems (D.A. Bruce) for US Department of Transportation, Federal Highway Administration.

FHWA-RD-99-167 (2001) An Introduction to the Deep Soil Mixing Methods as Used in Geotechnical Applications: Verification and Properties of Treated Soil, 434 p., Prepared by Geosystems (D.A. Bruce) for US Department of Transportation, Federal Highway Administration.

Filz, G. (2009). Design of deep mixing support for embankments and levees, *Proceedings from the International Symposium on Deep Mixing & Admixture Stabilisation*, May 19–21, 2009, Okinawa, Japan, pp. 37–59.

Fiorotto, R., Stötzer, E., Schöpf M. and Brunner, W. (2005). Cutter soil mixing (CSM). *Proceedings from the International Conference on Deep Mixing '05*, May 23–25, 2005, Stockholm, Sweden, pp. 521–526.

Forsberg, T. (2002). Oil tanks on Limix cement columns in Can Tho, Vietnam, *Proceedings of the Deep Mixing Workshop 2002*, Port and Airport Research Institute & Coastal Development Institute of Technology, Tokyo, Japan.

Fudo Construction Inc. (2011). *Deep Mixing Method with Large Diameter and Superior Quality CI-CMC*. Company bulletin.

Futaki, M. and Tamura, M. (2002). The quality control in deep mixing method for the building foundation ground in Japan, *Proceedings of the Deep Mixing Workshop 2002*, Port and Airport Research Institute & Coastal Development Institute of Technology, Tokyo, Japan.

Garbin, E., Hussin, J., and Kami, C.. (2010). Earth retention using the TRD method. *Proceedings of the Earth Retention Conference*, August 1-4, 2010, Bellevue, Washington, United States, ASCE Geotechnical Special Publication No. 208, pp. 318–325.

Geo-Con., Inc. (1998). Promotional information.

Herrmann, R., Hilmer, K. and Kaltenecker, H. (1992). Die Entwicklung des Bauverfahrens Mixed-in-Place (MIP) auf der Basis der Rotary-Auger-Soil-Mixing-Methode (RASM), *Vorträge der Baugrundtagung*, September 21–23, 1992, Dresden, Germany.

Hidetoshi, Y., Masato, U., Iwasaki, T. and Higaki, K. (1996). Removal of volatile organic compounds from clay layer. In: Yonekura R., Terashi M., and Shibazaki M. (eds.), *Grouting and Deep Mixing: Proceedings of the 2nd International Conference on Ground Improvement Geosystems*, May 14–17, 1996, Tokyo, Japan, pp. 787–792.

Hioki, Y. (2002). The construction control and quality control of dry method (DJM), *Proceedings of the Deep Mixing Workshop 2002*, Port and Airport Research Institute & Coastal Development Institute of Technology, Tokyo, Japan.

HITEC (Highway Innovative Technology Evaluation Center). (2002). *Evaluation of the Geo-Con Vert Wall System*, CERF Report: #40607, April 2002.

Holm, G. (2002). Nordic dry deep mixing method – execution procedure, *Proceedings of the Deep Mixing Workshop 2002*, Port and Airport Research Institute & Coastal Development Institute of Technology, Tokyo, Japan.

Holm, G. (2002a). Deep mixing – research in Europe, *Proceedings of the Deep Mixing Workshop 2002*, Port and Airport Research Institute & Coastal Development Institute of Technology, Tokyo, Japan.

Horpibiulsuk, S., Miura, N, Nagaraj, T.S., and Koga, H. (2002). Improvement of soft marine clays by deep mixing technique, *Proceedings of the 12th International Offshore and Polar Engineering Conference*, May 26–31, 2002, Kitakyushu, Japan, pp. 584–591.

Huiden, E.J. (1999). Soil stabilisation for embedment of Botlek Railwaytunnel in the Netherlands, *Proceedings from the International Conference for Dry Mix Methods for Deep Soil Stabilization*, October 13–15, 1999, Stockholm, Sweden, pp. 45–49.

Jasperse, B.H. and Ryan, C.R. (1992). *In situ* stabilization and fixation of contaminated soils by soil mixing, *Proceedings of the ASCE Geotechnical Division Specialty Conference Grouting, Soil Improvement and Geosynthethics*, February 25–28, 1992, New Orleans, Louisiana, United States.

Jelisic, N. and Leppänen, M. (2003). Mass stabilization of organic soils and soft clay, *Proceedings from the Third International Conference on Grouting and Ground Treatment*, February 10–12, 2003, New Orleans, Louisiana, United States, ASCE Geotechnical Special Publication No. 120, 552–561.

Kamon, M. (1996). Effect of grouting and DMM on big construction projects in Japan and the 1995 Hyogoken-Nambu earthquake. In: *Grouting and Deep Mixing: Proceedings of the 2nd International Conference on Ground Improvement Geosystems*, May 14–17, 1996, pp. 807–823.

Kawanabe, S. and Nozu, M. (2002). Combination mixing method of jet grout and deep mixing, *Proceedings from the Deep Mixing Workshop 2002*, Port and Airport Research Institute & Coastal Development Institute of Technology, Tokyo, Japan.

Kawasaki, K., Kotera, H., Nishida, K., and Murase, T. (1996). Deep mixing by spreadable wing method. In: *Grouting and Deep Mixing, Proceedings of the2nd International Conference on Ground Improvement Geosystems*, May 14–17, 1996, pp. 631–636.

Kim, H.Y., Shim, S.H., Shin, S.G., and Shim, M.B. (2009). Optimum design for the deep cement mixing in a soft soil with deep water depth, *Proceedings of the International Symposium on Deep Mixing & Admixture Stabilization*, May 19–21, 2009, Okinawa, Japan, pp. 333–336.

Kitazume, M., Okano, K., and Miyajima, S. (2000). Centrifuge model tests on failure envelope of column-type DMM-improved ground, *Soils and Foundations* 40(4):43–55.

Kitazume, M. (2002). Current design and future trends in Japan, *Proceedings of the Deep Mixing Workshop 2002*, Port and Airport Research Institute & Coastal Development Institute of Technology, Tokyo, Japan.

Kitazume, M., Maruyama, K., and Hashizume, H. (2009). Centrifuge model tests on failure mode of deep mixing columns. *Proceedings of the International Symposium on Deep Mixing & Admixture Stabilization*, May 19–21, 2009, Okinawa, Japan, pp. 337–342.

Kivelö, M. (1998). *Stabilisation of Embankments on Soft Soils with Lime/Cement Columns*, PhD Thesis 1023, Royal institute of Technology, Sweden.

Lambrechts, J. and Nagel, S. (2003). Coring soil-cement installed by deep mixing at Boston's CA/T project, *Proceedings from the 3rd International Conference on Grouting and Ground Treatment*, February 10–12, 2003, New Orleans, Louisiana, United States, ASCE Geotechnical Special Publication No. 120, pp. 670–680.

Larsson, S. (1999). The mixing process at the dry jet mixing method: dry deep mix methods for deep soil stabilization, *Proceedings from the International Conference for Dry Deep Mix Methods for Deep Soil Stabilization*, October 13–15, 1999, Stockholm, Sweden, pp. 339–346.

Larsson, S. (2001). Binder distribution in lime-cement columns, *Ground Improvement* 5(3): 111–122.

Lebon, S. (2002). Wet process soilmixing — a review of central European execution practice, *Proceedings from the Deep Mixing Workshop 2002*, Port and Airport Research Institute & Coastal Development Institute of Technology, Tokyo, Japan.

Matuso, O. (2002). Determination of design parameters for deep mixing, *Proceedings from the Deep Mixing Workshop 2002*, Port and Airport Research Institute & Coastal Development Institute of Technology, Tokyo, Japan.

Nagata, S., Azuma, K., Asano, M., Nishijima, T., Shiiba, H., Yang, D., and Nakata, R. (1994). Nakajima subsurface dam, water policy and management: solving the problems, *Proceedings of the American Society of Civil Engineers 21st Annual Conference of the Water Resources Planning and Management Division*, May 23–26, 1994, Denver, Colorado, United States., pp. 437–440.

Ohdaira, H., Hashimoto, H., Gotoh, K., and Nozu, M. (2002). Observation results for the embankment on soft ground improved by DJM, *Tsuchi-to-Kiso*, February 2002, pp. 31–33 (in Japanese).

Okumura, T. (1996). Deep mixing method of Japan. In: Yonekura R., Terashi M., and Shibazaki M. (eds.), *Grouting and Deep Mixing: Proceedings of the 2nd International Conference on Ground Improvement Geosystems*, May 14–17, 1996, Tokyo, Japan, pp. 879–887.

Pampel, A. and Polloczek, J. (1999). Einsatz des FMI- und HZV-Verfahrens bei der DB AG, *Der Eisenbahningenieurbau*, Nr. 3.

Paviani, A. and Pagotto, G. (1991). New technological developments in soil consolidation by means of mechanical mixing, implemented in Italy for the ENEL power Plant at Pietrafitta, *Proceedings of the Tenth European Conference on Soil Mechanics and Foundation Engineering*, Vol. 2, May 26–30, 1991, Florence, Italy, pp. 511–516.

Porbaha, A. (1998). State of the art in deep mixing technology: part I. Basic concepts and overview, *Ground Improvement* 2(2): 81–92.

Porbaha, A., Tanaka, H., and Kobayashi, M. (1998a). State of the art in deep mixing technology: Part II. Applications, *Ground Improvement* 2(2): 125–139.

Porbaha, A., Shibuya, S, and Kishida, T. (2000). State of the art in deep mixing technology. Part III: geomaterial characterization, *Ground Improvement* 3: 91–110.

Porbaha, A. (2000a). State of the art in deep mixing technology. Part IV: design considerations, *Ground Improvement* 3: 111–125.

Porbaha, A., Raybaut, J.-L., and Nicholson, P. (2001). State of the art in construction aspects of deep mixing technology, *Ground Improvement* 5(3): 123–140.

Porbaha, A. (2002). State of the art in quality assessment of deep mixing technology, *Ground Improvement* 6(3): 95–120.

Porbaha, A. and Puppala, A. (2003). *In situ* techniques for quality assurance of deep mixed columns, *Proceedings of the 3rd International Conference on Grouting and Ground Treatment*, February 10–12, 2003, New Orleans, Louisiana, United States, ASCE Geotechnical Special Publication No. 120, pp. 695–706.

Raju, V.R., Abdullah, A., and Arulrajah, A. (2003). Ground treatment using dry deep soil mixing for a railway embankment in Malaysia, *Proceedings of the 2nd International Conference on Advances in Soft Soil Engineering and Technology*, July 2–4, 2003, Putrajaya, Malaysia.

Rogbeck, Y., Jelisic, N., and Säfström, L. (1999). Properties of mass- and cell stabilization: Two case studies in Sweden, *Proceedings from the International Conference for Dry Mix Methods for Deep Soil Stabilization*, October 13–15, 1999, Stockholm, Sweden, pp. 269–274.

Rutherford, C.J., Biscontin, G., Koutsoftas, D., and Briaud, J.L. (2007). Design process of deep soil mixed walls for excavation support. *International Journal of Geoengineering Case Histories* 1(2): 56–72.

Saitoh, S. (1988). *Experimental Study of Engineering Properties of Cement Improved Ground by the Deep Mixing Method*, PhD Thesis, Nihon University.

Schmutzler, W., Nozu, M., Sakakibara, M., and Bartero, A. (2011). *Deep Mixing for the New Orleans East Back Levee (USA)*, ISSMGE Bulletin, Vol. 5, Issue 1.

Schwarz, W. and Seidel, A. (2003). Das mixed-in-place-Verfahren, *Symposium: 'Notsicherung von Dämmen und Deichen,'* Universität Siegen Institut für Geotechnik, July 2, 2003, Siegen, Germany.

Shiells, D.P., Pelnik III, T.W., and Filz, G.M. (2003). Deep mixing: an owner's perspective, *Proceedings from the 3rd International Conference on Grouting and Ground Treatment*, February 10–12, 2003, New Orleans, Louisiana, United States, ASCE Geotechnical Special Publ. No 120, pp. 489–500.

Sugiyama, K. (2002). The CDM-LODIC method – outline and applications, Proceedings from the *Deep Mixing Workshop 2002*, Port and Airport Research Institute & Coastal Development Institute of Technology, Tokyo, Japan.

Takemiya, H., Nishimura, A., Naruse, T., Hosotani, K., and Hashimoto, M. (1996). Development of vibration reduction measure wave impeding block. In: Yonekura R., Terashi M., and Shibazaki M. (eds.), *Grouting and Deep Mixing: Proceedings of the 2nd International Conference on Ground Improvement Geosystems*, May 14–17, 1996, Tokyo, Japan, pp. 753–758.

Taki, O. (2003). Strength properties of soil cement produced by deep mixing, *Proceedings from the 3rd International Conference on Grouting and Ground Treatment*, February 10–12, 2003, New Orleans, Louisiana, ASCE Geotechnical Special Publ. No. 120, pp. 646–657.

Taki, O. and Yang, D.S. (1991). Soil-cement mixed wall technique, *Proceedings of the ASCE Geotechnical Engineering Congress*, June 10–12, 1991, Boulder, Colorado, United States, pp. 298–309.

Tateyama, M., Trauma, H., and Fukuda, A. (1996). Development of a large diameter short reinforced anchor by cement mixing method. In: Yonekura R., Terashi M., and Shibazaki M. (eds.), *Grouting and Deep Mixing: Proceedings of the*

2nd International Conference on Ground Improvement Geosystems, May 14–17, 1996, Tokyo, Japan, pp. 759–765.

Terashi, M., Tanaka, H., and Kitazume, M. (1983). Extrusion failure of ground improved by the deep mixing method, *Proceedings of the 7th Asian Regional Conference on Soil Mechanics and Foundation Engineering*, August 14–19, 1983, Haifa, Israel, pp. 313–318.

Terashi, M. (2002a). Development of deep mixing machine in Japan, *Proceedings from the Deep Mixing Workshop 2002*, Port and Airport Research Institute & Coastal Development Institute of Technology, Tokyo, Japan.

Terashi, M. (2002b). Long-term strength gain vs. deterioration of soils treated by lime and cement, *Proceedings from the Deep Mixing Workshop 2002*, Port and Airport Research Institute & Coastal Development Institute of Technology, Tokyo, Japan.

Terashi, M. (2003). The state of practice in deep mixing methods, *Proceedings of the 3rd International Conference on Grouting and Ground Treatment*, February 10–12, 2003, New Orleans, Louisiana, United States. ASCE Geotechnical Special Publication No. 120, pp. 25–49.

Terashi, M., Ooya, T., Fujita, T., Okami, T., Yokoi, K., and Shinkawa, N. (2009). Specifications of Japanese dry method of deep mixing deduced from 4300 projects on land. *Proceedings of the International Symposium on Deep Mixing & Admixture Stabilisation*, May 19–21, 2009, Okinawa, Japan, pp. 647–652.

Tersahi, M. and Kitazume, M. (2009). Keynote lecture: current practice and future perspective of QA/QC for deep-mixed ground, *Proceedings of the International Symposium on Deep Mixing & Admixture Stabilization*, Okinawa, May 19–21, 2009, pp. 61–99.

Topolnicki, M. (2003). *Sanierung von Deichen in Polen mit dem Verfahren der Tiefen-Bodenvermörtelung (DMM)*, Hochwasserschutz, Ernst & Sohn Special 1/2003, pp. 45–53.

Topolnicki, M. (2009). Design and execution practice of wet soil mixing in Poland, *Proceedings of the International Symposium on Deep Mixing & Admixture Stabilisation*, May 19–21, 2009, Okinawa, Japan, pp. 195–202.

Topolnicki, M. and Sołtys, G. (2012). Novel application of wet deep soil mixing for foundation of modern wind turbines, *Proceedings of the 4th International Conference on Grouting and Deep Mixing*, February 16–18, 2012, New Orleans, Louisiana, United States.

Vähäaho, I. (2002). Deep mixing at Old City Bay in Helsinki, *Proceedings from the Deep Mixing Workshop 2002*, Port and Airport Research Institute & Coastal Development Institute of Technology, Tokyo, Japan.

Walker, A.D. (1992). Soil mixing and jet grouting on a hazardous waste site, Pittsburgh, PA, USA. In: Bell A.L. (ed.), *Grouting in the Ground: Proceedings of the Conference Organized by the Institution of Civil Engineers and Held in London on 25–26 November 1992*, London, England: Thomas Telford, pp. 473–486.

Walker, A.D. (1994). *A Deep Soil Mix Cut-off Wall at Lockington Dam in Ohio: In Situ Deep Soil Iimprovement*, ASCE Geotechnical Special Publication, No. 45, pp. 133–146.

Wehr, J., Topolnicki, M., and Sondermann, W. (2012). Design risks of ground improvement methods including rigid inclusions. *Proceedings of the ISSMGE-TC 211 International Symposium on Ground Improvement*, IS-GI Brussels, 31 May to 1 June, 2012.

Yang, D. and Takeshima, S. (1994). *Soil Mixing Walls in Difficult Ground: In Situ Deep Soil Improvement*, ASCE Geotechnical Special Publication, No. 45, pp. 106–120.

Yano, S., Tokunaga, S., Shima, M., and Manimura, K. (1996). Centralised control system of CDM method. In: Yonekura R., Terashi M., and Shibazaki M. (eds.), *Grouting and Deep Mixing, Proceedings of the 2nd International Conference on Ground Improvement Geosystems*, May 14–17, 1996, Tokyo, Japan, pp. 681–687.

Yu, Y., Pu, J., and Ugai, K. (1997). Study of mechanical properties of soil cement mixture for cutoff walls, *Soils and Foundations* 37(4): 93–103.

Zheng, G. and Liu, S. (2009). State of practice of deep mixing method in China: report for 2009 Okinawa Symposium, *Proceedings of the International Symposium on Deep Mixing & Admixture Stabilisation*, May 19–21, 2009, Okinawa, Japan, pp. 209–215.

Chapter 10

Dry soil mixing

Marcus Dahlström

CONTENTS

10.1 INTRODUCTION

Dry soil mixing (DSM) methods are the only techniques which directly mix soils with dry binder materials where the soil moisture is sufficient to hydrate the resulting in-situ mix. A range of binders can be used, but the most common are cement, lime-cement, and other cementitious blends which undergo beneficial chemical reactions with the soils into which they are mixed. These dry materials are fed into the ground using compressed air where they are comprehensively mixed, using purpose-designed tools, with the soils to be improved to the depth range identified.

Early development of in-situ dry mixing methods appears to have arisen simultaneously in Japan and Sweden during the late 1960s in order to deal with soft silts and clays. Since then the approaches have markedly diverged to address the differing technical and commercial demands of their respective markets. This chapter deals specifically with the Swedish method, sometimes dubbed the 'Nordic method' (Holm, 2002). By the end of the 1980s the method was widely established in Sweden, Finland, and Norway, and by the early 2000s was being used in the US and in several other European countries, notably Poland and the UK. At the time of writing the method is also being used in the Far East and Australia. The history and development of both Nordic and Japanese methods are covered in more detail in Chapter 9 (Section 9.2) of this book.

Single-axis column mixing is the simplest and most widely used form of construction using the Nordic method. The mixing tool is first rotated into the ground down to the depth previously identified for the toe of the column. The tool is then withdrawn at a high rate of rotation, during which the dry binder is injected through ports in the mixing tool at a controlled rate according to the design. This creates appropriate mixing of the binder with the soil contained within the blade diameter and shape of the tool. Typical equipment for column construction is a purpose-built hydraulic base machine with low ground pressure, suitable for traversing soft ground, carrying a mast, rotary head, hollow-stem Kelly bar and the mixing tool. A separate pressurised container is used to store and transfer the binder to the Kelly, and is either self-propelled or towed behind the base machine.

Dry soil mixing is applied in inorganic soft soils in which the natural moisture content is close to or above the liquid limit, and is increasingly being used in highly organic soils and peats. The main applications are for foundation bearing or settlement control and providing stability for slopes, embankments, and excavations. Other applications include reduction of vibrations, provision of liquefaction resistance, and solidification of contaminated soils and mud. The technique has continued to spread worldwide due to its advantages of avoiding the need to pre-mix materials with water before injection and the high rate of construction, leading to low costs per metre of column relative to other forms of soil mixing. Typical column diameters are 0.5–1.0 m and depths of treatment are often 5–17 m with deeper treatment available with special equipment.

10.2 THE DRY SOIL MIXING PROCESS

10.2.1 General

Dry soil mixing is a general term for mechanical in-situ mixing of soil by adding a dry additive (commonly called binder) distributed by air flow. The mechanical mixing is done by either vertical or horizontal mixing by rotating impellers of paddles see Figure 10.1 or by cylinders with cutting heads as in Figure 10.2.

The binder is transported from rig-mounted silos or from silos mounted on a separate unit (which could be either on a so-called shuttle or distributed directly from a bulk silo). The DSM process with the in-situ mixing of a dry additive into the soil by compressed air is referred to by several different terms, such as lime/cement columns, deep stabilisation, dry jet mixing

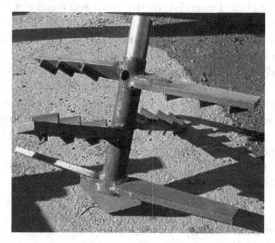

Figure 10.1 Vertical mixing tool (Pinnborr). (Courtesy of LCM.)

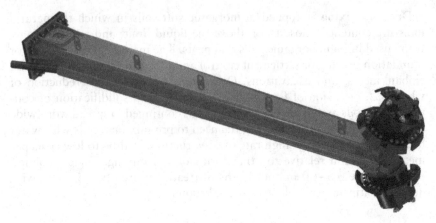

Figure 10.2 Horizontal mixing tool (Allu). (Courtesy of Allu.)

methods, dry deep mixing, column stabilisation, or mass mixing. These terms are explained in a later section.

The use of compressed air as the medium for transporting the binder is advantageous because it takes a relatively small amount of binding agent to achieve the requisite strength gain in the soil. Hence, soft soils contain large amounts of water, and by adding a dry binder instead of wet slurry (as is necessary in wet mixing or jet grouting processes) less binder is required.

10.2.2 The mixing process

The purpose of the mixing process is to distribute the binder into the soil efficiently in order to provide the best possible condition for the chemical reaction to take place. Depending on the depth and volume of the mixed soil, the mixing process is most frequently applied using columns of mixed material. Columns are typically 0.5–1.0 m in diameter and range 3–25 m in depth and are solely installed vertically (minor inclinations at 4:1 to 10:1 exists).

Alternatively, for shallower depths, mass mixing may be appropriate. Mass mixing can either be installed as interlocking columns into a block of vertical elements or as a mass-mixed soil volume with a horizontally rotating cylinder with cutter heads, see Figure 10.2. Mass mixing is generally performed in depth of 0.5–6 m; however, mass mixing using the interlocking column method has been performed up to 15 m.

The production process for DSM is similar for columns and mass mixing and can be divided into three principal phases:

(1) Penetration of the mixing tool to required depth
(2) In-situ mixing and dispersion of binder
(3) Molecular diffusion

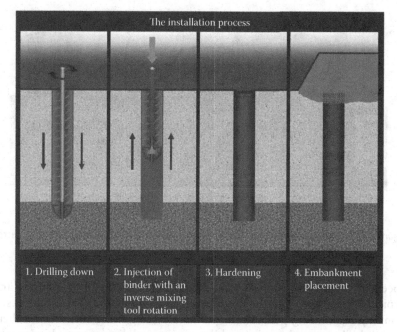

Figure 10.3 Installation sequence for column installation.

The installation of columns is presented in Figure 10.3, and the installation process of mass mixing is presented in Figure 10.4. For the mass mixing method with interlocking columns the installation sequence of columns is represented but with minor modifications.

In the European Standard EN 14679 2005 figure A.5, the installation sequence is described as a five step procedure:

(1) The mixing tool is correctly positioned
(2) The mixing shaft penetrates to the desired depth of treatment with simultaneous disaggregation of the soil by the mixing tool
(3) After reaching the desired depth, the shaft is withdrawn and at the same time the binder in granular or powder form is injected into the soil
(4) The mixing tool rotates in the horizontal plane and mixes the soil and binder
(5) Completion of the treated column

10.2.2.1 Phase 1: Penetration of the mixing tool

In the first phase the mixing tool penetrates the soil while it is rotated to the required depth. The soil structure is remoulded during this penetration. The magnitude of the remoulding depends on the penetration and rotation

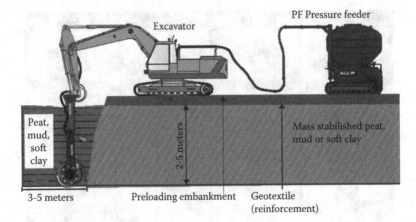

Figure 10.4 Installation sequence for mass mixing with rotated cylinders and cutter heads. (Courtesy of Allu Finland.)

speed of the mixing tool. Typical penetration speed is 100 mm/rev. and typical rotation speed of the mixing tool is <100 rev/min. The binder is seldom injected during penetration. In mass mixing using cutter heads there is no initial penetration—instead the mixing and dispersion of binder take place directly.

10.2.2.2 Phase 2: In-situ mixing and dispersion of binder

The formation of an in-situ mixed column takes place during withdrawal of the mixing tool. The process starts by delivering binder via compressed air from the powder tanks via hoses connected to the Kelly bar by a swivel. The mixing tool, connected to the lower end of the Kelly bar, has an outlet port or ports commonly at the top level of blades, see Figure 10.1, where the binder is distributed into the soil. The binder is spread through the cavities formed in the soil by rotation of the mixing tools upper level of blades. The spread of binder through the column diameter depends on the air pressure, the soil rheological properties, the geometry of the mixing tool and the diameter, and the rate of withdrawal and rotation. In order to increase mixing energy into the soil, tools with 2–4 levels of blade are commonly used.

Before the uplifting starts and the creation of a column begins, the mixing tool at the column base level is rotated just prior to lifting in order to reach the required rotation speed and to inject the required amount of binder in the toe of the column. This process takes approximately 10 seconds. Lifting of the mixing tool during high-speed rotation is then commenced. Typical rotation speeds are 120–180 rpm and lift speeds are 15–30 mm/rev. This means that production speed varies between 1.2–5.4

m of column per minute. The installation sequence for DSM columns is presented in Figure 10.3.

The distribution of binder through the column diameter significantly depends on the disposition of the outlet port in combination with the design of the upper level of blades on the mixing tool. Column installation commonly stops 0.3–1.0 m below the working platform. This means that the upper 1 m of a column should not be treated as a full-strength column.

In mass mixing using cutter heads, the distribution of binder takes place from an outlet port in the bottom of the mixing unit, see Figure 10.2. The cylinders (or drums) with the cutter head mounted rotate in a different sense to aid mixing of the soil. Mixing is carried out in a site-specific pattern, commonly in cells of 3 × 3 m, until the required depth is reached. Mixing energy in the soil, and distribution of binder evenly into the cell, is highly dependent on the skill of the operator. Mass mixing can be performed right to the ground surface. However, the top 0.5 m is often poorly mixed compared with the mixed soil at greater depth. Installation sequence for mass mixing is presented in Figure 10.4.

10.2.2.3 Phase 3: Molecular diffusion after installation

After the manufacture of a column or when mass mixing is completed, the molecular diffusion takes place in the mixed soil volume. Molecular diffusion of calcium ions from stabilised soil migrates into unstabilised surrounding soil, or from regions of stabilised soil with high concentration of calcium ions into regions with poor concentration. The increase in strength caused by the migration of calcium ions within a column has been poorly investigated. However, Axelsson and Larsson (2003) reported that observations on extracted lime-cement columns showed that the columns seem to heal a short time after column penetration tests, which may possibly be due to migration of calcium ions. Mitigation of calcium ions from column periphery has been the subject of many investigations (Rogers et al., 2000a, b; Hayashi et al., 2003; Rogers and Glendinning, 1996; Löfroth, 2005). These investigations show that calcium ions migrate approximately 20–30 mm within 1 year and 40–60 mm within 10 years.

10.2.3 Factors important in the mixing process

The mixing process is complex so the result and quality of an in-situ mixed column or the stabilised soil mass depends on a number of factors, after Larsson (2005). Significant issues include

- The rheological soil properties
- The type of binder and the dosage
- The mixing energy

- The mixing tool design and the drilling rig site–specific adjustments
- Air pressure delivery and amount of air entrained into the ground

10.2.3.1 Rheological soil properties

The rheological soil properties have a significant influence on the efficiency of the mixing and quality of a mixed soil volume. Cohesive soils with moderate to high water content have, by their nature, a considerable resistance to remoulding compared with cohesionless soil. In soft soils the natural water content is often near the liquid limit, and the incorporation of dry binder rapidly reacts with the soils' natural water and becomes more plastic and more resistant to remoulding. Incorporation of a dry powder will significantly change the rheological soil properties by dewatering the soil volume that is being mixed.

10.2.3.2 Type and amount of binder

The influence of type and quantity of binder affects the mixing process. The fact is that increasing the amount of binder will increase the strength of the soil, with some exceptions. On the other hand, increased amount of binder can decrease the production speed due to larger quantities of material to be transported in the feeding system and larger quantities needing dispersion into the soil. Very large amounts of binder can have a negative effect on the strength value due to inadequate water content. The relationship between water content and binder content, water/cement ratio (w/c), has been investigated in a number of studies (Åhnberg et al., 1995; Babasaki et al., 1997; Filz, 2012) and is an important factor to evaluate in the pre-design stage. The water/cement ratio and the relation to a certain strength gain are soil specific. Figure 10.5 shows the relation between shear strength and w/c-ratio.

The w/c ratio in soil mixed samples can be expressed as:

$$w/c = \frac{\left[\rho_{soil} - \dfrac{\rho_{soil}}{(1 + w_N)}\right]}{m_{binder}} \tag{10.1}$$

where
ρ_{soil} = bulk density of unstabilised soil (t/m³)
w_N = natural water content in unstabilised soil
m_{binder} = binder content (t/m³)

Graphs on the relationship between w/c ratio and strength from laboratory investigations and from literature studies are valuable information in the pre-design stage. However, mixing soil samples with different amounts

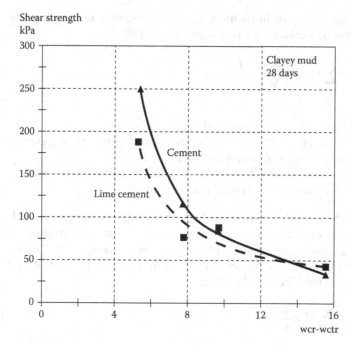

Figure 10.5 Relation between shear strength and water cement ratio (w/c) in a clayey mud from Sweden. (From Åhnberg, H., Johansson, S.E., Retelius, A., Ljungkrantz, C., Holmqvist, L., and Holm, G. (1995). *Cement och Lalk för Djupstabilisering av Jord – En Kemisk Fysikalisk Studie av Stabiliseringseffekter (Cement and Lime for Stabilisation of Soil at Depth – a Chemical Physical Investigation of Soil Improvement Effects)*, Report No. 48. Linköping, Sweden: *Swedish Geotechnical Institute.*)

of binder in laboratory is commonly performed in order to investigate sufficient amounts of binder. Increased binder content increases the strength gain to a certain limit; thereafter, strength gain is decreasing due to lack of water in the natural soil. On the other hand, if the binder content is decreased, that would mean that at a certain level there is no or very limited strength gain in the mixed soil (also known as a threshold condition).

10.2.3.3 Mixing energy

Mixing energy is a measure of the mixing work that the mixing tool causes during the installation process. The degree of mixing work implemented into a soil is related to the column strength and the dispersion of the binder across the column area (Muro et al., 1987a, b; Nishida et al., 1996; Larsson, 1999; Larsson et al. 2005a, c). Increased mixing work has an increasing effect of the column strength in the soil. However, it is not possible to predict column strength based on the strength of information of a certain mixing work.

Mixing energy is, in its more general sense, measured as mixing cycles per metre of a column T (Yoshizawa et al., 1997):

$$T = \sum M \times \left\{ \left(\frac{N_d}{V_d} \right) + \left(\frac{N_u}{V_u} \right) \right\} \qquad (10.2)$$

where:

$\sum M$ = number of mixing tool blades

N_d = rotation speed of mixing tool during penetration (rev/min)

V_d = mixing tool penetration velocity (m/min)

N_u = rotation speed of mixing tool during retrieval (rev/min)

V_u = mixing tool retrieval velocity (m/min).

For DSM, by the Nordic method, the uplift rate (mm/rev) of the mixing tool is used as a measure of the mixing time since mixing does not occur during penetration. The mixing energy is specified as the blade rotation number (BRN; see Figure 10.6) and the number of cycles per column metre T is calculated as:

$$T = \sum M \times \frac{1}{s} \times 1000 \qquad (10.3)$$

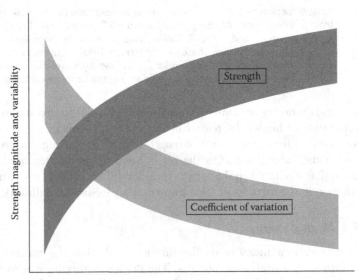

Blade rotation number, T: n/m

Figure 10.6 Principal changes in strength and coefficient of variation due to the variation in blade rotation number *T*. (From Larsson S. (2005). Keynote lecture: state of practice report – execution, monitoring and quality control. *Proceedings of the International Conference on Deep Mixing Best Practice and Recent Advances*, May 23–25, 2005, Stockholm, Sweden, pp. 732–785.)

where
   M = number of mixing tool blades
   s = retrieval rate of mixing tool during withdraw (mm/rev).

Increased mixing energy is favourable due to increased magnitude of strength and decreased variability in the mixed soil. However, a high degree of mixing energy reduces the production speed of a project. Negative effects of the column quality can also be a consequence of a high degree of mixing. For example, a high degree of mixing means that the mixing tool spends more time in the ground resulting in more compressed air into the soil. This can generate heaving and a phenomenon causing 'craters'.

The magnitude of mixing energy needs to be evaluated for each specific site and the site-specific type of soil. Laboratory mixing tests gives an indication of the required magnitude of mixing energy, which will be verified (or adjusted if necessary) during the installation of test columns before the production begins.

10.2.3.4 Mixing tool design and drilling rig site–specific adjustments

Adjustments of the drilling rig to suit the site-specific soil properties are important in DSM projects. This includes evaluation of mixing tools, adjustments of rotation speed, evaluation of mixing energy (testing different retrieval rates), and adjustments of air pressure in the feeding system. These parameters and the adjustments are discussed in Section 10.3.

10.2.4 Binders and soil properties

10.2.4.1 Binders

Binders used for DSM are cement (standard Portland cement), lime (quicklime), slag (granulated blast furnace slag), fly ash, and gypsum. There have also been some investigations of other binders, primarily in laboratory testing; however, these additives are not used extensively. Indeed, the most commonly used binders are cement and quicklime. These are often pre-blended to gain the most suitable binder for the site-specific soil properties. The local access to different additives and regional industrial manufacturers (access to cost-effective material) also influences the choice and blend for site-specific binders. Notwithstanding general experience available, laboratory mixing tests on soil samples from the site to be treated are an essential component in the selection of binder and binder blends.

In the Nordic countries, pre-blended binders of quicklime and cement are by far the most commonly used binders. Pre-blended mixtures of lime/cement and fly ash are moderately used as are pre-blended mixtures of cement and slag. In other European countries, the UK, and the US, the most commonly used binders are cement only or cement and slag. However, blended mixtures of cement and quicklime are moderately used in Poland.

10.2.4.2 Applicable soil types

Dry soil mixing is applicable in soft soils with high moisture content. As the dry binders need in-situ water for the chemical reaction, a certain amount of evaporable water is necessary. As a guideline, a minimum moisture content of 30% is necessary for DSM. Table 10.1 provides a summary of binders that are commonly used in different soil types.

The presented amounts of binders are for soft to medium soft columns with a design undrained shear strength of 150 kPa or less.

The described binder in Table 10.1 is one, two, or three component blends. In almost every blend cement is the basic additive and lime, slag, fly ash and gypsum are secondary additives. Historically, in the Nordic countries lime is the basic binder and other admixtures have been developed from there.

10.2.4.3 Binders and reaction process

The reaction process and the function of different binders have been investigated in numerous reports. The investigation by Janz and Johansson (2002) is recommended for a detailed description of the reaction process. In this chapter the most common used binders—cement, lime, blast furnace slag (GGBS) and fly ash—are described.

10.2.4.3.1 Cement

Cement used as binder in a DSM process is predominately standard Portland cement. When cement is mixed with a soil it reacts with the water content and instantly the hydration process commences in which a hard cement paste forms of calcium silicate hydrate $Ca_3Si_2H_4$ (CSH-gel). The CSH-gel is formed on the cement particles and increases in size, filling the voids between the particles. With time the porosity decreases, the particles bind together, and the mass becomes stronger and more dense. Initially, the rate of strength gain is controlled by the temperature; the higher the temperature, the more reactions that take place leading to better strength gain (Timoney et al., 2012). The reaction process of 100g cement and 25g H_2O can be expressed as:

$$Cement + H_2O \rightarrow CSH\text{-gel} + Ca(OH)_2$$
$$100\,g \quad 25\,g \rightarrow 100\,g \quad 25\,g$$

Table 10.1 Soil types and binders

Soil type	Binder	Amount of binder	Commentary
Clay	Lime, Cement, Lime/ Cement, Lime/ Cement/Fly Ash	70–110 kg/m³ *Pure cement* 90–150 kg/m³	
Quick clay	Lime, Lime/Cement, Lime/Cement/ Fly Ash	70–100 kg/m³	Quick reaction, especially with high lime content.
Silty clay	Lime/Cement, Cement, Cement/ Slag	70–110 kg/m³	High degree of cementation with cement.
Organic clay	Cement, Lime/ Cement, Cement/ Slag	100–200 kg/m³	Slow reaction, minor part with lime commonly speeds up the reaction.
Sludge	Cement, Cement/Slag	120–250 kg/m³	Slow reaction. Difficult to predict strength increase.
Clay with high sulphide content	Cement, Lime/ Cement, Cement/ Slag	120–250 kg/m³	Slow reaction. Large variations in strength gain. Local knowledge important.
Silt	Cement, Cement/ Slag, Lime/Cement	100–150 kg/m³	
Sandy silt	Cement, Cement/Slag	60–110 kg/m³	Natural moisture content needs to be larger than 30%.
Peat	Cement, Slag/Cement	150–>300 kg/m³	Very important with field and laboratory tests.
Dredged material (Mud)	Cement, Cement/ Slag, Cement/Fly Ash	70–110 kg/m³	Tests necessary, especially due to contaminations.
Contaminated soils		70–110 kg/m³	Tests necessary, especially due to type of contamination and leakage tests.

10.2.4.3.2 Lime

Lime used as binder is mainly calcium oxide (CaO), known as *quicklime* or *burnet lime*. When quicklime is mixed with water slaked lime or hydrated lime, $Ca(OH)_2$, is formed:

$$CaO + H_2O \rightarrow Ca(OH)_2$$
$$100g \quad 32g \quad \rightarrow 132g$$

The reaction is instant, and reaches its maximum within 5 minutes while generating a great deal of heat. The reaction also results in the pH increasing to ~12.5, which is a condition for the secondary pozzolanic reaction. During soil stabilisation, hydration dewaters the soil giving a rapid gain in stability (Janz and Johansson, 2002). The primary reaction is not gaining any strength. The strength gain in stabilised soil is due to the secondary pozzolanic reactions with other additives or with the surrounding soils:

$$Ca(OH)_2 + pozzolana + H_2O \rightarrow CSH \ (CASH)$$

Humic acid inhibits strengthening reactions, which leads to poor strength gain performance in organic soils.

10.2.4.3.3 Granulated blast furnace slag (GGBS)

Ground GGBS is a by-product of the steel manufacturing processes. GGBS is a latent hydraulic material, which means it contains lime but requires an activation before it can react with water. GGBS is used as a secondary binder in soil stabilisation. The temperature created during the reaction is low, which results in slow strength gains. In the Nordic countries GGBS blended with cement has been used successfully in mass mixing projects (Jelisic and Leppännen, 1999).

10.2.4.3.4 Fly ash

Pulverised fly ash (PFA), similar to GGBS, is also a latent hydraulic material that either can be obtained from flue gas in the coal-fired power industry or from the paper mill industry. The flue gas from the power generation plants is a fine-grained material with pozzolanic qualities. The pozzolanic reaction in PFA requires an activator as cement or lime to obtain strength gain. PFA, lime, and cement in the proportions 33%/33%/33% is commonly used in Sweden as a binder alternative to lime/cement 50%/50% in marine clays.

10.2.4.4 Strength gain

Strength gain and the magnitude of strength gain in stabilised soil depends on several factors, such as soil characteristics, type of and amount of binder, temperature, time aspects, and stress situation. The influences of these factors in the strength gain process differ between binder combinations and soil types.

10.2.4.4.1 Soil characteristics

As shown in Table 10.1, different types of soil are more suitable for certain types of binders or binder combinations. The required amount of binder to gain a certain magnitude of strength depends predominately on the natural moisture content and the organic content in the soil. Soils with high organic content (>6%) require large amounts of binder (not necessarily higher w/c ratio compared to inorganic soils).

In the mixing process, some general soil characteristics can be identified; soils with high sensitivity (for example quick clay) are easier to mix than soils with low sensitivity; soils with high organic content (sludge, mud) often have a slow increase in strength compared to inorganic clay.

Timoney et al. (2012) reviewed published data on a wide range of peats and highly organic soils and found that the 28-day unconfined compression strength (UCS) showed some correlation with the ratio of water to binder by weight in the mix. Significantly, the amount of humification was also highly relevant with a clear inverse relationship found between UCS and humification using the von Post* classification.

10.2.4.4.2 Type of binder and amount of binder

As described, different binders react differently in the same type of soil. An outline of the strength gain in clay and silty clay is shown in Figure 10.7. Typical amount of binders in different type of soils are presented in Table 10.1. Local experience and results from laboratory mixing tests are important parts in determining a sufficient amount of binder. Site-specific requirements shall also be evaluated; for instance, the need for rapid strength gain in order to use the installed column shortly after installation; and the interaction with unstabilised soil especially when used to increase the stability in natural slopes.

10.2.4.4.3 Temperature

Soil masses stabilised with lime and/or cement generate heat during the cement and pozzolanic reactions; Halkola (1999) reports a temperature of 70°C in lime and Åhnberg et al. (1995) reported that cement increases the ground temperature by 5–10°C and lime can increase the ground temperature 40–50°C, and locally it can even generate temperatures up to 100°C. Binders like GGBS produce less heat during exothermic reactions, and are therefore more susceptible to temperature changes of the soil being stabilised, thus often resulting in less reaction and lower initial strength (Timoney et al., 2012). Admixtures with lime or cement increase the ground temperature, which activates any secondary binder such as GGBS or fly ash.

---

* von Post is a classification system for peat on a scale between H1 to H10 based on its physical properties.

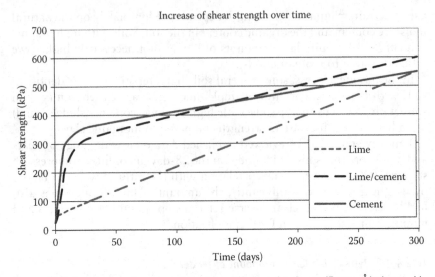

Figure 10.7 Strength gain over time for different binders. (From Åhnberg, H., Johansson, S.E., Retelius, A., Ljungkrantz, C., Holmqvist, L., and Holm, G. (1995). *Cement och Lalk för Djupstabilisering av Jord – En Kemisk Fysikalisk Studie av Stabiliseringseffekter (Cement and Lime for Stabilisation of Soil at Depth – a Chemical Physical Investigation of Soil Improvement Effects)*, Report No. 48. Linköping, Sweden: Swedish Geotechnical Institute.)

10.2.4.4.4 Pre-stress loading

Strength gain in peat is dependent on the pre-stress level during curing. Laboratory tests carried out by Åhnberg (2001) investigated the effect on pre-stress loading on stabilised peat samples. Samples were loaded with 0, 9, and 18 kPa at 45 minutes, 4 hours, and 24 hours after mixing, respectively. It was observed that pre-stress loading within 45 minutes increased the strength in the samples up to several times compared to unloaded samples. Compression that occurs during preloading reduces the distance between the binder grains and the particles in the peat and facilitates building bonds in the stabilised soil mass. Hebib and Farrell (2003) showed from tests on Irish peat that the permeability of the stabilised samples were also reduced by pre-stressing, whereas the permeability of samples not subject to pre-stress were the same as for unstabilised peat.

In the field, strength gain of stabilised peat (especially mass mixed) needs to be pre-stressed by approximately 1 m of fill in order to compress the loosened soil volume, which has been remoulded by the mixing tool and the compressed air. Pre-stress load is recommended to be placed on the mixed soil as soon as possible after mixing; however, the common requirement is that the load shall be applied within 24 hours after mixing. Pre-stressed loading on DSM columns in clay has also have a positive effect on the strength gain, but less than in peat.

10.2.4.4.5 Prediction of strength gain

The time dependency of strength gaining and prediction of strength increase according to type of binder and soil conditions has been investigated in a number of reports, such as Nagaraj et al. (1996), Porbaha et al. (2000), Horpibulsuk et al. (2003), and Åhnberg (2006). For cement-stabilised soils, Åhnberg (2006) investigated strength increase in clay and sludge samples tested between 7 and 800 days, see Figure 10.8.

Prediction of strength gain for cement-stabilised clay as:

$$\frac{q_t}{q_{28}} = 0.3 \times \ln t \tag{10.4}$$

where
q_t = UCS after t days
q_{28} = UCS at 28 days
t = time (days).

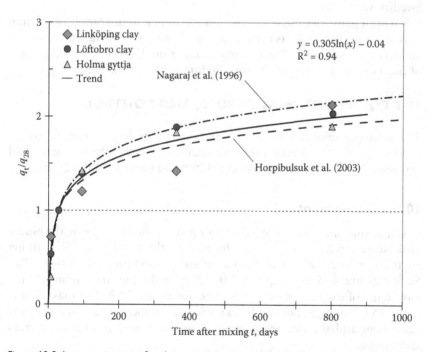

Figure 10.8 Increase in unconfined compression strength over time for cement-stabilised clay and sludge. (From Åhnberg, H. (2006). *Strength of Stabilized Soils – A Laboratory Study on Clays and Organic Soils Stabilised with Different Types of Binder.* PhD Thesis, Lund University, Sweden.)

Prediction on strength gain for cement-stabilised soil has also been investigated by Horpibulsuk et al. (2003) by using a modification of Abram's law[*].

Predictions of strength gain by time for cement-stabilised clays are difficult and the presented methods are empirical methods that can be used as an indication of the strength gain over time. However, these can only provide indications while site-specific tests are the most reliable method.

10.2.4.5 Long-term strength

Long-term strength and durability have been investigated by field tests and laboratory tests. For Japanese experience, Terashi (2003) summarised the results from a number of studies. The results demonstrate that significant increases of strength have occurred from the 28 days' strength to 10-year-old lime and cement-treated soils. Minor leaching of calcium (CaO) ions that can decrease strength has been observed at the periphery of the columns. However, the leaching was slow and small in scale, and the author's conclusion was that the leaching is much less than the strength gained over time. Löfroth (2005) reported similar results on 9.5- to 11-year-old lime/cement columns installed in organic clay and highly sensitive clays on the Swedish west coast.

Strength gain continues over the long-term due to migration of calcium ions and pozzolanic effects in the columns. The above investigations showed an increasing strength 10 years after installation. However, the magnitude of the increase depends on the site conditions as well the binder type.

10.3 EQUIPMENT, MONITORING, AND CONTROL

The development of new equipment and the modification of existing equipment is ongoing. Description of the execution procedure of the Nordic method was recently presented by Bredenberg (1999) and Larsson (2003, 2005).

10.3.1 Equipment

One machine unit consists of a drilling rig and a shuttle carrying the binder tank (some machines have the binder tank on the drilling rig). The drilling rig is constructed with wide tracks for low ground pressure (40–60 kPa). Such rigs are 4–5 m long and 3.0–3.5 m wide and are commonly the base units of excavators or a piling rig, see Figure 10.9. The mast is normally 15–17 m high, but can be extended up to about 27 m. The machine equipment and shuttles are commonly manufactured by the contractors themselves.

---

[*] Abram's law is an empirical model to predict strength gain extensively used in concrete technology.

Figure 10.9 DSM machine unit, drilling rig, and shuttle. (Courtesy of dmixab.)

The shuttles carrying binders have a storage capacity of 10–15 tons and are pressurised to 5–10 bar. The pressures applied to the ground by the shuttle are similar to those from drilling rigs. The binder is fed from the shuttle to the top of the drilling shaft and blown out by the outlet hole at the mixing tool via air pressure, see Figure 10.1. Drilling can generally, with equipment today, be carried out to approximately 25 m depth; however, in the majority of projects drilling length is 15 m or less. Rotation speed of the drilling shaft is in the interval of 100–200 rpm, depending on the ground conditions. Downward penetration is usually controlled manually by the operator and corrected due to ground conditions and the risk of hitting obstacles. Typical downward movement is 100 mm/rotation. During downward movement, the shaft is pressurised by air to prevent water and soil from entering. During upward movement, the binder is fed out by air pressure. The amount of binder is regulated by feed-out valves in the shuttle. The upward movement is generally 15–30 mm/rotation, depending on the required mixing work. The blowout of binder into the surrounding soil is stopped at 0.5 m beneath the ground surface to prevent release into the open air. In urban areas a protection head can be used to avoid clay splash and blow outs of binder into open air.

Mass mixing equipment is similar to DSM equipment with a drilling unit (rig and shuttle). The drilling rigs are commonly smaller than a DSM rig and have lower ground pressure (30–40 kPa), see Figure 10.10. Shuttles are in many cases not used for binder delivery due to limited storage capacity

(a)   (b)

Figure 10.10 (a) Mass mixing drilling rig and shuttle. (Courtesy of LCM.) (b) Storage supply in bulk silos. (Courtesy of LCM.)

(large amount of binder is commonly the case in mass mixing). The binder
is delivered directly from bulk silos by compressed air in tubes or supplied
by bulk trucks behind the drilling rig.

The mixing tool is an important part of the equipment. There are some
well-established shapes of mixing tools. The three most commonly used
are (1) standard mixing tool (the first developed mixing tool, Figure 10.11);

Figure 10.11 Standard mixing tool, Pinnborr (shown in Figure 10.1), and Allu rotary head
(shown in Figure 10.2).

(2) Pinnborr, a three-level blade mixing tool (the most common in Nordic countries today); and (3) Allu rotary head (mass mixing tool).

Since the turn of the century, industrial mixing has made progress as a new technique based on knowledge gained from DSM experience. Industrial mixing is predominately used for dredge material pumped into the machine where it is mixed with different types of binder and thereafter transported to a land reclamation area (a lagoon) for hardening.

10.3.2 Monitoring and control

Monitoring and control of the installation process are very important since the stabilisation process itself seldom leads to direct inspection. The amount of binder injected as well as the geometry and homogeneity of the stabilised soil volume, whether it is columns or mass-mixed soil, must be evaluated by indirect measurements of binder use, retrieval rate, rotation speed, and so on. The technical problems related to monitoring are more pronounced for the dry than for the wet method since the binder is distributed by a compressed air stream in the dry method. Therefore, the weight loss of binder in the storage tank is used as a measure of the injected amount of binder. Load transducers continuously monitor (20 readings per second) the weight in the storage tank and transfer the information to the machine computer.

Each contractor has their own monitoring system for control, recording, inspection, and documentation of the improvement work, but these parameters at a minimum should be recorded for every DSM project:

- Column name
- Machine name
- Responsible operator
- Used mixing tool
- Time and date of installation
- Drilling time
- Retrieval rate (mm/r)
- Speed of rotation (rpm)
- Drilling length and/or stabilising length
- Amount of binder along the column
- Weight of the binder tank
- Position of the column (GPS)

Manual registration by the machine operator parallel to the computer registration is recommended. The following is registered on manual logs:

- Machine name
- Responsible operator
- Date of installation
- Column name

- Drilling length and/or stabilising length
- Deviations and other obstacle
- Service, change of mixing tool, and so on

Monitoring data is part of the quality assurance (QA) and quality control (QC) for the stabilisation work. The information is commonly delivered from the contractor to the client on a daily or weekly basis. Adjustments to the installation process and decisions for in-situ testing can be made from the printouts of the installation data. It is important not to focus solely on performing in-situ tests on apparently less good columns, as it is the mass properties that will often be most relevant.

The production capacity is dependent on a number of factors and site-specific conditions such as stiffness in the unstabilised soil, magnitude of obstacles, length of columns, rotation speed and restrictions in vertical movement (special installation pattern etc.), access to the area and transportations roads, distance to binder storage, and amount of binder in the columns. Typical production capacity for 10-m long columns is 40–80 stabilised metres per hour per unit.

10.4 APPLICATION, DESIGN AND TESTING

10.4.1 Applications

There are a number of applications using DSM columns and mass mixing for permanent and temporary works either on land or in marine environment. Some of the main applications include

- Reduction of settlement
- Improvement of stability
- Reduction of ground vibration
- Foundations of structures and houses
- Mass mixing of organic soils and dredged sediments
- Solidification and stabilisation of contaminated soils

Secondary applications include the following:

- Increase passive earth pressure for sheet pile walls in soft clay
- Reduce active earth pressure on retaining/sheet-piled walls in soft clay
- Preventing liquefaction in seismic hazard areas
- Creating geohydrological barriers

DSM columns and mass mixing is sometimes combined with other soil improvement techniques in order to design the most technical and economical solution:

- DSM columns combined with light fill aggregates or expanded polystyrene (EPS) material in the embankment. This combination is commonly used for embankments in the transit zones of bridges and piled structures.

- DSM columns and wick drains are occasionally used, especially in areas with deep deposits of clay. In these applications DSM columns are mainly installed to improve the stability of the embankment and wick drains are installed to reduce settlements at depth.
- DSM columns and vibro replacement have also been combined, (Dahlström, 2012). DSM columns increase the confinement and shear strength in the soil. The improved ground thereby increases the load capacity of the vibro replacement columns.

Depending on the application, column spacing (or improvement ratio), layout, diameter of column, and length of column all require careful attention.

10.4.1.1 Reduction of settlements

Reduction of settlements for road and railroad embankments, parking areas, and areas around structures are the most common application of DSM columns and mass mixing. In soft soils with shear strength greater than 8 kPa, columns are installed as single elements with a spacing of 1.3–3.0 times the column diameter. In extremely soft soils (shear strength less than 8 kPa) the confinement of the columns are limited, hence the columns need to be installed in panels or grids to support each other. In these extremely soft soils mass mixing is an alternative to interlocking columns in panels or grids. A combination of DSM columns and mass mixing is commonly used in areas with extreme soft soil in the top 1–5 m followed by soft clay deposits (e.g, Dahlström and Eriksson, 2005).

Columns are mainly installed into a firm soil layers in order to distribute the load from the embankment. In very deep soft soil deposits, and a low to moderate height of the embankment, columns are installed to a predesigned depth. The improvement ratio can also vary by depth (e.g., every second column is installed to greater depth) in order to provide the most economical solution.

Arching between single elements such as columns needs to be checked, especially in the case of low embankments and column spacing greater than twice the column diameter. Geogrids or load transfer platforms can be used to secure arching between the elements.

The magnitude of reduction of settlements mainly depends on the improvement ratio and column strength. Generally, settlements are reduced 2–5 times compared with unimproved soil.

10.4.1.2 Improving stability

Columns can be connected with each other to create a panel or a grid of interlocking columns (see Figure 10.12) for improving the stability of road

Figure 10.12 Interlocking columns in panels at Bärbyleden, Uppsala, Sweden.

and railroad embankments, slopes, and temporary excavations. The spacing between columns in a panel is generally in the range of 0.75–0.85 times column diameter. The connection between columns is the most critical part of a panel, and shearing between columns is the most common cause of failure. Panels are installed perpendicularly to the most critical failure plane and have a length along the top to ensure that the total area of improved soil does not slide as a mass block. Panel length must be enough to mobilise sufficient shear resistance in the panel. In slopes, panels installed with inclinations of 10:1 to 5:1 are advantageous due to the increased axial load on the panels. Panel depth depends on the most critical failure plane in stabilised and unstabilised conditions, and both cases therefore need to be investigated. Overlapping of columns with great depth is extremely difficult to perform with the equipment available at the time of writing. A recommendation by the Swedish Road Authorization is that overlapping of columns deeper than 8 m shall, without investigations, be used only with restrictions.

The distance between panels depends mainly on three parameters: the width of the panel, the stability of the unstabilised soil between panels (squeezing), and the interaction between stabilised and unstabilised soil. Typical spacing between panels is 1.0–3.5 times column diameter. Panels can be installed as single panels, which is the most common design, or as two interlocking panels. Interlocking panels are sometimes used in areas where the function of the panel is very critical, especially at depth.

Construction of columns that will form a panel requires higher requirements of precision and inclination during installation. Tests using vertical inclinometers to control the declination after installation of columns in a panel have been undertaken in some projects. Installation of panels in area with low stability or for live railway embankments (Pye et al., 2012a, b) needs to consider the effects of increasing pore pressure and movements during installation.

10.4.1.3 Reduction of vibrations

Reduction of vibrations is predominately associated with high-speed trains (>180 km/h) travelling over soft clay deposits. The train-induced ground vibrations generate different waves (P-, S-, and R-waves) with different velocity propagating in the soil. When the train speed exceeds any of these wave velocities, the character of the propagation of the waves is dramatically changed. This phenomenon is called a shock front and will give rise to high levels of vibrations in the soil with large displacements as a secondary effect.

Since the late 1990s, DSM columns have been used as soil improvement to reduce the ground vibrations and displacements connected to the high-speed phenomenon of a train passing over soft clay on low embankment. Columns are installed in a specific pattern, see Figure 10.13, which was first developed within a Swedish research project (Holm et al., 2002). In their report, the measurements of displacements on the trackbed showed a reduction of approximately 5 times at low speed and of approximately 15 times at high speed (200 km/h) compared with measurements taken before installation of DSM columns.

The specific layout was adopted in order to develop stiff ground below the ballasted track structure, and also to create a barrier for the surface waves travelling along the track. As shown in Figure 10.13, the longitudinal panels are centred underneath the rails in order to create a stiffened base. These columns are 6–8 m long (commonly 7 m) and are the most important part of the structure. Therefore, these columns are installed first in order to secure a sufficient interlocking. Thereafter the perpendicular panels connecting the longitudinal panels are installed to minor depth, commonly 4–6 m. Finally, the single columns for reducing settlements are installed. The critical part in this system is the contact between column and track ballast. Hence columns are exposed and visually examined before back filling of track ballast.

This application is today a standard method, adopted in Scandinavian countries for construction of new high-speed railways on soft soils. Barriers of DSM columns installed in 2–4 interlocking panels for reduction of ground vibrations have also been used as protection of built-up areas.

10.4.1.4 Foundations of structures and houses

Buildings, warehouses, and smaller bridges are structures for which DSM columns are used to improve the sub soil layers. Concerning smaller bridges

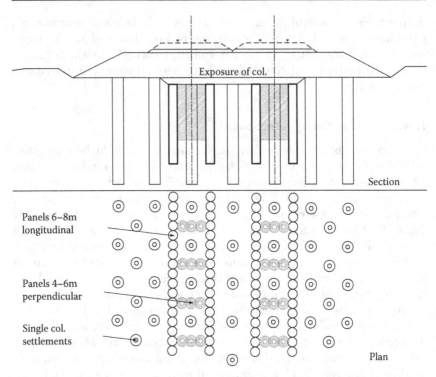

Section

Panels 6–8m longitudinal

Panels 4–6m perpendicular

Single col. settlements

Plan

Figure 10.13 Column layout for reduction of vibration connected to high-speed train phenome.

and culverts, the adjoining embankments are also usually stabilised with DSM columns. By improving the soil underneath the structure, a system with small differential settlements is obtained. An example of column layout is shown in Figure 10.14. The ratio of improvement is increased underneath the structure and at the adjacent embankment, but further from the embankment the ratio of improvement decreases. Columns in panels are installed to increase the stability toward the passway. Panels along the road are installed to increase the stability during excavation and foundation works. This type of layout takes care of the permanent situation as well as the temporary excavation and foundation works.

Buildings, warehouses, and residential homes built in soft clay and silt areas have been supported by DSM columns as an economical alternative to piling. When it comes to warehouses, a combination of DSM and piling has been used in areas with soft clay deposits. The warehouse framework structure is supported on piles taken to a firm bearing soil layer, and the floor is supported by DSM columns.

Designing DSM for supporting structures means that small settlements and restrictions of differential settlements are critical. Hence the static load

(a)

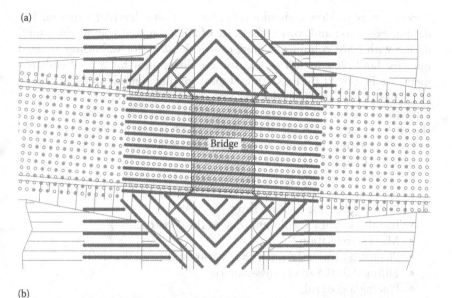

(b)

Figure 10.14 (a) Column layout applied on small bridges and culverts. (b) Interlocking panels for slope stability at Skepplanda, Sweden (Courtesy of Johnny Wallgren.)

(permanent load) on a column shall be limited to 0.4–0.6 times the bearing capacity of a column.

10.4.1.5 Mass mixing and stabilisation of highly organic soils and dredged muds

Mass mixing is used in extremely soft soils or soils with high compressibility and organic content as peat, mud and in mangrove (see case history for Jewfish Creek). The majority of mass mixing projects are for roads and parking areas and low embankments (0.5–3 m) to a depth of 1 to 5 m. The design, execution and working scheme of a mass mixed volume, in combination with DSM columns, is more or less the same for stabilising dredged mud for land reclamation as it is for stabilising peat for a road embankment, see Figure 10.15. Execution of mass mixing remoulds a large volume of soil and mixes in a considerable amount of air from the process into the mixed soil volume. To obtain the target strength, it is necessary to allow the air to dissipate into the atmosphere, which is best achieved by starting preloading within 24 hours. Installation is executed in square or rectangular cells of 15–25 m² at a time.

The performance will be carried out in the following steps:

- Working bed for the first front
- Installation of LC columns (due to the construction)
- Mass stabilisation
- Placing a geotextile and/or a geogrid on the stabilised soil
- Filling 0.3–0.5 m of cross-material
- Placing a geogrid
- Filling 0.3–0.5 m of cross-material
- After approximately 1 month (28 days curing time), the fill to final level
- The surcharge 3–6 months before the finalising the embankment

Dahlström and Eriksson (2005) have reported two projects using mass stabilisation and DSM columns for road embankments in Sweden. Jelisic

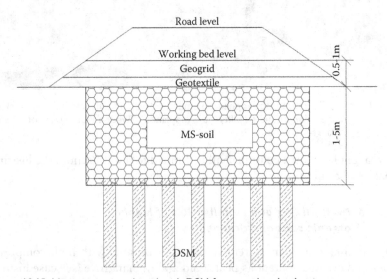

Figure 10.15 Mass mixing combined with DSM for a road embankment.

and Leppännen (2005) and Forsman et al. (2008) have reported experiences of mass stabilisation in contaminated dredge mud.

10.4.1.6 Solidification and stabilisation of contaminated soil

Solidification and stabilisation of contaminated soils (the so-called S/S method) has been an increasingly common application in the last decade. The S/S method is predominately executed by mass-mixing equipment in limited cells or barge. DSM columns have also been used as barriers to enclose the contaminated soil. Industrial mixing is a new up-and-coming technique based on the knowledge from the DSM experience. The dredged material is usually pumped into the machine where it is mixed with different types of binder, and thereafter transported to the land reclamation area (a lagoon) for hardening. This application is under development and stems from the experiences with DSM.

10.4.2 Design

The design of the Nordic method and the Japanese method differ in their basic philosophy. This chapter only presents the Nordic method as it has been developed alongside the DSM processes.The original design was developed for lime columns and was first presented by Boman and Broms in 1975 at the Nordic Geotechnical Conference, NGM-75, in Copenhagen. Only minor changes and complementary theories have been developed since the first paper regarding the design method was published. In 2005 Álen et al. (2005a) presented a new design method to determine the compression strength in the columns. Today, both the original design philosophy developed by Boman and Broms and the new design model for determining the compression strength and bearing capacity of a column is used in the Nordic countries. The original design philosophy is still the most common design method. Design guidelines such as EuroSoilStab (2002) and SGF (2000) are based on Boman and Broms' design philosophy. Stability failure has been investigated and presented in numerous papers, including Kitazume et al. (1996), Kivilö and Broms (1999), Terashi (2005), and Filz et al. (2012).

DSM columns and mass mixing are inhomogeneous to varying degrees, with an irregular structure and varying properties. The columns and mass mixing are primarily intended to interact with unstabilised soil at axial loading. For other load situations such as horizontal loading (direct shearing) or uplifting (tensile stresses), shear strength can be significantly lower than measured. Columns subjected to tensile stresses shall be avoided. DSM columns and mass mixing structures are designed for ultimate limit state (ULS) and serviceability limit state (SLS).

10.4.2.1 Design models

The design models are based on the assumption that the DSM column inter-acts with the unstabilised soil, which implies that the design is based on semi-hard columns. The assumption of interaction between columns and unstabilised soil is based on the assumption that deformation in a column is equal to deformation in unstabilised soil. The characteristic properties of the stabilised soil volume can therefore be calculated using the ratio of improvement. The ratio of improvement is defined as the area of a column divided by the spacing between the columns (total area), see Figure 10.16.

The ratio of improvement for single elements can be expressed as:

$$a = \frac{A_{col}}{A_{tot}} \tag{10.5}$$

where A_{col} = area of the column; A_{tot} = total area, which can be calculated from the effective diameter (D_e) according to column layout and spacing between columns (s); $A_{tot} = \pi D_e^2/4$.

Soil characteristics such as the shear strength, settlement modulus, and permeability of the improved soil volume are calculated based on the ratio

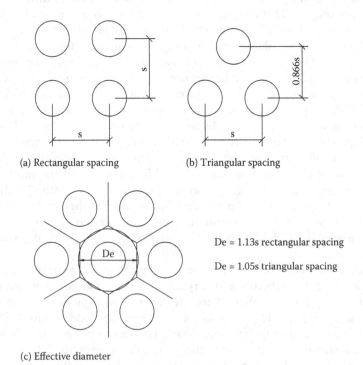

(a) Rectangular spacing   (b) Triangular spacing

$D_e = 1.13s$ rectangular spacing

$D_e = 1.05s$ triangular spacing

(c) Effective diameter

Figure 10.16 Ratio of improvement.

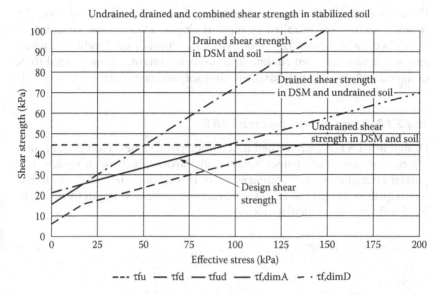

Undrained, drained and combined shear strength in stabilized soil

Figure 10.17 Critical shear strength due to in-situ stress situation in the stabilised soil volume; the bold line is the calculated shear strength used in the stability analysis.

of improvement. The improved soil volume can be divided into three zones, A, B and C, see Figure 10.17 (Álen et al., 2005a), which are:

A. Upper zone of 0.5–2 m, transition zone. In the top part of the columns, the homogeneity and strength can vary considerably. Hence, the column shall not be treated as a full-strength column. Spacing between columns due to arching in the embankment fill needs to be evaluated. Column spaces greater than 3 times the column diameter should be used with caution.

B. Stabilised volume. Column and soil are assumed to have full interaction.

C. Unimproved soil underneath the stabilised volume. Soft soil underneath the columns is considered according to basic soil mechanics. For time-settlement analysis the drainage length has to be adjusted due to the fact that DSM columns have higher permeability than unimproved soil.

10.4.2.1.1 Column layout

The layout and the application of the DSM columns have significant influence on the design and design calculations (Table 10.2). Columns and mass mixing underneath embankments are predominately axially loaded. In these situations full interaction between column and soil can be assumed. When columns are installed in natural slopes, excavations, or embankments with a very low factor of safety in unstabilised conditions where full interaction between column and soil is uncertain, columns need to be designed and checked in undrained, drained, and combined shear strength.

10.4.2.1.2 Design ultimate limit state (ULS)

Design in ULS is recommended for characteristic values. However, according to new design guidelines adopted in Europe (Eurocode), design with partial safety factors is recommended. In this section, designs with characteristic values are presented. The characteristic shear strength of the stabilised soil volume is calculated as:

$$\tau_f = \tau_{f,col} + (1 - a) \times \tau_{f,soil} \tag{10.6}$$

where

a = ratio of improvement, see Equation 10.5
$\tau_{f,col}$ = shear strength in a column
$\tau_{f,soil}$ = shear strength in unstabilised soil.

Characteristic shear strength in undrained conditions is calculated as:

$$\tau_{fu} = C_{u,col} + (1 - a) \times C_{u,soil} \tag{10.7}$$

where

$C_{u,col}$ = UCS/2, (unconfined compression strength).

Undrained shear strength greater than 150 kPa is not recommended for use in the design guidelines.

Table 10.2 Recommendations of column layout for certain applications

Design/Application	Recommended layout			
	Single columns	Panels	Grid	Block/Mass mix
Embankments with FOS ≥ 1.0	X	(x)	(x)	(x)
Columns in shear and passive zone	-	X	X	X
Natural slopes	-	X	(x)	(x)
Excavations, temporary works	-	X	X	(x)
Railways, high-speed trains	-	(x)	X	X

FOS, factor of safety for the unstabilised soil condition; X, primary use for the application; (x) moderate use for the application; -, not recommended for the application.

Characteristic shear strength in drained conditions is calculated as:

$$\tau_{fd} = \tau_{fd,col} + (1 - a) \times \tau_{fd,soil} \tag{10.8}$$

where

$$\tau_{fd,col} = c'_{col} + \sigma' \times \tan(\phi'_{col}) \tag{10.9}$$

$$c'_{col} = \beta \times C_{u,col} \tag{10.10}$$

$\beta = 0$ to 0.5
$\varphi'_{col} = 30\text{--}35$ deg.

$$\tau_{fd,soil} = c'_{soil} + \sigma' \times \tan(\phi'_{soil}) \tag{10.11}$$

Characteristic shear strength in combined conditions is calculated as:

$$\tau_{f,comb} = \tau_{fd,col} + (1 - a) \times C_{u,soil} \tag{10.12}$$

Combined analysis means that the drained shear strength in the columns and undrained shear strength in unstabilised soil are used in calculations of shear strength in the stabilised soil volume.

Combined shear strength is often the most critical factor in the stability analysis for slopes and excavations, due to limited axial load on the columns. In stability analysis the most critical shear strength in the stabilised soil is due to the stress situation on the columns and the shape of the slip surface (evaluation of columns in active, direct and passive shear zone). Figure 10.17 shows an evaluation of the critical shear strength due to the in-situ stress for the active shear zone and direct shear zone.

10.4.2.1.3 Seismic design and dynamic loads

Design for dynamic loads, such as with high-speed trains and mitigation of earthquake ground deformations, has been performed successfully. Studies from both Japan and Scandinavia shows that cyclic loads with a shear stress level less than 0.4–0.6 times the undrained shear strength (measured from laboratory triaxial and UCS tests) had no reduction of the undrained shear strength in the samples. Japanese studies showed increased undrained shear strength, Bengtsson and Karlsson (2006).

Triaxial cyclic load tests on laboratory mix samples are valuable information in the design procedure for dynamic loads. Example of design for mitigation of earthquake ground deformations is presented by Martin et al. (1999) and an example of a design and field study of the mitigation of track and ground vibrations by high-speed trains is presented by Holm et al. (2002).

10.4.2.1.4 Design serviceability limit state (SLS)

Design in SLS involves models for settlement calculations, bearing capacity, and dynamic loading required in the reduction of vibrations and deformations arising from dynamic loads. The ultimate bearing capacity of a column was originally proposed by Broms (1984) as a function of the undrained shear strength in a column and the effective horizontal stress. The guideline EuroSoilStab and SGF's Report 2:2000 is based on these functions:

$$\sigma_{f,col} = q_{u,col} + K_P(\sigma_{v0} + m_{soil} \times \Delta\sigma_v - u_{col}) + u_{col} \qquad (10.13)$$

where:

$q_{u,col}$ = UCS (unconfined compression strength) $\sim 2^*C_{u,col}$

K_P = Coefficient of passive earth pressure, $K_P = \dfrac{1 + \sin\phi'_{col}}{1 - \sin\phi'_{col}}$

σ_{v0} = Initial total overburden pressure

m_{soil} = Factor of stress increase into the unstabilised soil due to applied weight from unit loads (e.g., embankment). The ratio of stress increase m_{soil} = 0 to 0.5. The stress increase depends on the load distribution on the unstabilised soil

$\Delta\sigma_v$ = Applied unit load on the columns (e.g., embankment)

u_{col} = Pore water pressure.

The ultimate bearing capacity increases with depths and the UCS in a column.

Creep strength is a commonly used term in design of DSM columns. The creep strength is a function of the expected homogeneity of the column depending on the mixing work, the virgin soil properties and type of binder. For instance, organic soils and peats are more difficult to mix to achieve high homogeneity in the column. Creep strength of a column is estimated at 65%–90% of the ultimate bearing capacity.

The creep strength can be calculated as:

$$\sigma_{creep} = m_{creep} \times \sigma_{f,col} \qquad (10.14)$$

where:

m_{creep} = 0.8–0.9 (columns in clayey silt, silty clay and sandy clay)

m_{creep} = 0.7–0.8 (columns in clay)

m_{creep} = 0.65–0.7 (columns in organic clay, peat and contaminated soils).

If a small settlement is required, then a safety factor on the creep strength should be applied. The safety factor is 0.7–0.8 times the creep strength (σ_{creep}).

10.4.2.1.5 Settlement calculations

The settlement of a column group is governed by the weighted average modulus of elasticity of the column and the compression modulus of the unstabilised soil.

$$s_{group} = \sum \frac{\Delta h \times \Delta \sigma_v}{a \times E_{col} + (1 - a) \times M_{soil}} \tag{10.15}$$

where:

$E_{col} = 50\text{–}300 * C_{u,col}$

M_{soil} = compression modulus of the unstabilised soil and depends on the stress applied on the soil and the preconsolidation pressure in the unstabilised soil between the columns.

In mass mixing settlement calculations are governed by the modulus of the mass mixed soil. Due to the requirement of preloading (in order to gain strength in the mixed soil volume) considerable settlement, up to 30%–35% of the stabilised volume, can occur during curing.

$$s_{MS} = \sum \frac{\Delta h \times \Delta \sigma_v}{M_{MSl}} \tag{10.16}$$

where:

$M_{MSl} = 50\text{–}100 * C_{u,MS}$

Young's modulus in a column E_{col} depends on the soil properties, shear strength, and the stress level. For example, columns in organic soils shall assume low values, and columns in inorganic clay and silty sandy soils can be assumed to have high values. For mass mixing, compression modulus M_{MS} is in the range of $50\text{–}100 * C_{u,MS}$.

Due to large variations of Young's modulus in the columns and the differences and uncertainties between samples prepared in the laboratory and in-situ strength in a column, settlement calculations are recommended for probable maximum and minimum values of Young's modulus in a column. Hence, monitoring of settlements and consolidation are of great importance in a DSM project.

New models for calculation of settlement have been developed by Álen et al. (2005a) based on field studies on trial embankments on the Swedish west coast and finite element calculations. Álen et al. (2005a) commented, 'The material behavior of DSM columns can be regarded as similar to a highly over consolidated clay or maybe a very low strength concrete. Both descriptions highlight that it is the drained strength properties that governs the behavior of the DSM column.' The compression strength (bearing capacity) can therefore be expressed with Mohr-Coulomb failure criteria as:

$$\sigma'_{col} = \frac{2 \times \cos(\phi')}{1 - \sin(\phi')} \times c' + \frac{1 + \sin(\phi')}{1 - \sin(\phi')} \times \sigma'_h - \sigma'_{v0} \tag{10.17}$$

The horizontal stress situation in the soil can be expressed as:

$$\sigma'_h = \sigma'_{h0} + \Delta\sigma'_{h,soil} = \sigma'_{h0} + 0{,}5 \times \Delta\sigma'_{v,soil} \tag{10.18}$$

where:

σ'_{col} = maximum increase of stress in a column

$\sigma'_{h,0}$ = horizontal effective stress in-situ conditions

$\Delta\sigma'_{h,soil}$ = increased horizontal stress in unstabilised soil

$\Delta\sigma'_{v,soil}$ = increased vertical stress in unstabilised soil.

Stress distribution from the columns to the unstabilised soil has been reported by Álen et al. (2005a). The investigation showed that the load distribution from trial embankments on floating columns (columns not installed to firm ground) was similar to load distributions in unstabilised soil. Álen et al. (2005a) presented a modified Boussinesq's stress distribution with depth. The authors also presented a load split model (see Figure 10.18) where part of the total applied load is divided into two loads, q = q1 + q2. Load (q2) is applied at top of the columns and load (q1) is transferred to the toe of the columns. Stress distribution by depth is thereafter calculated for the two cases. In applications with floating columns and large deposits of soft clay (commonly the case in Sweden) this model has been successfully used.

10.4.2.1.6 Consolidation rate

When the effective stress in the soil is less than the preconsolidation pressure, settlement will develop rapidly. When the effective stress in the soil exceeds the preconsolidation pressure, then the rate of consolidation settlement in the stabilised soil volume is calculated similar to vertically drained soil. The permeability of the stabilised soil is 200–600 times the permeability of the soil (EuroSoilStab, 2002). The rate of consolidation can be calculated by Equation 10.17, after Barron (1948) and Hansbo (1979) and modified by Åhnberg et al. (1986):

$$U = 1 - \exp\left[\frac{-2 \times c_{vh} \times t}{R^2 \times f(n)}\right] \tag{10.19}$$

where:

U = degree of consolidation

c_{vh} = coefficient of consolidation in unstabilised soil with respect to lateral drainage, normally assumed to be equal to $2{*}c_{vv}$

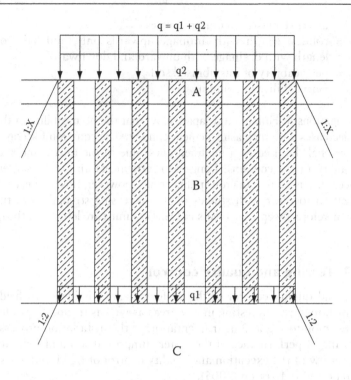

$$q = q1 + q2$$

Figure 10.18 Conceptual zones, principle of load split model and stress distribution from a stabilised soil volume.

c_{vv} = coefficient of consolidation in unstabilised soil with respect to vertical flow

t = period of consolidation

R = radius of influence.

When columns are installed at distance c between the centre in a square grid, the influence radius can be expressed as $R = 0.56^*c$. If the columns are installed in a triangular grid, the influence radius can be expressed as $R = 0.53^*c$.

$$f(n) = \frac{n^2}{n^2 - 1} \times \left[\ln(n) - 0.75 + \frac{1}{n^2} \times \left(1 - \frac{1}{4n^2} \right) \right] + \left[\frac{n^2 - 1}{n^2} \times \frac{1}{r^2} \times \frac{k_{soil}}{k_{col}} \times L_D^2 \right]$$

(10.20)

where:

$n = R/r$

r = column radius

c = distance between column centres
L_D = column length with drainage upwards only, and half column length with drainage both upward and downward
k_{soil} = permeability of unstabilised soil
k_{col} = permeability of column.

Investigations in Finland and Japan show that the permeability in the columns decreases with increasing cement content. The equation for consolidation time shall be used as a qualified guess due to the uncertainties of the permeability in the columns. Long-term settlements and creep settlements have been discussed in numerous publications; however, this area needs more investigation together with studies of the increasing strength over time in DSM. Therefore, creep settlements in DSM columns are left out of the design today.

10.4.3 Testing and quality control

Testing and quality control of the performance is divided into field testing, which involves pretesting in test areas as well as testing of production columns, monitoring and instrumentation of the installation process, and monitoring of performance of the system (improved soil and structure). A recent review of the execution and quality control of DSM and mass mixing is reported in Larsson (2005).

10.4.3.1 Field test methods

The mechanical properties of the stabilised soil are controlled in situ with various types of penetration test methods. Sampling of stabilised soil samples in fresh columns (MOSTAP) or in hardened columns (coring) can also be performed. The samples can afterwards be tested in the laboratory. Visual inspection of the homogeneity of columns can be executed through trial pits and exposure of columns. Samples to analyse the chemical composition can also be collected from trial pits or by coring. Extraction of an entire column and subsequent testing may also be performed using large split-tube samplers. However, this method is expensive and has only been used for research projects and special situations. Plate load tests and test embankment are also used to determine the elasticity module in the column and the combined compression module in the stabilised soil volume.

10.4.3.1.1 Column penetration tests (KPS or SCPT)

Column penetration tests or the lime column probe is the most common test method in Scandinavia. The method is a penetration test method using a vane with a diameter of 400–500 mm (see Figure 10.19).

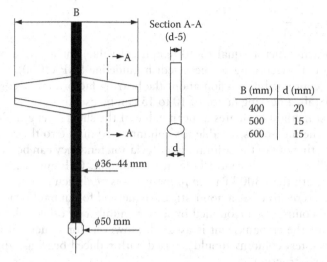

B (mm)	d (mm)
400	20
500	15
600	15

Figure 10.19 Column penetration test (the lime column probe). (From Svenska Geotekniska Förening (Swedish Geotechnical Society). (2000). *Lime and lime-cement columns. Guide for design, construction and control,* Report 2. , Linköping, Sweden (in Swedish).)

The vane is pressed down into the column with a penetration rate of 20 mm/sec and the push-down force is recorded. The method was developed from the Iskymeter method from the late 1930s and the mid-1950s. The Iskymeter, which was developed for penetration tests in very soft clays, was calibrated with respect to vane tests and fall cone tests. A semi-empirical relation was developed where the undrained shear strength τ_{fu} can be evaluated according to Equation 10.21. In 1979 Boman presented a simplified equation to evaluate the undrained shear strength in a lime column.

The equation to determine the shear strength in soft soils using the Iskymeter method:

$$\tau_{fu} = \frac{0.092 \times P}{\left(1 + \frac{2}{S_t}\right) \times A} + \frac{0.06 \times \gamma \times h \times \left(1 - \frac{1}{S_t}\right)}{\left(1 + \frac{2}{S_t}\right)} \quad (10.21)$$

where:
P = penetration force
γ = density of the soil
A = area of the probe
S_t = sensitivity of the soil.

Boman proposed a simplified evaluation to evaluate the undrained shear strength in a lime column.

$$\tau_{fu} = \frac{F}{N \times A} \tag{10.22}$$

N is a bearing factor equal to 10 (empirical value) for a probe with the area 100 cm^2, according to the Swedish guideline SGF (2000). There is, however, continued discussion about the bearing factor and, according to EuroSoilStab, a bearing factor of 10 to 15 can be used.

The test method evaluates a mean value of the shear strength along the column. The method is suitable for columns <10 m due to the risk of the vane deviating out of the column. This deviation tendency can be overcome by predrilling a small centre hole in the column. Columns with greater shear strength than 300 kPa can in many cases be difficult to penetrate.

The test is regarded as a nondestructive method for embankments where the tested column is surrounded by a great number of other columns and support for the embankment is assured. However, for structural foundation works, test columns should be used with reduced bearing capacity in the final construction.

For evaluation of the test result, the mantle friction along the probe shall be taken into account. In recent years the method has been improved by attaching a CPT (Cone Penetration Test) device to measure the penetration resistance without the friction. In the CPT device an inclinometer has also been added for verticality control (Forsgren and Ekström, 2002). A minimum of 5%–10% of the total tested columns shall be taken in unstabilised soil in order to compare the shear strength and to evaluate the mantle friction along the probe.

10.4.3.1.2 Reverse column penetration test (PORT)

The reverse column penetration test (PORT; Figure 10.20) is a pull-out test. A vane is installed below the bottom of the column, with a wire up to ground surface. The vane can be installed at the same time as the columns or directly after the column has been installed. The vane shall be installed a minimum 1 m deeper than the column tip.

The vane is pulled out through the column with a penetration rate of 20 mm/sec and the pull-out force is measured. The shear strength of the column can then be evaluated according to Equation 10.22. The area of the vane shall be 100 cm^2 and a bearing factor 10 is recommended in the Swedish guideline SGF (2000). As previously noted, however, EuroSoilStab recommends a bearing factor of 10 to 15.

The PORT method evaluates the shear strength similar to the column penetration test. With PORT testing there is no theoretical limitation of column length and columns with shear strength of up to 600 kPa can be tested. For evaluation of the test result, the mantle friction along the wire shall be taken into account. A minimum of 5%–10% of the total tested columns shall be installed only with the wire to evaluate the mantle friction along the wire.

(a)

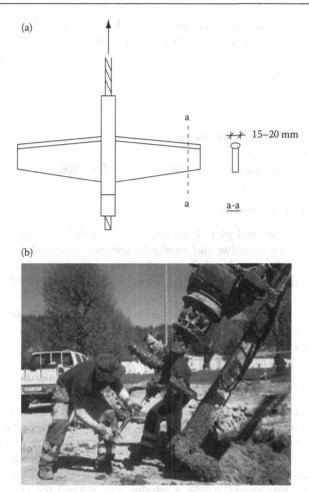

15–20 mm

a-a

(b)

Figure 10.20 (a) Reverse column penetration test. (From Holmqvist, L. (1992). The lime column method. *Bygg and Teknik,* 7–8:40–44, in Swedish.) (b) Picture of installation of the probe. (Courtesy of LCM.)

10.4.3.1.3 Cone penetration tests (CPT)

Cone penetration testing is frequently used in Norway and Finland and is a primary test method in other European countries. However, in Sweden CPT is used as a complementary test method. The method tests a small area of the column. Therefore, small local weak zones can have a major influence on the test result and the apparent shear strength of the column cross-section may not be representative from CPT tests. When CPT testing is used, a larger number of tests are recommended, typically 1%–4% of the installed columns in order to make a statistical evaluation from which a mean value and standard deviation can be identified. The shear strength in a column can be evaluated according to Equation 10.23:

$$\tau_{fu} = \frac{q_c - \sigma_{v0}}{N_c}$$

(10.23)

where:

q_c = measured cone resistance
σ_{v0} = total overburden stress
N_c = Bearing factor 10 to 25 (according to EuroSoilStab N_c = 10–13)

10.4.3.1.4 Test embankments and plate load tests

Test embankment and plate load tests are suitable methods for evaluating the elasticity modulus and combined compression modulus M_{comb} for a composite column/soil stabilised zone. Test embankments are expensive methods and take a long time but are very valuable, especially for large projects.

Plate load tests can be performed as traditional load tests or as special compression tests using a plate under the bottom of the column as counterforce, with a wire to the column top (Baker, 2000).

10.4.3.1.5 Visual inspection

Visual inspection cannot be used as a test method but is valuable in understanding variation in the product and in the interlocking zone in panels and grids. Visual inspection can be performed in trial pits down to 2 or 3 m. In the trial pits, it is possible to take soil samples from the columns and perform chemical analysis in laboratory. While providing qualitative information, this would not be a reliable test method for DSM columns (Figure 10.21).

10.4.3.1.6 Other test methods

Other test methods used for DSM columns and mass stabilisation are the Finnish vane test, core sampling, extraction of whole column, and rock sounding (or total sounding).

The Finnish vane test (Figure 10.22) was developed from the ordinary vane test method. The method is suitable in soft columns and in mass stabilised soil to evaluate the shear strength. The method has limitations in stiffer columns and has a tendency to disturb the stabilised soil during penetration. The shear strength can be evaluated according to Equation 10.24.

$$\tau_{fu} = N_v \times M_{max}$$

(10.24)

Figure 10.21 Exposed DSM column.

where:

N_v = Vane factor (345)
M_{max} = Maximum torque.

Extraction of the whole column (Figure 10.23) has been used in some large projects and in research projects. Extraction is performed by a special designed casing, pressed down around the column and extracted by a mobile crane. This is an expensive test and seldom used.

Core sampling has frequently been used as a test method for wet mixing columns. In Japan core sampling is a standard test method and the samples are collected after 5 or 6 days. Typical sample equipment is a Denison sampler, double core tube, and triple core tube. Diameters of the sample tubes are 86–150 mm. For DSM columns, undisturbed core sampling is difficult to perform because of the risk of cracks in the samples. It is recommended that the samples be tested with consolidated-undrained triaxial tests.

MOSTAP sampling is a core sampling method and has been used in a number of projects in the UK. The method uses standard CPT equipment and consists of a cone and cutting shoe at the base of the sampling tube. The sampling method is easy to use, and samples of 36–65 mm in diameter can be obtained. The samples lengths are 1.20 m and are retained within a stocking in a UPVC liner. The liner is sealed within airtight end caps. Samples are

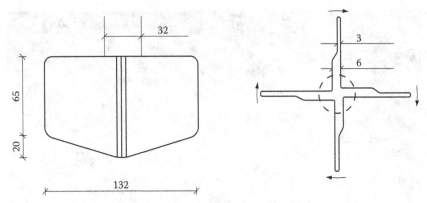

Figure 10.22 Finnish vane test. (From Halkola, H. (1983). In-situ investigation of deep stabilized soil. *Proceedings of the 8th European Conference on Soil Mechanics and Foundation Engineering,* May 23–26, 1983.)

preferably taken in fresh columns (soon after installation) and stored in laboratory for hardening. After hardening the sample tube is split and laboratory tests can be performed on the samples. This method has shown good results especially for visual inspection as well as for chemical testing.

Soil/rock sounding and total sounding are common Swedish and Norwegian methods. These methods are commonly used for predrilling a centre hole in the columns before the column penetration tests are performed. The penetration resistance can be roughly estimated by adding a bearing capacity or a correlation factor to the results from column penetration tests, which identifies the undrained shear strength. However, this method alone is not reliable.

10.4.3.2 Performance of penetration tests

The performance of the testing is crucial to the outcome of the test results. Here is a simple guideline for test procedure:

(1) Exposure of column head and survey of the column position and level.
(2) Documentation of the column head (e.g., photo documentation).
(3) Predrilling with a soil/rock sounding. During drilling, registration of penetration force, rotation speed (if necessary), and torque. Spoil water shall not be used.
(4) Execute column penetration test, with or without the CPT device.
(5) Taking short notes (e.g., if the probe tends to deviate out of the column) or other observations that could be of interest to the designer.

The test result is best presented graphically for each column as well as for the penetration force of the predrilling. All the tests from one test session

Figure 10.23 Extracted column. (From Axelsson, M. (2001). *Djupstabilisering med Kalkcementpelare-Metoder for Produktionsmassig Kvalitetskontroll i Falt (Deep Stabilization with Lime Cement Columns – Methods for Quality Control in the Field)*, Report No. 8. Linkoping, Sweden: Swedish Deep Stabilization Research Centre.)

should be summarised and presented graphically with average value, median value and bound values.

Example requirements for field tests with the column penetration method are shown in Table 10.3. Minimum required shear strength and average required shear strength at different levels and time scales are valuable information.

Table 10.3 Example of a test table for evaluation of undrained shear strength with the column penetration test method

Depth below working platform	Average shear strength (kPa)	Lower 15% percentile (kPa)	Minimum strength of local value (kPa)
0–0.5	-	-	-
0.5	≥100	≥50	-
0.5–2.0	Straight line interpolation	Straight line interpolation	-
>2.0	≥150	≥120	90

If the shear strength is not fulfilling the requirements, then additional testing in a nearby location should be the next step.

Data from the penetration force is commonly presented as a floating average of z − 200 mm to z + 200 mm. This gives an average of 400 mm. The floating average is today practiced in many projects in order to reduce the influence of locally large variations in the columns. For soil improvement with DSM columns, a locally weaker or stiffer zone <0.5 m has no significant effect on the construction, hence the boundary of the floating average.

10.5 CASE HISTORIES

Dry soil mixing also referred to as the Nordic method was rarely used outside the Nordic countries until the late 1990s. After this the number of projects has significantly increased and today the method is well accepted and used successfully in numerous countries around the world. Four case histories from different countries are presented in this chapter.

10.5.1 Thames estuary, UK

In the UK, a large number of projects have been carried out during the last few years. The projects include DSM improvement for roads, increasing of stability in harbour areas (Lawson et al. 2005), improvement of stability for railway embankments (Pye et al. 2012a), land reclamation, mass mixing, and support for deep excavations.

Soil improvement by using DSM columns (Figure 10.24) was carried out to provide a renewed access road for a large new 'Energy from Waste' facility near London on the Thames estuary. An existing route to the new plant was sited over soft alluvial deposits and was in very poor condition and quite unsuitable for trafficking by trucks during construction, as well as for the future operation of the new plant. The soil profile underneath the existing road was soft clay overlying 2–3 m of peats overlying very soft clay with peat layers in the clay. The soft deposit was 6–9 m deep overlying a dense sand deposit (see Figure 10.25). The moisture content varied between 160%–400% with an undrained shear strength of 6–21 kPa in the peat layer.

DSM columns were selected to increase the stiffness of the soft soil in order to reduce settlements and provide a new foundation for the access road. DSM columns with diameters of 800 mm were selected. These were formed in a rectangular grid with 1.1 m spacing, giving a replacement ratio of 0.41. The selected binder was Cement CEM I with a dosage rate of 200 kg/m^3. The amount and type of binder to be used was based upon evaluation of the soil characteristics from the site investigations together with the geotechnical engineer's local knowledge of soil mixing. Soil samples from the site were collected for laboratory mixing tests in order to verify the stabilisation effect. A pretest was conducted to demonstrate the performance of the improvement. Column penetration testing was carried out after 5–7 days and 14–16 days according to the field test scheme (Figures 10.25 and 10.26). The preselected binder and amount of binder demonstrated good increase in shear strength, and the design shear strength of 150 kPa (28 days' strength) was demonstrated within 7 days test.

Installation of production columns started after finalising the pretest, which demonstrated the performance of the DSM columns, with binder content of 100 kg/m column (200 kg/m^3) and mixing energy of 350 BRN. In total, some 7,500 columns were installed at an average depth of 7.8 m.

Testing was performed using CPT, column penetration tests (KPS tests) and reverse column penetration tests (pull-out test, FOPS). The column penetration tests showed the most reliable test results and were selected as the main test method. Evaluation of column shear strength at 7–14 days testing showed a wide range of strength. The evaluated average shear strength was 2–3 times design shear strength.

This case study demonstrates that DSM columns can be used in peaty soils with high organic content and high moisture content, which reportedly are difficult to improve.

Figure 10.24 Installation of DSM. (Courtesy of Keller Géotechnique.)

Figure 10.25 Field test with column penetration tests. (Courtesy of Keller Géotechnique.)

Figure 10.26 Exposed columns for visual examination. (Courtesy of Keller Géotechnique.)

10.5.2 Jewfish Creek, Key Largo, Florida

In the United States, improvement by DSM columns and mass mixing has been used since the early 1990s. However, in the last decade there has been a major increase in the use of soil improvement techniques, especially

Table 10.4 Jewfish Creek project-specific information

Source	Hayward Baker, Mann, J.A, Sehn, A., Burk, G.
Location	Highway between Florida City and Key West in Key Largo, Florida, USA
Construction site	8 km (5 miles) long, 12 m (40 ft) wide (see Figure 10.27)
Soils	Very soft organic mangrove with peat, muck, and marl, organic content 40%–60%, moisture content w = 50%–650%
Design requirements	Design in-place minimum shear strength of 75 kPa (1,500 lbs/ft^2). Long-term settlement maximum 50 mm in 5 years.
Applied DM method	Mass mixing with rotary head shaft (see Figure 10.28)
Installation data	Volume to be mixed 270,000 m^3 (350,000 yd^3)
Design shear strength	250 kPa under the rails, 150 kPa remaining area
Binder type and factor	Slag 75% and cement 25%, 140–160 kg/m^3
Field test methods	Sampling by coring (see Figure 10.29), Penetration test by KPS-vane, PORT, test embankment (see Figure 10.30)

for supporting flood levees and floodwalls after Hurricane Katrina and Hurricane Rita devastated parts of the southern United States.

The project Jewfish Creek is located in Key Largo, Florida (Table 10.4). The existing road was extended by 12m in width and 8km in length, see Figure 10.27. A typical mass-mixing setup is illustrated in

Figure 10.27 Site location of the Jewfish creek project.

Figure 10.28. Field test methods are illustrated in Figures 10.29 and 10.30. As an alternative to traditional solutions, such as removing and replacing the soft mangrove material, surcharge or installation of piles with structural platform in-situ treatment with mass mixing (Figure 10.31) was selected.

Conversion between psi and kPa, 1psi ≈ 6.89 kPa

(a)   (b)

Figure 10.28 Mass mixing equipment. (a) Mixing rig. (b) Blending station. (Courtesy of Hayward Baker.)

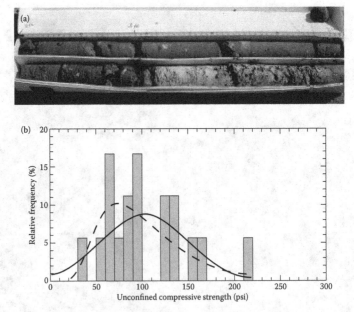

Figure 10.29 (a) Typical core samples. (b) Compression strength from core samples taken. (Courtesy of Hayward Baker.)

Figure 10.30 (a) Test embankment. (b) Settlement measurements. (Courtesy of Hayward Baker.)

Figure 10.31 Production of mass mixing on site. (Courtesy of Hayward Baker.)

10.5.3 Road 45 and Norway/Väner Route, Sweden

Road 45 and Norway/Väner route is one of the largest infrastructure projects in Sweden, which started in 2007 and will be finalised in 2012 (see Table 10.5). The project expands the European Road E45 between Gothenburg and Trollhättan (80 km) from a two-lane to a four-lane highway. The railway system is expanded from an extreme one-lane track to a two-lane high-speed railway (train speed 250 km/h).

Table 10.5 Project-specific information for road 45 and Norway/Väner route

Source	Swedish Road Authorities, www.banavag.se
Location	Road- and railway (high-speed trains) between Gothenburg and Trollhättan, Sweden.
Construction site	80 km highway and railway (see Figure 10.32)
Soils	Very soft clay to large deposit more than 100 m locally. Typical undrained shear strength 8–25 kPa, moisture content w = 40%–150%, sensitivity up to St < 400.
Design requirements in DSM columns	Design undrained shear strength of 150 kPa. Long-term settlement maximum 250 mm in 40 years, stability; factor of safety 1.5–1.65 (1.65 in quick clay areas).
Applied DM method	Lime/cement columns with diameter 600 mm
Installation data	Approximately 9.2 million linear metres installed up to 25-m depth. Interlocking columns in panels and grids for stability and high-speed trains
Binder type and factor	Lime 50% and cement 50%, 90–30 kg/m³, some areas lime 33%, cement 33%, and fly ash 33%.
Field test methods	Test embankments, Álen et al. (2005b), column penetration tests with CPT devise and inclinometer device.

Location of the project is in the Göta Älv River Valley (see Figure 10.32). The geological formation in the valley is characterized by large clay deposits with highly sensitive clays (quick clays) overlying bedrock. Slope stability is a major issue along the Göta Älv River, and two major slip failures with human loss have occurred in the area (the Surte failure 1950 and the Göta failure 1957). In addition, there have been a large number of failures resulting in material losses.

The soft clay deposits are stabilised by approximately 9 million linear DSM columns (Figure 10.33). Columns are installed to increase stability, reduce settlements, and prevent mitigation of ground vibrations connected to high-speed trains. Other applications are foundations of structures such as bridges (see Figures 10.14a, b and 10.15).

Before the project started (5–7 years), test embankments were installed and monitored. Field and laboratory tests were performed in order to determine binder combination, amount of binder, and mixing parameters for the total project before construction started. This is, however, an unusual way of designing because all contractors are restricted to certain parameters. In construction, pretesting is performed for each part of the project. Pretesting of DSM columns is carried out in test areas to verify the predesigned installation concept. Field testing is performed on two occasions, 12–16 days and 28–34 days. No other time periods are allowed for testing.

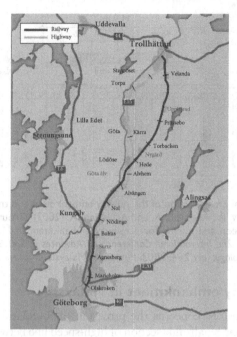

Figure 10.32 Site map of the project. (From Banaväg in Väst website, http://www. trafikverket.se/banavag.)

(a)

(b)

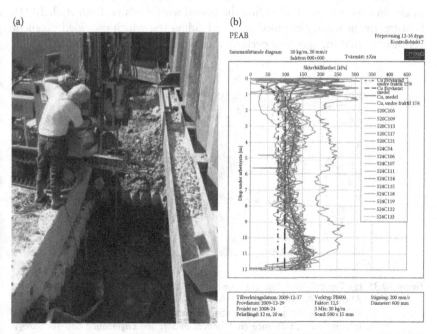

Figure 10.33 (a) Testing of DSM columns. (b) Printout of KPS tests. (Courtesy of dmixab.)

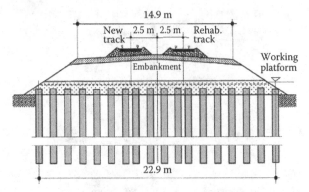

Figure 10.34 Typical cross-section of the railway embankment and treated zone. (From Raju, V. R., Abdullah, A. and Arulrajah, A. (2003). Ground treatment using dry deep soil mixing for a railway embankment in Malaysia, *Proceedings of the 2nd International Conference on Advances in Soft Soil Engineering and Technology*, July 2–4, 2003, Putrajaya, Malaysia.)

10.5.4 Railway embankment in Malaysia

The final case history concerns the provision of foundations for a railway embankment along a half-mile section of high-speed line between Rawang and Ipoh in Malaysia (Figure 10.34). Ground conditions consisted of very soft alluvial deposits (Figure 10.35 and Table 10.6), and dry soil mixing was selected as an effective means of providing the design requirements (Raju et al. 2003). Post treatment testing included an area loading test which provided excellent verification that the specified performance was achieved (Figure 10.36).

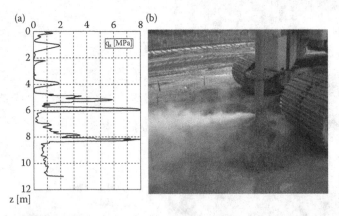

Figure 10.35 (a) Typical CPT log. (b) The LCM machine at work. (From Raju, V. R., Abdullah, A. and Arulrajah, A. (2003). Ground treatment using dry deep soil mixing for a railway embankment in Malaysia, *Proceedings of the 2nd International Conference on Advances in Soft Soil Engineering and Technology*, July 2–4, 2003, Putrajaya, Malaysia.)

Table 10.6 Railway embankment in Malaysia

Source	Raju, Abdullah, and Arulrajah (2003)
Location	Railway line between Rawang and Ipoh, Malaysia
Construction site	800 m long, 20–25 m wide (see Figure 10.34)
Soils	Very soft silty clay or clayey silt to loose silty clayey sand, typical CPT log see Figure 10.35a, moisture content w = 50%– 70%, groundwater ca. 1 m below ground surface
Embankment height and load	1.5–3 m, equivalent traffic load 30 kPa
Design requirements	Train speed 160 km/h, max. settlement 25 mm in 6 months of operation, max. differential settlement 0.1% along the centreline, safety factor for slope failure 1.5 (long term)
Applied DM method	(Lime-cement column), single shaft (Figure 10.35)
Column data	Diameter 0.6 m, length 7–14 m, overall 50,000 lin. m
Column pattern	Detached columns, square/rectangle, 1–1.3 m c/c under the rails (a_p = 28 to 17%), 1.4–1.5 m c/c remaining area (a_p = 14 to 13%)
Design shear strength	250 kPa under the rails, 150 kPa remaining area
Binder type and factor	Portland cement 100%, 100–150 kg/m³
Embankment reinforcement	Geotextile 100/50 kN/m (longitudinal/transverse direction)
Observed performance	Settlement below 10 mm for embankment 1–1.5 m, lateral displacement below 15 mm, loading test (see Figure 10.36).

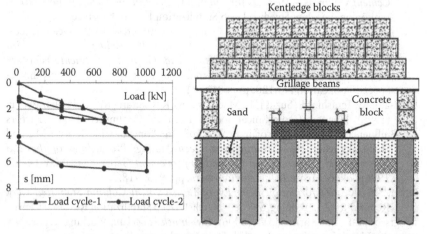

Figure 10.36 Control static loading test over an area of 3 × 3 m, 4 columns. (From Raju, V. R., Abdullah, A. and Arulrajah, A. (2003). Ground treatment using dry deep soil mixing for a railway embankment in Malaysia, *Proceedings of the 2nd International Conference on Advances in Soft Soil Engineering and Technology*, July 2–4, 2003, Putrajaya, Malaysia.)

REFERENCES

Åhnberg, H. and Holm, G. (1986). *Kalkpelarmetoden – Resultat av 10 Års Forskning och Praktisk Användning samt Framtida Utveckling (The Lime Column Method – Results from Research and Practical Applications during 10 Years and Future Development)*, Report No. 31. Linköping, Sweden: Swedish Geotechnical Institute.

Åhnberg, H., Johansson, S.E., Retelius, A., Ljungkrantz, C., Holmqvist, L., and Holm, G. (1995). *Cement och Lalk för Djupstabilisering av Jord – En Kemisk Fysikalisk Studie av Stabiliseringseffekter (Cement and Lime for Stabilisation of Soil at Depth – a Chemical Physical Investigation of Soil Improvement Effects)*, Report No. 48. Linköping, Sweden: Swedish Geotechnical Institute.

Åhnberg, H., Bengtsson, P-E. and Holm, G. (2001). Effect of initial loading on the strength of stabilised peat. *Ground Improvement*, 7(1):9–23.

Åhnberg, H. (2006). *Strength of Stabilized Soils – A Laboratory Study on Clays and Organic Soils Stabilised with Different Types of Binder*. PhD Thesis, Lund University, Sweden.

Álen C., Baker, S., Bengtsson, P-E., and Sällfors, G. (2005a). Lime/cement column stabilised soil – a new model for settlement calculation. *Proceedings from the International Conference on Deep Mixing Best Practice and Recent Advances*, May 23–25, 2005, Stockholm, Sweden, pp. 205–212.

Álen C., Baker, S., Ekström, J., Hallingberg, A., Svahn, V. and Sällfors, G. (2005b). Test embankments on lime/cement stabilized clay. *Proceedings from the International Conference on Deep Mixing Best Practice and Recent Advances*, May 23–25, 2005, Stockholm, Sweden, pp. 213–220.

Axelsson, M. (2001). *Djupstabilisering med Kalkcementpelare-Metoder för Produktionsmässig Kvalitétskontroll i Fält (Deep Stabilization with Lime Cement Columns – Methods for Quality Control in the Field)*, Report No. 8. Linköping, Sweden: Swedish Deep Stabilization Research Centre.

Axelsson, M. and Larsson S. (2003). Column penetration tests for lime-cement columns in deep mixing – experiences in Sweden, *Proceedings from the Third International Conference on Grouting and Ground Treatment*, February 10–12, 2003, New Orleans, Louisiana, United States, ASCE Geotechnical Special Publication No.120, pp. 681–694.

Babasaki, R., Terashi, M., Suzuki, T., Maekawa, A., Kawamura, M., and Fukazawa, E. (1996). Japanese Geotechnical Society Technical Committee Report: Factors influencing the strength of improved soil. In: Yonekura R., Terashi M., and Shibazaki M. (eds.), *Grouting and Deep Mixing: Proceedings of the 2nd International Conference on Ground Improvement Geosystems*, May 14–17, 1996, Tokyo, Japan: Balkema Publishers, pp. 913–918.

Baker, S. (2000). *Deformation Behavior of Lime/Cement Column Stabilized Clay*, Report No. 7. Linköping, Sweden: Swedish Deep Stabilization Research Centre.

Bengtsson, P-E. and Karlsson, M. (2006). *Dynamisk Påverkan*, Working Report No. 39. Linköping, Sweden: Swedish Deep Stabilization Research Centre.

Boman, P. and Broms, B. (1975). Stabilization of cohesive soil by lime columns. *Proceedings of the Nordic Geotechnical Meeting*, May 22–24, 1975 Copenhagen, Denmark, pp. 265–279 (in Swedish).

Boman, P. (1979). *Quality Control of Lime Column. Part 2: Results from Investigations at Sollentunaholm.* Stockholm, Sweden: Royal Institute of Technology (in Swedish).

Bredenberg, H. (1999). Keynote lecture: equipment for deep soil mixing with the dry jet mix method. *Proceedings of the International Conference on Dry Methods for Deep Soil Stabilization,* October 13–15, 1999, Stockholm, Sweden, pp. 323–331.

Bredenberg, H., Holm, G., and Broms, B., eds. (1999). *Dry Mix Methods for Deep Soil Stabilization: Proceedings of the International Conference on Dry Methods for Deep Soil Stabilization,* October 13–15, 1999, Stockholm, Sweden.

Broms, B. (1984). *Stabilization of Soil with Lime Columns – Design Handbook,* 3rd ed. Kungsbacka, Sweden: Lime Column Method AB.

Broms, B. (1999). Keynote lecture: Design of lime, lime/cement and cement columns. *Proceedings of the International Conference on Dry Methods for Deep Soil Stabilization,* October 13–15, 1999, Stockholm, Sweden, pp. 125–154.

Dahlström, M. and Eriksson, H. (2005). Mass stabilisation in Smista Allé (2002) and Moraberg (2003) by cell and block stabilisation methods. *Proceedings from the International Conference on Deep Mixing Best Practice and Recent Advances,* May 23–25, 2005, Stockholm, Sweden, pp. 425–429.

Dahlström, M. and Wiberg, D. (2012). Dry soil mixing and vibro replacement in combination for a high embankment. *Proceedings from the International Conference on Grouting and Deep Mixing,* February 16–18, 2012, New Orleans, Louisiana, United States.

EuroSoilStab. (2002). *Development of Design and Construction Methods to Stabilise Soft Organic Soils: Design Guide Soft Soil Stabilisation,* CT97-0351, European Commission Project BE 96-3177.

EN14679:2005 (2005). *European Standard, Execution of Special Geotechnical Works – Deep Mixing.* Publisher information

Filz, G., Adams, T., Navin, M., and Tempelton, A.E. (2012) Keynote lecture: Design of deep mixing for support of levees and floodwalls. *Proceedings from the International Conference on Grouting and Deep Mixing,* February 16–18, 2012, New Orleans, Louisiana, United States.

Forsgren, I. and Ekström, J. (2002–2012). Personal communication.

Forsman, J., Maijala, A., and Järvinen, K. (2008). Case stories, harbours – mass stabilisation of contaminated dredging mud in Sörnäinen, Helsinki. *Proceedings of the International Mass Stabilisation Conference 2008,* October 8–10, 2008, Lahti, Finland.

Halkola, H. (1983). In-situ investigation of deep stabilized soil. *Proceedings of the 8th European Conference on Soil Mechanics and Foundation Engineering,* May 23–26, 1983, Helsinki, Finland.

Halkola, H. (1999). Keynote lecture: quality control for dry mix method. *Proceedings of the International Conference on Dry Methods for Deep Soil Stabilization,* October 13–15, 1999, Stockholm, Sweden, pp. 285–294.

Hansbo, S. (1979). Consolidation of clay by bandshaped prefabricated drains, *Ground Engineering* 12(5): 21–25.

Hayashi, H., Nishikawa, J., Ohishi, K. and Terashi, M. (2003). *Field Observation of Long-Term Strength of Cement Treated Soil,* ASCE Geotechnical Special Publication No. 120, pp. 598–609.

Hebib, S. and Farrell, E.R. (2003). Some experience on the stabilisation or Irish peats, *Canadian Geotechnical Journal* 40:107–120.

Holm, G. (1999). Keynote lecture: applications of dry mix methods for deep soil stabilization. *Proceedings of the International Conference on Dry Methods for Deep Soil Stabilization,* October 13–15, 1999, Stockholm, Sweden, pp. 3–14.

Holm, G. (2002). Nordic dry deep mixing method – execution procedure, *Proceedings of the Deep Mixing Workshop 2002,* Port and Airport Research Institute & Coastal Development Institute of Technology, Tokyo, Japan.

Holm, G., Andréasson, B., Bengtsson, P-E., Bodare, A., and Eriksson, H. (2002). *Mitigation of Track and Ground Vibrations by High Speed Trains at Ledsgård, Sweden,* Report No. 10. Linköping, Sweden: Swedish Deep Stabilization Research Centre.

Holmqvist, L. (1992). The lime column method. *Bygg and Teknik,* 7–8:40–44 (in Swedish).

Horpibulsuk, S., Miura, N. and Nagara, T. (2003). Assesment of strength development in cement-admixed high water content clay with Abrams' law as a basis. *Géotechnique* 53(4): pp. 439–444.

Janz, M. and Johansson, S.E. (2002). *The function of different binding agents in deep stabilization,* Report No. 9. Linköping, Sweden: Swedish Deep Stabilization Research Centre.

Jelisic, N. and Leppännen, M. (1999). Mass stabilization of peat in road and railroad construction. *Proceedings of the International Conference on Dry Methods for Deep Soil Stabilization,* October 13–15, 1999, Stockholm, Sweden, pp. 59–64.

Jelisic, N. and Leppännen, M. (2005). Remediation of contaminated land of Sörnäinen, Helsinki, by using mass stabilization. *Proceedings of the International Conference on Deep Mixing Best Practice and Recent Advances,* May 23–25, 2005, Stockholm, Sweden, pp. 353–356.

Kitazume, M., Tabata, T., Ishiyama, S. and Ishikawa, Y. (1996a). Model tests on failure pattern of cement treated retaining wall, In: Yonekura R., Terashi M., and Shibazaki M. (eds.) *Grouting and Deep Mixing: Proceedings of the 2nd International Conference on Ground Improvement Geosystems,* May 14–17, 1996, Tokyo, Japan, pp. 509–514.

Kivilö, M. and Broms, B. (1999). Mechanical behavior and shear resistance of lime/cement columns. *Proceedings of the International Conference on Dry Methods for Deep Soil Stabilization,* October 13–15, 1999, Stockholm, Sweden, pp. 193–200.

Lawson, C.H, Spink, T.W, Crawshaw, J.S. and Essler R.D. (2005). Verification of dry soil mixing at Port of Tilbury, UK. *Proceedings of the International Conference on Deep Mixing Best Practice and Recent Advances,* May 23–25, 2005, Stockholm, Sweden, pp. 453–462.

Larsson, S. (1999). The mixing process at the dry jet mixing method. *Proceedings of the International Conference on Dry Methods for Deep Soil Stabilization,* October 13–15, 1999, Stockholm, Sweden, pp. 339–346.

Larsson, S. (2003). *Mixing Process for Ground Improvement by Deep Mixing,* Doctoral Thesis, Royal Institute of Technology, Stockholm, Sweden.

Larsson, S. (2005). Keynote lecture: state of practice report – execution, monitoring and quality control. *Proceedings of the International Conference on Deep Mixing Best Practice and Recent Advances,* May 23–25, 2005, Stockholm, Sweden, pp. 732–785.

Larsson, S., Dahlström, M. and Nilsson, B. (2005a). Uniformity of lime-cement columns for deep mixing: a field study. *Ground Improvement* 9(1):1–15.

Larsson, S., Stille, H. and Olsson, L. (2005b). On horizontal vaiability in lime-cement columns in deep mixing. *Géotechnique* 55(1):33–44.

Larsson, S., Dahlström, M. and Nilsson, B. (2005c). A complementary field test on the uniformity of lime-cement columns. *Ground Improvement* 9(2): 67–77.

Löfroth, H. (2005). Properties of 10-year-old lime-cement columns, *Proceedings of the International Conference on Deep Mixing Best Practice and Recent Advances*, May 23–25, 2005, Stockholm, Sweden, pp. 119–127.

Martin, G.R., Arulmoli, K., Yan, L., Esrig, M.I., and Capelli, R.P. (1999). Dry mix soil-cement walls: An application for mitigation of earthquake ground deformations in soft or liquefiable soils, *Proceedings of the International Conference on Dry Methods for Deep Soil Stabilization*, October 13–15, 1999, Stockholm, Sweden, pp. 37–44.

Muro, T., Fukagawa, R. and Mukaihata, K. (1987a). Fundamental study of mixing conditions affecting the mechanical properties of soil-cement piles. *Doboku Gakkai Ronbunshu* 385:98–195 (in Japanese).

Muro, T., Fukagawa, R. and Mukaihata, K. (1987b). Effect of mixing condition on deformation and strength characteristics of cement treated soil. *Ehime Daigaku Kogakubu Kioy* 11(2):393–403 (in Japanese).

Nagaraj, T.S., Muira, N. Yaligar, P.P. and Yamadera, A. (1996). Predicting strength development by cement admixture based on water content. In: Yonekura R., Terashi M., and Shibazaki M. (eds.), *Grouting and Deep Mixing: Proceedings of the 2nd International Conference on Ground Improvement Geosystems*, May 14–17, 1996, Tokyo, Japan, pp. 431–436.

Nishida, K., Koga, Y., and Muira, N. (1996). Energy consideration of dry jet mixing method. In: Yonekura R., Terashi M., and Shibazaki M. (eds.), *Grouting and Deep Mixing: Proceedings of the 2nd International Conference on Ground Improvement Geosystems*, May 14–17, 1996, Tokyo, Japan, pp. 643–648.

Porbaha, A., Shibuya, S. and Kishida, T. (2000). State of the art in deep mixing technology. Part III geomaterial characterization. *Ground Improvement* 4(3):91–110.

Pye, N., O'Brien, A., Essler, R.D. and Adams, D. (2012a). Deep dry soil mixing to stabilize a live railway embankment across Thrandestone Bog. *Proceedings from the International Conference on Grouting and Deep Mixing*, February 16–18, 2012, New Orleans, Louisiana, United States.

Pye, N., O'Brien, A., Essler, R.D. and Adams, D. (2012b). Laboratory and field testing for deep dry soil mixing to stabilize a live railway embankment across Thrandestone Bog. *Proceedings from the International Conference on Grouting and Deep Mixing*, February 16–18, 2012, New Orleans, Louisiana, United States.

Raju, V. R., Abdullah, A. and Arulrajah, A. (2003). Ground treatment using dry deep soil mixing for a railway embankment in Malaysia, *Proceedings of the 2nd International Conference on Advances in Soft Soil Engineering and Technology*, July 2–4, 2003, Putrajaya, Malaysia.

Rogers, C.D.F. and Glendinning, S. (1996). The role of lime migration in lime pile stabilization of slopes. *Quarterly Journal of Engineering Geology* 29:273–284.

Rogers, C.D.F., Glendinning, S. and Holt, C.C. (2000a). Slope stabilization using lime piles – a case study, *Ground Improvement* 4(4):165–176.

Rogers, C.D.F., Glendinning, S. and Troughton, V.M. (2000b). The use of additives to enhance the performance of lime piles. *Proceedings of the 4th International Conference on Ground Improvement Geosystems,* June 7–9, 2000, Helsinki, Finland, pp. 127–134.

Svenska Geotekniska Förening (Swedish Geotechnical Society). (2000). *Lime and lime cement columns. Guide for design, construction and control,* Report 2, Linköping, Sweden (in Swedish).

Terashi, M. (2003). *The State of Practice in Deep Mixing Methods,* ASCE Geotechnical Special Publication No. 120, pp. 25–49.

Terashi, M. (2005). Keynote lecture: design of deep mixing in infrastructure applications. *Proceedings of the International Conference on Deep Mixing Best Practice and Recent Advances,* May 23–25, 2005, Stockholm, Sweden, pp. K25–K45.

Timoney, M.J., McCabe B.A., and Bell, A.L. (2012). Experiences of dry soil mixing in highly organic soils, *Proceedings of the Institution of Civil Engineers – Ground Improvement* 165(1):3–14.

Yoshizawa, H., Okumura, R., Hosya, Y., Sumi, M. and Yamada, T. (1997). Japanese Geotechnical Society Technical Committee Report: factors affecting the quality of treated soil during execution of DMM. In: Yonekura R., Terashi M., and Shibazaki M. (eds.) *Grouting and Deep Mixing: Proceedings of the 2nd International Conference on Ground Improvement Geosystems,* May 14–17, 1996, Tokyo, Japan, pp. 931–937.

Index

Printed in the United States
by Baker & Taylor Publisher Services

Printed in the United States
by Baker & Taylor Publisher Services